DISCARDED

Communist Trade in Oil and Gas

PRAEGER SPECIAL STUDIES IN
INTERNATIONAL ECONOMICS AND DEVELOPMENT

Communist Trade in Oil and Gas

AN EVALUATION OF THE
FUTURE EXPORT CAPABILITY
OF THE SOVIET BLOC

Robert E. Ebel

PRAEGER PUBLISHERS
New York • Washington • London

The purpose of Praeger Special Studies is to make specialized research in U.S. and international economics and politics available to the academic, business, and government communities. For further information, write to the Special Projects Division, Praeger Publishers, Inc., 111 Fourth Avenue, New York, N.Y. 10003.

PRAEGER PUBLISHERS
111 Fourth Avenue, New York, N.Y. 10003, U.S.A.
5, Cromwell Place, London S.W.7, England

Published in the United States of America in 1970
by Praeger Publishers, Inc.

All rights reserved

© 1970 by Praeger Publishers, Inc.

Library of Congress Catalog Card Number: 77-95699

Printed in the United States of America

PREFACE

The study of the economics of the Soviet Union and in particular its oil and gas industry is pursued by two distinct groups of observers: those who possess an adequate language background and are able to utilize primary source material, and those who, lacking this facility, to a large degree must rely on analyses and conclusions of other researchers. This study is addressed to both groups.

The emergence in the middle 1950's of the Soviet Union as an oil exporter of international significance has been accompanied by a veritable flood of original-language technical and economic publications, of varying quality, issued by a host of publishing houses in the Soviet Union. The listing of newspapers and periodicals given in Appendix B is ample testimony to the often frustrating task which faces the serious student, the task of searching endless pages of material in an attempt to construct a reasonable appraisal of these industries. Indeed, just the periodicals noted constitute a monthly output of more than one thousand pages. At the same time, the researcher is aware that certain aspects of the Soviet oil and gas industry--crude oil reserves, refinery capacity and output, and domestic demand, to name a few--are rarely discussed in the open press, and then only in terms of often meaningless link relatives or percentages. He can attempt to put these pieces together in the hope that somehow the final picture may make sense, or he can invite the attention of his reader to the impossibility of the task at hand and proceed to the next sector of the industry.

In attempting to project the future export capability of the Soviet bloc, one should first examine realistically the oil and gas industries as an integral part of the primary energy economy as a whole and then undertake a careful autopsy of each sector of these industries--reserves, exploration, drilling, production, refining, transportation, and consumption. But this is not always possible, nor is it always feasible.

This study illustrates the methodology employed by the author in evaluating the export capability of the Soviet bloc without necessarily having to resort to a painstaking analysis of each and every detail that somehow might influence export levels. Understandably, this evaluation is made against a continuing review of pertinent information on all aspects of the oil and gas industry as it becomes available in the Soviet press; but in developing forward-looking estimates of exports, only certain guidelines will be at hand. These guidelines will point to planned ranges in crude oil and natural gas production, to broad pipeline-construction programs, and to cumulative drilling effort scheduled for the forthcoming plan period and perhaps may hint at possible levels of exports to other Communist countries. But the most important criterion for oil, the forecast of future domestic demand, will be missing, and without this guideline any projection of probable future exports can reflect nothing more than a subjective analysis of what information is available.

Similarly, projections of future gas sales to the West are exceedingly tenuous at best and are based on the acceptance of a number of assumptions. Exports of gas are signaled well in advance by the initialing of long-term trade agreements and the construction of pipelines and/or natural gas liquefaction facilities for the delivery of the gas to the ultimate consumer. Until such contracts are signed, estimates of gas sales can serve as little more than broad guidelines and as indications of what might be expected under special circumstances.

Over the years a number of important studies in English relating to energy and to oil and gas in the Soviet Union have become available. These studies, also listed in Appendix B, are highly recommended reading for anyone wishing to gain an appreciation of how the oil and gas industry of the Soviet Union functions. Each successive publication has been able to offer a more penetrating insight and critical review of the subject matter than has its predecessors, simply because of the improvement which has taken place in the depth of the material made public by the Soviet Union.

Yet, the energy industry of the Soviet Union has been so dynamic in recent years, centered around important new finds of oil and gas in western Siberia and Central Asia, with attendant reformulations of development programs and concomitant re-evaluations of prospects for the export of oil and gas, that the results of research, no matter how conscientious, may be out of date even before publication. In addition, the other Communist countries are continually re-assessing their need for imports of energy from the Soviet Union, a re-assessment made against changing political and economic backgrounds, within and without. Such changes greatly influence those quantities of Soviet oil and gas remaining for possible sale to non-Communist buyers.

These factors in concert place a dubious value on the validity of any attempts to project future levels of exports of oil and gas from the Soviet bloc as a whole and from the Soviet Union in particular. Yet such projections are of more than passing interest to the policy-maker who, in his general assessment of probable future courses of action by the Communist world, must be aware of what might be expected of the Soviet bloc in the international oil and gas markets where the bloc is seeking enlarged economic and political gain. But moreover, such projections are of immediate concern to the international oil industry which must compete with Soviet oil and gas on a day-to-day basis. And to be able to compete profitably depends to a large extent upon success in anticipating the forthcoming moves of the competition.

The estimates of probable future levels of exports of oil and gas by the Soviet bloc developed in this study are believed to be reasonable and acceptable in the light of information currently available. They are not, of course, definitive estimates, and are subject to revision for the reasons stated. It is hoped that further research into the factors which seem to influence the export capability of the Soviet bloc will be stimulated by the discussions contained herein. Other researchers, utilizing the same material, but approaching the subject with a fresh point of view, may arrive at completely differing conclusions from those presented by the author.

To assist those Western observers who are interested in attempting to evaluate the export capability of the Soviet bloc and who may not have access to primary source material or who may lack the facility to utilize such material, Part II presents a selection of translated readings from original-language sources. These readings are by no means an exhaustive presentation of what is available. They are designed, however, to provide the reader with quick insight and perhaps a better understanding of the subject and, it is hoped, will encourage independent investigation by others.

CONTENTS

	Page
PREFACE	v
LIST OF TABLES	xi
LIST OF FIGURES	xvi
GLOSSARY	xviii
CONVERSIONS	xix

PART I: HISTORY AND FORECASTS

Chapter

1	OIL IN CZARIST RUSSIA	5
2	EXPANSION AND CONTRACTION: 1918-40	14
3	WORLD WAR II	25
4	RECONSTRUCTION: 1946-55	30
5	THE POSTWAR DEVELOPMENT OF OIL EXPORTS FROM THE SOVIET BLOC	39
6	THE OIL PRICING POLICY OF THE SOVIET UNION	57
7	CONFLICT WITH THE ARAB WORLD	74
8	THE FUTURE OF COMMUNIST TRADE IN OIL	81

Recapitulation	81
Problems of Forecasting	84
Out of Oil?	87
Communist Oil Through 1970	91
East European Search for Oil	91
Impact of 1967 Middle East Crisis	100
Soviet Oil Supplies, 1970	103
A Look Ahead to 1975	111

Chapter   Page

9   SOVIET TRADE IN NATURAL GAS   127

      Exports   134
         Before 1970   135
         After 1970   137
            Italy as a Market   140
            Japan as a Market   149
            West Germany as a Market   153

      Imports   155
         From Iran   155
         From Afghanistan   160
         Summary Imports   163
      Net Trade in Gas   164

   NOTES   **171**

           PART II: SELECTED READINGS

   Introduction   191

   Soviet Trade in Oil   193

   Soviet Attitude Toward Western Oil Holdings
      in the Middle East and North Africa   298

   Major Pipeline Developments   317

                  APPENDIXES

A   SELECTED STATISTICS   **385**

B   PERTINENT CURRENT SOURCE MATERIALS
     ON THE COMMUNIST OIL INDUSTRY   **442**

   ABOUT THE AUTHOR   449

LIST OF TABLES

Tables		Page
1.	Dominance of Russian Oil at Turn of Century	7
2.	Production of Crude Oil in Russia, 1889-1917	9
3.	Russian Oil Exports as Share of Oil Production, Selected Years 1890-1914	10
4.	Russian Oil Exports, 1873-1917	12
5.	Exports of Oil from the Soviet Union, 1918-40	19
6.	Production of Crude Oil in the Soviet Union, 1918-40	20
7.	Comparison of Soviet Oil Exports with Oil Production, 1926-37	21
8.	Oil Stockpiling in the Soviet Union, 1938 and 1940	23
9.	Rate of Growth in Soviet Oil Output, 1928-42	25
10.	Regional Concentration of Oil Production in 1940	27
11.	Soviet Crude Oil Production During World War II	28
12.	Soviet Imports of Oil from the United States, 1941-45	29
13.	Production of Crude Oil in the Soviet Union, 1946-55	31
14.	Trade in Oil by the Soviet Union, 1946-55	32
15.	Soviet Oil Trade with Non-Communist Countries, 1946-55	34

Tables		Page
16.	Soviet Oil Trade with Non-Communist Countries, by Country, 1946-54	35
17.	Soviet Net Oil Exports as Percentage of Crude Oil Production, 1955-68	40
18.	Export of Oil from the Soviet Union, 1955-68	42
19.	Import of Oil by the Soviet Union, 1955-68	43
20.	Distribution of Soviet Oil Exports Among Communist and Non-Communist Buyers, 1955-68	44
21.	Soviet Exports of Crude Oil and Petroleum Products to Other Communist Countries, 1955-68	47
22.	Soviet Exports of Crude Oil and Petroleum Products to Non-Communist Countries, 1955-68	48
23.	Simplified Method for Estimating Crude Oil and Petroleum Product Exports from the Soviet Union	49
24.	Soviet Imports of Oil from Non-Communist Sources, 1955-67	50
25.	Exports of Oil from Eastern Europe to Non-Communist Countries, 1955-68	52
26.	Average Prices Paid for Soviet Crude Oil by Other Communist Countries, 1955-67	58
27.	Average Prices Paid for Soviet Crude Oil by Non-Communist Countries, 1955-67	59
28.	Apparent Average Annual Prices Paid for Soviet Crude Oil, 1955-67	61

Tables		Page
29.	Soviet Oil Price Differentials, East and West Germany, 1959-67	63
30.	Foreign Exchange Earnings from the Sale of Soviet Oil to Non-Communist Countries, 1955-67	70
31.	Shifts in Earnings from the Sale of Soviet Oil to Non-Communist Countries, 1961-67	71
32.	Computation of Per Capita Consumption of Crude Oil in Soviet Union, 1965	90
33.	Fluctuations in Short-Term and Long-Term Crude Oil Production Planning in the Soviet Union	96
34.	Revisions in Soviet Coal and Gas Production Plans, 1965 and 1970	97
35.	Soviet Crude Oil and Petroleum Product Exports to Other Communist Countries, 1965, 1970, and 1975	98
36.	Probable Supply and Demand for Oil in East Europe, 1970	99
37.	Probable Supply and Demand for Oil in the Soviet Union, 1970	105
38.	Allocation of Growth in Soviet Crude Oil Production, 1965-70	108
39.	Planned and Actual Soviet Coal Extraction, 1959-65	110
40.	Siberian Output as Percent of Total, 1965-1975	113
41.	Probable Supply and Demand for Oil in the Soviet Union, 1975	115
42.	Comparative Growth in Demand for Oil in the Soviet Union and in Eastern Europe, 1955-75	116

Tables		Page
43.	Probable Supply and Demand for Oil in East Europe, 1975	117
44.	Actual and Estimated Soviet Bloc Trade in Oil with Non-Communist Countries, 1965, 1970, and 1975	122
45.	Trends in the Extraction of Natural Gas in the Soviet Union	128
46.	Trends in Manufactured Gas in the Soviet Union	130
47.	Failure of the Soviet Natural Gas Industry During the Five Year Plan 1966-70	131
48.	Soviet Exports of Natural Gas, Selected Years 1950-68	138
49.	Probable Exports of Natural Gas by the Soviet Union, 1970	139
50.	Probable Exports of Natural Gas by the Soviet Union, 1975	156
51.	Imports of Natural Gas by the Soviet Union, 1970 and 1975	165
52.	Probable Soviet Net Trade in Natural Gas, 1970 and 1975	166
53.	Soviet Exports of Oil to the Free World, by Country, 1955-63	386
54.	Exports of Petroleum from the Soviet Union, 1964	389
55.	Exports of Petroleum from the Soviet Union, 1965	400
56.	Exports of Petroleum from the Soviet Union, 1966	410

Tables		Page
57.	Imports of Petroleum by the Soviet Union, 1966	421
58.	Exports of Petroleum from the Soviet Union, 1967	423
59.	Imports of Petroleum by the Soviet Union, 1967	437
60.	Apparent Consumption of Petroleum in the Soviet Union, Selected Years 1955-75	440
61.	Derivation of Apparent Consumption of Petroleum in the Soviet Union, 1964 and 1966-68	441

LIST OF FIGURES

Figure		Page
1.	Dominance of Russian Oil at Turn of Century	8
2.	The Export of Oil From Russia, Selected Years 1890-1910	11
3.	Soviet Oil Exports as a Share of Crude Oil Production, 1926-37	22
4.	Net Oil Exports From the Soviet Union, 1955-68	41
5.	Allocation of Soviet Oil Exports Between Communist and Non-Communist Countries, 1955-68	45
6.	Average Prices Paid for Soviet Crude Oil, 1955-67	60
7.	The Friendship Crude Oil Pipeline	65
8.	Proliferation of Oil Pipelines in East Germany	66
9.	Fluctuations in Earnings from Sale of Soviet Oil to Non-Communist Countries, 1961-67	72
10.	The Israeli Python and U.S. Oil Monopolies	78
11.	Fluctuations in Increments to Crude Oil Production and Oil Consumption in the Soviet Union, 1960-69	88
12.	Oil and Gas Deposits in Western Siberia	94
13.	Oil and Gas Deposits in Western Kazakhstan and Western Turkmen	95
14.	The Export of Oil from the Black Sea	106

Figure		Page
15.	Probable Allocation of Oil Supply in the Soviet Union, 1970 and 1975	118
16.	Actual and Estimated Increments in Soviet Oil Sales to Non-Communist Areas, 1955-75	124
17.	Failure of the Soviet Natural Gas Industry During the Five Year Plan 1966-70	132
18.	Pipeline Supply of Soviet Gas to Czechoslovakia, Poland, Austria, and Hungary	136
19.	Planned Pipeline for Delivery of Soviet Natural Gas to Bulgaria	143
20.	Gas Pipelines to Originate in Northern Tyumen Oblast	144
21.	The "Northern Lights" Natural Gas Pipeline System: Route of the Vuktyl-Ukhta-Torzhok Sector	145
22.	Central Asia--Center Natural Gas Pipeline System	147
23.	Proposed Export of Soviet Natural Gas to Japan: Sakhalin, Three Proposals	151
24.	Proposed Export of Soviet Natural Gas to Japan: Yakutsk	152
25.	The Iran-Soviet Union Natural Gas Pipeline	159
26.	Tie-In of Gas Pipeline from Iran with Transcaucasus and North Caucasus Pipeline Systems of the USSR	161
27.	The Afghanistan-Soviet Union Natural Gas Pipeline	162
28.	Soviet Trade in Natural Gas, 1970 and 1975	167

GLOSSARY

Communist Countries - The Soviet Bloc plus Albania, Cuba, Yugoslavia, Communist China, North Korea, North Vietnam and Mongolia.

Soviet Bloc - The Soviet Union and Eastern Europe.

Eastern Europe - Bulgaria, Czechoslovakia, East Germany, Hungary, Poland and Rumania.

The East - A synonym for Communist countries.

The West (or Free World) - A synonym for non-Communist countries.

Western Europe (or Free Europe) - Continental Europe plus Scandinavia and the United Kingdom, less the Communist countries noted above.

CEMA - The initials for the Council for Economic Mutual Assistance (Sovyet Ekonomicheskoy Vzaimopomoshchi - SEV), to which belong the Soviet Union, Bulgaria, Czechoslovakia, East Germany, Hungary, Poland, Rumania, and Mongolia, with Yugoslavia as an associate member.

CONVERSIONS

standard fuel — The term standard fuel (uslovnoye toplivo), sometimes referred to as conventional fuel, is the measurement employed by the Soviet Union to express various forms of primary energy in a common denominator. Standard fuel has a heat content of 7,000 kilocalories per kilogram. To convert from natural units to units of standard fuel, the following factors may be applied (data for 1965):

Form of Energy	Tons of Standard Fuel Per Actual Ton[a]
Coal	0.72
Crude oil	1.43
Natural gas	1.19[b]
Peat	0.37
Shale	0.35

[a] Derived from data in Narodnoye khozyaystvo SSSR v 1965 g. (Moscow, 1966).

[b] Tons of standard fuel per 1,000 cubic meters.

metric ton — All tons in this study are given in metric tons. There are approximately 7.3 barrels of crude oil in a metric ton, and about 7.5 barrels of petroleum products. As an acceptable conversion from the measure generally applied in the oil industry, that is, barrel per day, to metric ton, simply multiply by 50. To illustrate, a measure of 1,000 barrels per day (multiplied by 50) would translate into 50,000 metric tons per year. Similarly, to convert from metric ton per year to barrel per day, reverse the procedure and divide metric tons by 50 (50,000 metric tons/50 is the equivalent of 1,000 barrels per day). To convert metric tons to short tons, multiply by 1.102.

ruble	Multiply by	1.11	to convert to	$U.S.
meter	Multiply by	3.281	to convert to	feet.
kilometer	Multiply by	0.621	to convert to	miles.
cubic meter	Multiply by	35.314	to convert to	cubic feet.

PART I: HISTORY AND FORECASTS

CHAPTER 1   OIL IN CZARIST RUSSIA

In 1973 the Soviet Union will celebrate one hundred years of history as an exporter of oil. These have been rather tumultuous years, marked by movement in and out of the market and by a general acceptance of the Soviet product among prospective buyers, but to be remembered more for the controversy which these sales have generated than by the amounts sold.

History has never been kind to the relationships between the Russians and the international oil industry. The beginnings of this antagonism can be traced back more than half a century, to 1917 and the Communist revolution, and to the subsequent nationalization of the Russian oil industry and the heavy financial losses which this expropriation inflicted upon foreign groups.

Prior to the Communist revolution in 1917, Western oil interests and investments in Russia were considerable, under the majority control of the three Nobel brothers, the Rothschilds, Henri Deterding (Royal Dutch) and Marcus Samuel (Shell), the latter two companies merging into the Royal Dutch-Shell Group in 1907. Under their direction the Russian oil industry, centered around Baku and Groznyy in the Caucasus, was developed rapidly and the oil was used to meet the challenge of the Standard Oil Company in Europe and later in the Far East. Russian kerosine became a familiar sight around the world, warmly regarded for its high quality and competitive price. It has been said that Standard would have dominated Europe as thoroughly as the United States if not for the arrival of Russian oil.[1]

At the turn of the century Russia actually was outproducing the United States, drawing upon highly prolific wells, closely concentrated,[2] and benefiting from geographic convenience to the expanding European market. However, this leadership (see Table 1) was but a passing thing, for with the discovery of oil in Texas in early 1901, which quickly led to other finds in Louisiana and Oklahoma, the United States soon stood far in the lead, in a position never relinquished. Nevertheless, the surge in production (traced in Table 2) provided oil in amounts well in excess of those required domestically by an economy yet to be stimulated by industrialization and by mechanization of agriculture. In fact, production was increasing so rapidly that the Russian oil industry in 1900 and 1901 was faced with what has become a perennial world-wide anathema to the oil industry -- oversupply. And as always, overproduction was followed by a sharp decline in prices. It is possible that the collapse in prices attendant on supply exceeding demand could have been avoided had appropriate steps been taken to enlarge the means to move the oil to points of export, but such expansion was slow in coming. By the time oil producers were able to envision a return to some degree of prosperity, the industry was faced with a new crisis, that of labor disturbances. The first major strike occurred in 1903; it was settled to the satisfaction of the workmen, but continued agitation and rumor, together with the embarrassment of the military reversals suffered in the Russian-Japanese War of 1905, finally led to a general insurrection.

Despite the absence of a broad industrial base, the Russian economy nevertheless absorbed the preponderant share of indigenous oil production. The demands for kerosine for illumination and cooking, and for fuel oil for use by industry, the railroads, and inland waterway navigation represented a major claim on local supply. In addition, the transportation and freight policy of the Russian government, which exacted crippling tariffs on those volumes of oil moving from producing areas around Baku to the port of Batumi on the Black Sea, deliberately diverted the flow of oil so that it remained within Russia.[3] These factors in concert greatly limited the quantities of oil which could be freed for export. Even in the peak year of exports from Russia, 1904, only 17.1 per cent of the supply was sold abroad. The export of oil from Russia in selected years from 1890 to 1914 is compared with production in Table 3.

TABLE 1

Dominance of Russian Oil at Turn of Century

(in million metric tons)

Year	Crude Oil Production Russia[a]	World[b]	Russia as Percent of World Production	U.S. as Percent of World Production
1898	8.3	17.1	48.5	44.3
1899	9.0	18.0	50.0	43.5
1900	10.4	20.4	50.9	42.7
1901	11.7	22.9	51.1	41.4
1902	11.0	24.9	44.2	48.8
1905	7.5	29.5	25.4	62.6

[a] Data from Table 2.

[b] American Petroleum Institute, Petroleum Facts and Figures, Centennial Edition, 1959 (New York, 1959), p. 437. Converted from barrels to metric tons at the rate of 7.3 barrels per ton.

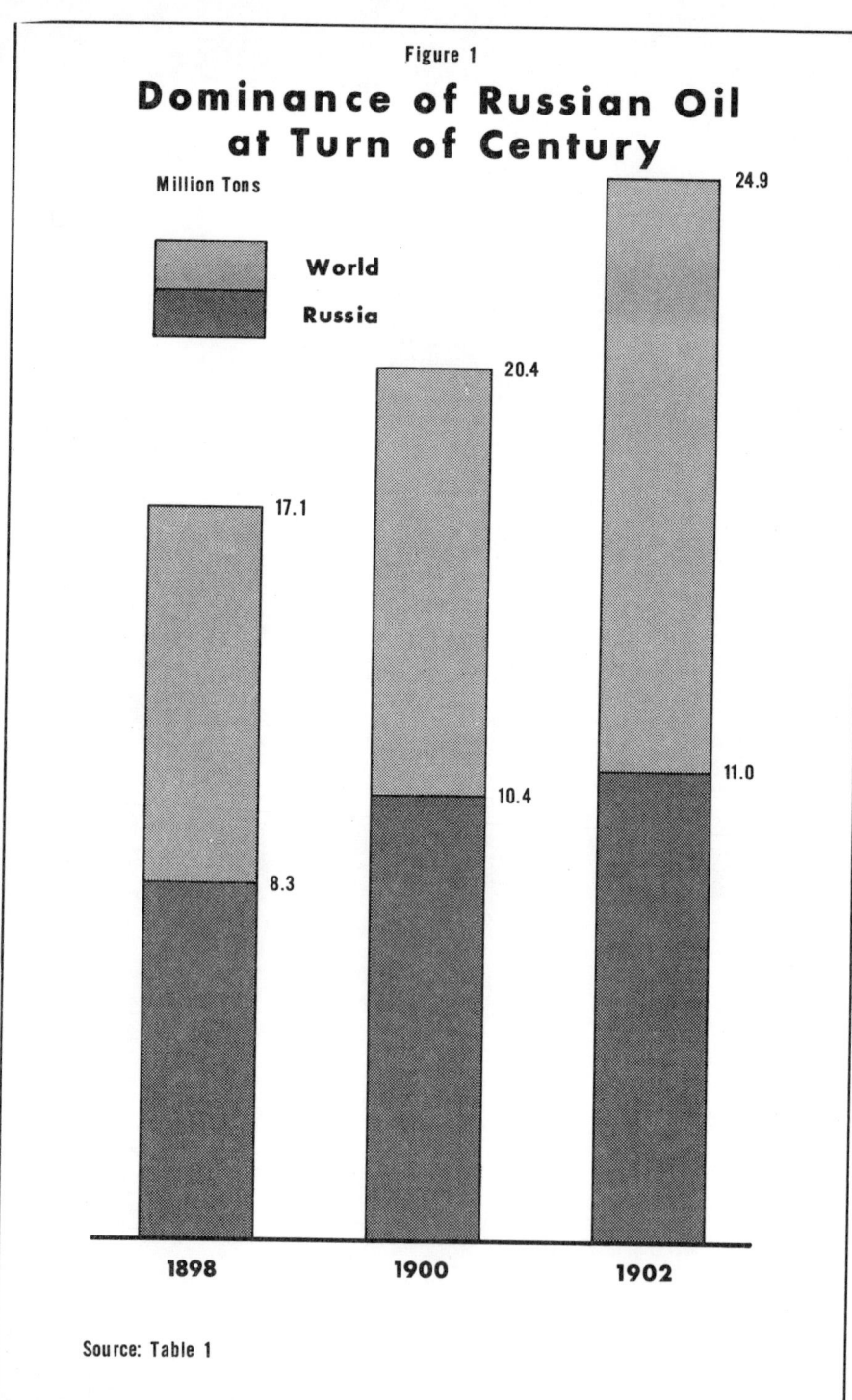

TABLE 2

Production of Crude Oil in Russia, 1889-1917

Year	Million Metric Tons	Year	Million Metric Tons
1889	3.170	1904	10.760
1890	3.750	1905	7.470
1891	4.520	1906	8.060
1892	4.720	1907	8.590
1893	5.450	1908	8.670
1894	4.960	1909	9.240
1895	6.650	1910	9.660
1896	6.690	1911	9.160
1897	7.220	1912	9.340
1898	8.270	1913	9.230
1899	9.020	1914	9.020
1900	10.350	1915	9.320
1901	11.680	1916	9.880
1902	10.990	1917	8.720
1903	10.320		

[a] For data on crude oil production prior to 1889, see Robert E. Ebel, The Petroleum Industry of the Soviet Union (Washington, D.C.: American Petroleum Institute, 1961), p. 74.

Source: Amtorg Trading Corporation, Soviet Oil Industry (New York, 1927), p. 26.

TABLE 3

Russian Oil Exports as Share of Oil Production, Selected Years 1890-1914

Year	Quantity (Thousand Metric Tons) Production[a]	Exports[b]	Exports as Percent of Production
1890	3,750	788.3	21.0
1895	6,650	1,076	16.2
1900	10,350	1,442	13.9
1903	10,320	1,783.6	17.3
1904	10,760	1,837.1	17.1
1905	7,470	945.2	12.7
1910	9,660	859.1	8.9
1914	9,020	529.1	5.9

[a] Data from Table 2.
[b] Data from Table 4.

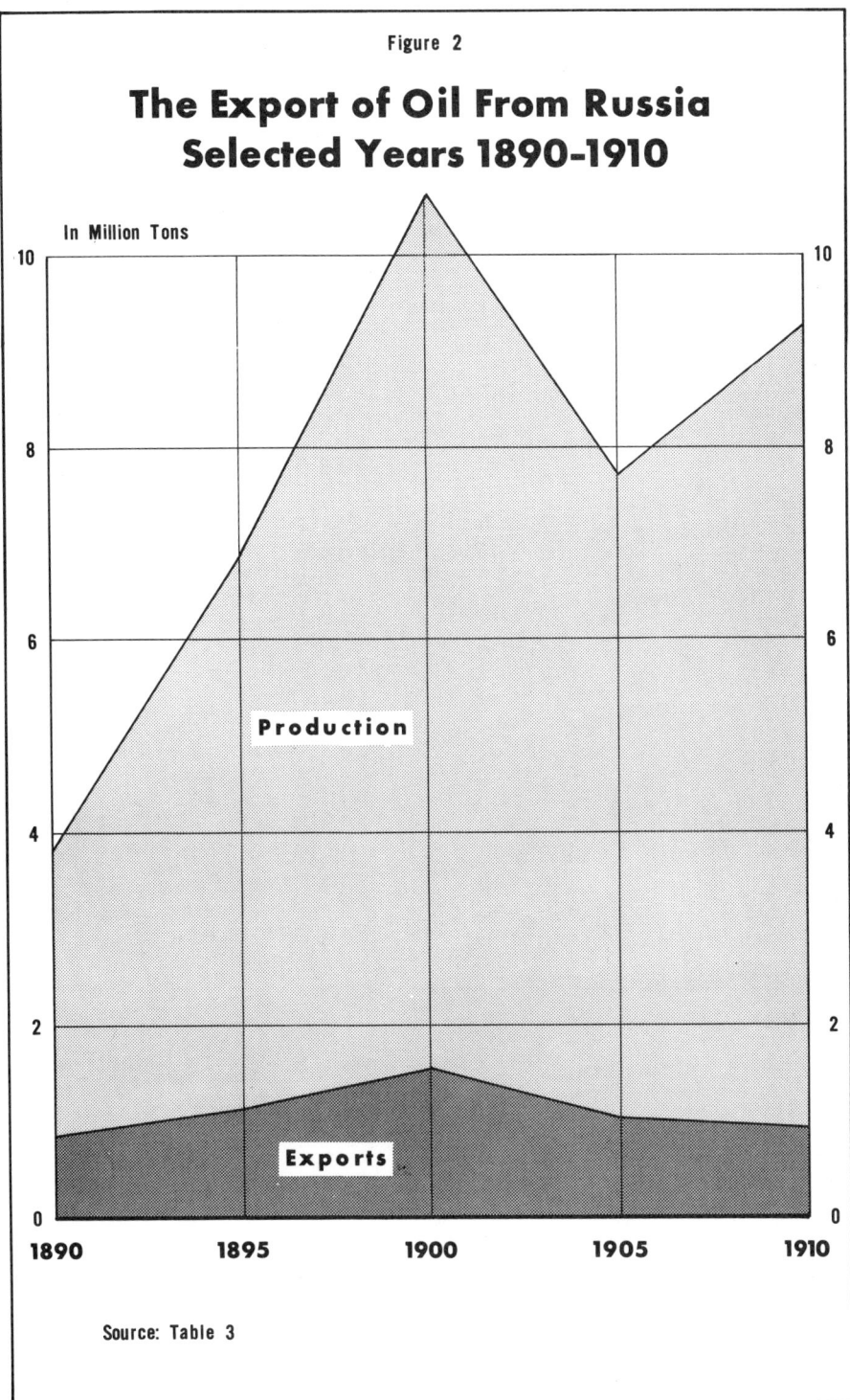

TABLE 4

Russian Oil Exports, 1873-1917

Year	Quantity (Metric Tons)	Year	Quantity (Metric Tons)
1873	500	1896	1,058,400
1874	1,400	1897	1,046,000
1875	2,100	1898	1,115,100
1876	1,500	1899	1,392,000
1877	1,000	1900	1,442,000
1878	1,300	1901	1,558,700
1879	4,500	1902	1,534,800
1880	3,400	1903	1,783,600
1881	4,100	1904	1,837,100
1882	1,800	1905	945,200
1883	59,100	1906	661,300
1884	113,300	1907	733,000
1885	177,500	1908	796,700
1886	246,000	1909	796,000
1887	311,200	1910	859,100
1888	572,800	1911	825,100
1889	734,400	1912	839,100
1890	788,300	1913	947,700
1891	889,000	1914	529,100
1892	938,800	1915	77,700
1893	986,900	1916	no data
1894	880,300	1917	no data
1895	1,076,000		

Source: Amtorg Trading Corporation, Soviet Oil Industry, (New York, 1927), p.28.

In 1905 the production of crude oil in Russia fell off very sharply, to 7.5 million tons, compared with an output of about 10.8 million tons in the preceding year, as labor disputes and armed uprisings in the Baku area led to constant shut-downs of operations and the destruction of oil producing, refining, and transport facilities.[4]

A. Beeby Thompson, in the preface to the second edition of his <u>The Oil Fields of Russia,</u> (1908) had this to say about these labor disturbances:

> A series of lamentable disturbances, which commenced in 1906, have continued almost without intermission, causing consternation amongst the producers, and the complete disorganization of the great Russian industry, which has always been such a rich source of revenue to the Russian Government. Apart from the direct pecuniary loss sustained by producers, the lives of leading producers have been in constant danger, and . . . some of [the leading employers of labor]have been murdered. Social Democratic bodies, supported by armed bands of desperadoes, have dictated terms of labor, and in the absence of Government aid producers have been compelled to make outrageous concessions to their workmen.[5]

Supplies of oil available for export were reduced with the fall in production, concomitant and exports in 1905 (945,000 tons) were roughly half of exports in 1904. Indeed, the Russian oil industry was never to regain the heights it had once held at the turn of the century, for World War I was soon upon the country, to be followed by the revolution of 1917 and the ravages of foreign intervention and civil war. Supplies of fuel were strained even prior to World War I and became chaotic in the early months following the Communist revolution. The disruptions of civil war and foreign intervention did little to improve the situation. By the summer of 1920 the shortage had reached unheard-of proportions. A leading Soviet historian, S. M. Lisichkin, has noted that in Central Asia, which had just been freed from the White Guard forces, dried fish were being used to fire the boilers of steam locomotives--coal and fuel oil simply were not available.[6]

CHAPTER 2 — EXPANSION AND CONTRACTION: 1918-40

As Louis Fischer has described in his Oil Imperialism, when victory in the World War failed to give the Russian oil prize to any of the Allied nations, they fell to quarreling for it among themselves.[1] This scramble involved Great Britain, France, Belgium, and the United States. A decree of the Soviet government on June 1, 1918 nationalized the oil industry and all of the Western oil interests in the country were expropriated.[2] Baku first fell into the hands of Turkey in September, 1918; two months later the English took command, reinstated the former property-owners, and immediately took steps to renew the export of oil from Batumi. Some eighteen months later, in May, 1920, the Bolshevik forces, victorious in the civil war, came into control of Baku and again the oil properties were nationalized.

Western property-owners were understandably reluctant to give up their claim to the Russian oil industry without a struggle.[3] In April of 1922 the major powers met at Genoa; the main point of interest was the question of restoring private property to its former owners. However, representatives on the Soviet delegation refused to consider denationalization of the property they had appropriated or to recognize any claims upon it. The conference carried over into May, in excited anticipation, for Moscow was now considering the granting of limited concessions over its oil lands, but the proposal was lost because each major power was anxious to preclude any participation in the Soviet oil industry by the other powers. The conference came to an end with the Russians rejecting the demands with respect to private property; furthermore, there would be no agreement with and no concessions from the Soviet Union.

EXPANSION AND CONTRACTION: 1918-40

Disappointed by their failure to gain recognition of the property claims, the major oil interests joined together later that year in a boycott of any dealings relative to oil with the Soviet government. Behind the boycott was the reasoning that if no one would buy Soviet oil nor seek concessions with the Soviet government, the Soviet Union would soon be forced to come to terms with the former property-owners. But boycotts are traditionally broken, and oil boycotts, particularly when evaluated in the light of cold commercial considerations, are no exception. The united front against the Soviet Union lasted roughly five months after its formalization and came to an end with the purchase of certain quantities of Soviet oil products by Royal Dutch. Not to be outdone by its major competitor, the Standard Oil Company set out not to buy oil from the Russians, but to secure concession rights.[4] These actions brought the ill-fated cordon sanitaire to an abrupt end, the decision of the Soviet government not to recognize the claims of the former property-owners remained unchallenged and unchanged, the control of the Soviet government over its oil industry was solidified, and international oil firms were excluded from any further participation in Soviet oil.

In the interim, Soviet oil products once again began to appear in the world markets, at first in quite modest quantities, for three obstacles confronted its return to international trade: (1) a seven-year absence from the foreign market during which contact was lost with former customers; (2) the lack of storage and distribution facilities abroad; and (3) the entrenchment of new and powerful distributors in the former Russian marketing areas.[5]

But sales were gradually enlarged as crude oil production was able to be increased at rates exceeding that minor growth in demand generated by an economy laid waste by a succession of wars. In addition, as pointed out in defense of expanding exports, the population of the Soviet Union in the 1920's was nearly 20 per cent smaller than the population of the Russian Empire in 1913.[6] This loss of potential demand had its own particular influence on the export surplus. Soviet oil products were competitive in quality and in price and the refinery yields were fitted to approximate the requirements of the world oil market. A second pipeline was laid between Baku and Batumi in 1928 and later the port of Tuapse, also on the Black Sea, was linked by pipeline with the producing fields of Groznyy.

All the requirements for a successful re-entry into the market were now at hand.

That this re-entry was indeed successful is underscored by the subsequent actions of the competitors. In 1925 the Standard Oil Company offered to sign an agreement with the Soviet trading organizations which would permit Standard to purchase every barrel of Soviet crude oil and petroleum product allocated for export.[7] When this proposal was rejected, Standard submitted the following: it was ready to offer credit to the Soviet oil industry in the amount of 24 million dollars and, in addition, would build a pipeline from Groznyy to Tuapse. These were also rejected.

By 1927 Soviet oil exports exceeded 2 million tons, surpassing peak annual exports by Czarist Russia. An All-Russian Oil Trade Syndicate (Vserossiyskiy nefyanoy torgovyy sindikat) was established, storage facilities were erected in several key market areas, and branch offices and agents were available in such widely separated cities as Milan, Vienna, and Harbin. In addition, the Oil Syndicate was a shareholder in joint-stock companies in England, Germany, France, Italy, and Iran. Trading with the Communists always seems to arouse the emotions of those who perceive political danger in such trade, and the resurgence of Soviet oil exports in the late 1920's and early 1930's was no exception. There was a wide divergence of opinion among the international oil companies with regard to trade contacts with the Soviet Union. The opposition was forcefully led by Sir Henri Deterding, chairman of the Royal Dutch-Shell Group,[8] who summed up his apprehensions very clearly by stating that "the plain fact is that the Soviets are trying to use this oil to bring about the bankruptcy of honest trading companies."[9] At the other end of the spectrum, the Vacuum Oil Company believed it good policy to draw supplies for various markets from the most economic source, giving preference at all times to American supplies. But in this philosophy it saw a place for purchases of Soviet oil, if such supplies were available on proper terms and of proper quality, for those markets where the Soviet Union was indisputably the natural and economic source. The Standard Oil Company of New York justified its purchases of Soviet oil with much the same reasoning, pointing to the obvious advantages of supplying its markets in India with oil obtained at Black Sea ports, a shorter route by about 5,000 miles than if American oil were to be used for the same purpose.

EXPANSION AND CONTRACTION: 1918-40

During the late 1920's and early 1930's the world oil industry was plagued by a growing imbalance between production and consumption. A good share of the blame for the resultant decline in prices, and in profits, was attributed to the current Soviet oil export drive which, in the eyes of its competition, seemed bent upon the placement of the largest volumes of oil possible at whatever prices could be obtained.[10]

The attraction of high quality petroleum products, offered at comparatively low prices,[11] combined with supplies available for loading at Black Sea ports--a considerable geographic advantage compared with American or Persian Gulf oil--had during this period served as a powerful stimulus for oil exports from the USSR. And as the Soviet Union embarked on the first of its five year plans (1928-32), the oil industry became a particularly important source of earnings of foreign exchange. By the close of this first five year plan in 1932, the export of oil had increased to more than 6 million tons, having almost tripled within five years.[12]

But whatever the cause, leading spokesmen for the industry recognized that drastic steps were in order to return the industry to sound financial footing. The most publicized plan for restoring the health of the world oil industry was that formulated by I. B. Aug. Kessler, managing director of Royal Dutch-Shell. His plan, outlined in the fall of 1931, was simple: world conservation through restricted output and restricted drilling and an allocation of limited export quotas among the United States, Venezuela, and Rumania.[13] Although a good share of the industry found considerable merit in the plan, it was unacceptable to the United States because of anti-trust laws.

In the interim, world production continued to soar, demand remained stagnant, competition heightened, and prices fell even further. At the same time exports from the Soviet Union continued to expand, precipitating a growing anxiety among the major international oil companies who in looking ahead could see only further increases in Soviet oil exports and a concomitant reduction in company profits. Fears for larger Soviet exports were somewhat justified. There had been announcement of the discovery of rich new fields in the eastern regions of the USSR, and Professor I. M. Gubkin, the chief architect of these discoveries, proposed that the new areas be rapidly developed and that the Groznyy and Baku areas, the historic bulwark of Soviet oil, be reserved for export.[14] Earlier, the chairman of Russian Oil Products, Ltd., in reviewing the growth in oil exports from his country, had stated: "We expect shortly to export 8 million tons per annum"[15] compared to the 5.2 million tons exported in 1931. Although these statements may have been made as much for anticipated impact upon Western oil company posture toward the USSR as for an objective assessment of the future, they did serve to underscore, in the estimation of these companies, the need to find some means of controlling Soviet oil sales.

In the end, after having exhausted attempts to cartelize the industry, an offer to buy the Russians out seemed to be the only remaining feasible alternative. Subsequently, representatives of the major international oil companies, with a revised Kessler plan in hand,[16] met with Soviet officials in New York during May and June of 1932, in a "world oil conference," with C. E. Arnott, chairman of the Socony-Vacuum Corporation, presiding.[17] At that time the British-American group proposed that it take over all Soviet exports for the next ten years, at the 1931 level of about 5.2 million tons. As an adjunct, the Soviet Union would be required to dispose of all of its distributing facilities held in foreign countries, to insure the effective elimination of the Soviet Union from the world oil market. The Russian counter-offer, while not rejecting this proposal out of hand, expressed a firm desire to retain the distribution facilities in question and, more important, limited fixed sales to the group to a period of only three years. Although the Soviet delegation indicated the USSR was anxious to cooperate in any manner which gave promise of improving the conditions then prevailing, these differences were too great to be reconciled and the conference dissolved without agreement of any sort having been reached.

There is much similarity to be found between the Soviet oil export campaign of the 1920's and early 1930's and that which began in the late 1950's and is still under way. This similarity extends to the philosophy behind the oil exports, the tactics employed to find markets for the oil, and of course the competition's reaction, but it ends at a very important point. World concern over Soviet oil reached a peak in 1932, at a time when these exports were also at a peak and, as noted above, comments by responsible Soviet officials pointed to likely increases in the coming years.

TABLE 5

Exports of Oil from the Soviet Union, 1918-40

(in thousand metric tons)

Year	Quantity
1918	negligible
1919	no data
1920	negligible
1921	31.3
1922	169.7
1923	382.9
1924	815.1
1925	1,505.4
1926	1,685.3
1927	2,097.1
1928	2,787
1929	3,625
1930	4,712
1931	5,224
1932	6,011
1933	4,894
1934	4,315
1935	3,368
1936	2,665
1937	1,930
1938	1,400
1939	500
1940	900

Sources: Discrepancies are considerable among the available sources with respect to the export of oil from the Soviet Union during 1918-40, particularly for 1921-29. In these latter years, official statistics were reported by fiscal rather than calendar year. The fiscal year for the Soviet Union then ran from October 1 to September 30. Data have been selected from what are believed to be the most authoritative references:

For 1918, 1920, 1938-40, Vneshnyaya torglovlya SSSR, statistecheskiz sbornik 1918-1966 (Moscow, 1967), pp. 18-19.

For 1921-27, S.M. Lisichkin, Ocherki razvitiya neftedobyvayushchey promyshlennosti SSSR (Moscow, 1958), pp. 213, 217, 399.

For 1928-29, 1937, World Petroleum, April, 1939, p. 62.

For 1930-36, U.S. Department of State, Despatch No. 1200, March 29, 1938, from the Legation, Riga, Latvia, subject: "The Petroleum Industry of the Soviet Union." See pp. 48-51 for a discussion of the export of crude oil, and Annex 19, which provides data on the export of crude oil and petroleum products for 1913 and 1930-36.

TABLE 6

Production of Crude Oil in the Soviet Union, 1918-40
(in thousand metric tons)

Year	Quantity
1918	4,146
1919	4,448
1920	3,851
1921	3,781
1922	4,658
1923	5,277[a]
1924	6,064
1925	7,061
1926	8,318
1927	10,285
1928	11,625
1929	13,684
1930	18,451
1931	22,392
1932	21,414
1933	21,489
1934	24,218
1935	25,218
1936	27,427
1937	28,501
1938	30,186
1939	30,259
1940	31,121

[a] The recovery in the production of crude oil, once the Communists were able to establish firm and complete control over the country, is not a particular tribute to the restorative qualities of this form of government. Rather, as a degree of order was returned to the oil-producing areas, idle wells were gradually brought back into production. In the summer of 1921, only 700 wells were in operation in Baku; 3,000 stood idle. Similarly, in Groznyy, out of 388 wells only 65 were producing. The return of these idle wells to active production provided the major basis for the growth in crude oil output during the middle 1920's.

Source: Narodnoye khozyaystvo SSSR v 1958 godu: Statisticheskiy yezhegodnik (Moscow, 1959), p. 185.

TABLE 7

Comparison of Soviet Oil Exports[a]
with Oil Production, 1926-37

Year	Quantity (Thousand Metric Tons) Production[b]	Exports[c]	Exports as Percent of Production
1924	6,064	815	13.4
1925	7,061	1,505	21.3
1926	8,318	1,685	20.3
1927	10,285	2,097	20.4
1928	11,625	2,787	24.0
1929	13,684	3,625	26.4
1930	18,451	4,712	25.5
1931	22,392	5,224	23.3
1932	21,414	6,011	28.1
1933	21,489	4,894	22.8
1934	24,218	4,315	17.8
1935	25,218	3,368	13.4
1936	27,427	2,665	9.7
1937	28,501	1,930	6.8

[a] Imports were negligible during these years.
[b] Data from Table 6.
[c] Data from Table 5.

Figure 3

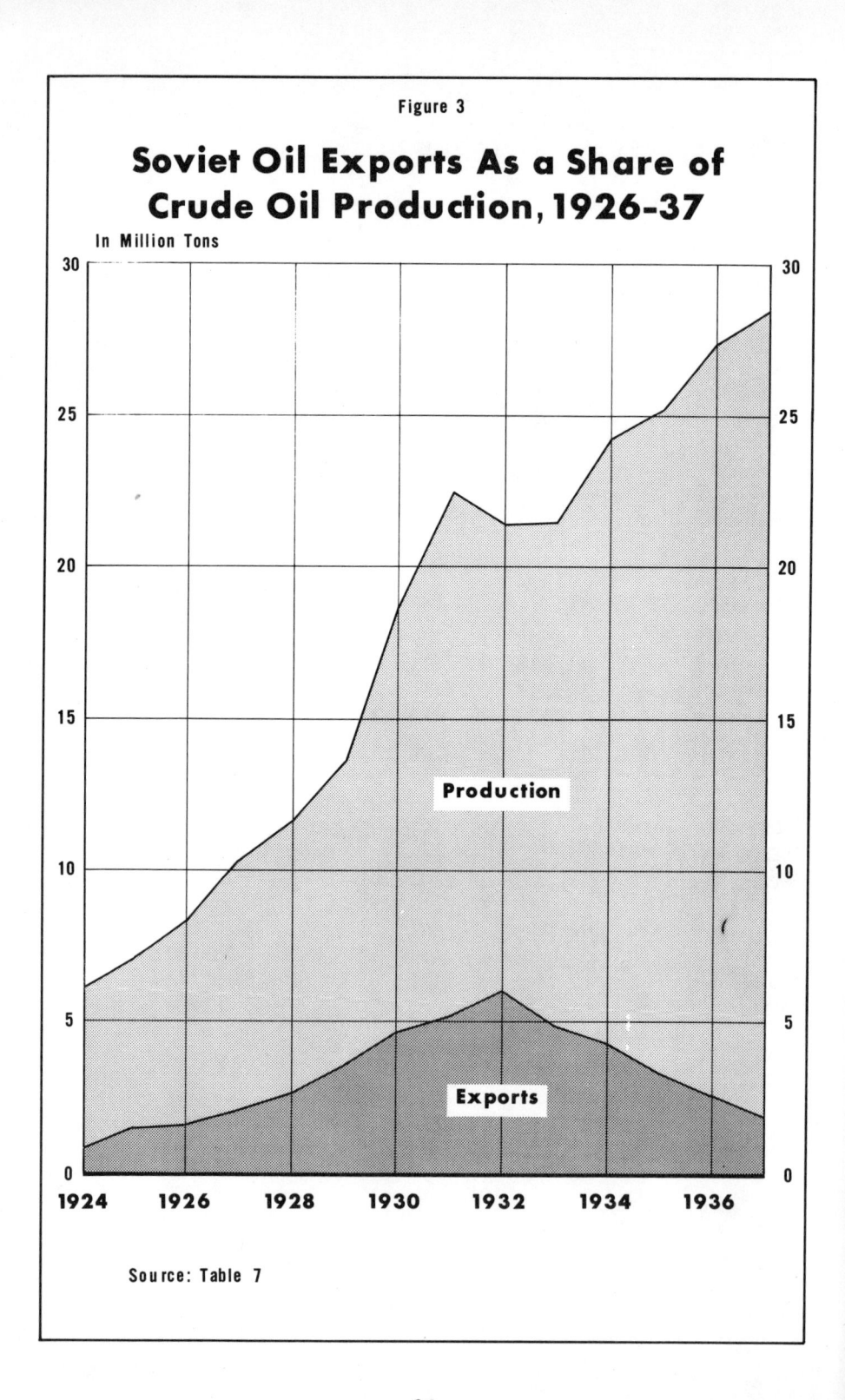

Soviet Oil Exports As a Share of Crude Oil Production, 1926-37

In Million Tons

Production

Exports

Source: Table 7

TABLE 8

Oil Stockpiling in the Soviet Union, 1938 and 1940
(in million metric tons)

Indicator	1938	1940
Crude oil		
Production[a]	30.186	31.121
Imports[b]	0	0
Supply of crude oil	30.186	31.121
Exports[b]	0.200	0
Losses in field, etc.	1.586[c]	1.649[d]
Charge to refining	28.400[e]	29.472[f]
Petroleum products		
Yield from refining[g]	26.128	27.114
Exports[b]	1.200	0.900
Imports[b]	0.142	0.068
Supply of petroleum products	25.070	26.282
Consumption	23.000[h]	24.656[i]
Surplus available for allocation to stocks[j]	2.070	1.626
Equivalent days supply	33	24

[a] Robert E. Ebel, The Petroleum Industry of the Soviet Union (Washington, D.C.: American Petroleum Institute, 1961), p. 70.
[b] Vneshnyaya torgovlya SSSR, statisticheskiy sbornik, 1918-66 (Moscow, 1967), pp. 18-10, 26-27.
[c] Derived as the difference between reported availability and charge to refining.
[d] This category in 1938 represented 5.3 per cent of total crude oil supply. It has been held constant for 1940.
[e] Heinrich Hassman, Oil in the Soviet Union (Princeton, 1953), p. 53.
[f] Calculated as total supply less losses, etc.
[g] Calculated as 92 per cent of the charge to refining.
[h] Hassman, op. cit, p. 56.
[i] Neftyanaya promyshlennost SSSR (Moscow, 1958), p. 275.
[j] Total supply of petroleum products less consumption.

In 1933 oil sales by the USSR fell off to about 4.9 million tons, a decline of 19 per cent in one year, and by 1935 exports had been reduced to less than 3.4 million tons. This decline was brought about by the failure of crude oil production to meet plan targets and by substantial increases in domestic demand as the equipment paid for by earlier exports began to generate petroleum requirements of its own. The oil industry could do little more than match this growth in requirements. By 1938 exports had declined to less than 1 million tons, the lowest level in about fifteen years.

With the withdrawal of Soviet oil from the market, the relations between the Soviet Union and the international oil companies became less strained. The open animosity and distrust disappeared, or at least was well hidden, to be renewed about twenty years later, as the Soviet Union once again turned to oil exports to finance the technical re-equipping of its economy.

Granted that consumption of oil increased tremendously in the Soviet Union during the late 1930's, owing to the increase in population and the gigantic investment in the production of capital goods. But another factor was perhaps equally responsible for the disappearance of Soviet oil from the market place. It became quite clear that the Soviet government during these years was stockpiling oil for war. The bulk of the crude oil produced in the country continued to originate in the Caucasus, in the areas of Baku and Groznyy, as did the capacity to refine. These areas lay extremely vulnerable to an enemy sweep from the south or to a pincer movement from the west. New oil had been found in the interior of the country, between the Ural Mountains and the Volga River, but development of these fields was only in the initial stages. Caucasus oil would still have to bear the brunt of fueling whatever military actions might be called for.

In 1938, allocations of more than 2 million tons of products to stocks exceeded exports by 600 thousand tons, and represented the equivalent of about 33 days' supply. (For derivation, see Table 8.) Exports continued to decline, to only 500,000 tons in 1939, but improved slightly in 1940, increasing to 900,000 tons. Nevertheless, in 1940 allocations to stocks still were far in excess of exports, if representing only 24 days' supply that year.

CHAPTER 3 WORLD WAR II

In the years immediately preceding the outbreak of World War II, the rate of growth in the production of crude oil in the USSR had declined sharply in comparison with the rates achieved in the early 1930's. Production of crude oil in 1937 had reached 28.5 million tons but by 1940 had increased to only 31.1 million tons. The bulk of the national output of crude oil originated with the Baku fields of Azerbaydzhan; in 1937 these fields yielded 21.4 million tons, or more than 75 per cent of total USSR production. Difficulties in expanding crude oil producing capacity at Baku therefore had an immediate effect on national output. In the succeeding three years, output in Azerbaydzhan increased by less than 1 million tons and, in addition, the Groznyy oil fields in the North Caucasus, which had been a major contributor for a number of years, entered into a decline, and during 1937-40, output fell by 19 per cent, to 2.2 million tons.

TABLE 9

Rate of Growth in Soviet Oil Output, 1928-42

Period	Average Annual Rate of Growth in Crude Oil Output (Percent)
First Five Year Plan, 1928-32	15.8
Second Five Year Plan, 1933-37	5.9
Third Five Year Plan, 1938-42 (interrupted by World War II)	2.9

Source: Tables 6 and 11.

At both Baku and Groznyy, drillers were unable to prove-up reserves adequate to support continued high growth in output. Drillers were forced to attempt to increase output through the development of low-yield strata, and the new crude oil was barely sufficient to offset declining yields from old wells.

At the same time, the search for oil in the eastern regions of the country, particularly in the Urals-Volga, offered little reason for enthusiasm. Drilling crews in the Urals-Volga were poorly equipped and available techniques and technology were inadequate for any rapid progress in drilling the hard-rock deep-lying strata being encountered. Moreover, the search for oil then was concentrated in the wrong areas (the Arctic, for example) and in the wrong formations (Carboniferous instead of Devonian). The first commercial show of oil in the Urals-Volga, in 1929, was an accidental discovery. Expanded exploratory efforts in the next several years failed to produce any substantial new finds and most geologists favored a re-emphasis on prospecting and exploration in the Caucasus. Only one--I. M. Gubkin--held out for further search in the Urals-Volga.[1]

A turning point in attitude came in May, 1932, when Permian oil was found at Ishimbay. Stimulated by the discovery at Ishimbay, particular emphasis was given during the second Five Year Plan (1933-37) to the creation of a new oil-producing region in the Urals-Volga. Drillers continued to probe the Permian and Carboniferous but the results were less than satisfactory and in the first year of the third Five Year Plan, 1938, Urals-Volga oil accounted for only 4.3 per cent of the national total.

When the third Five Year Plan (1938-42) was drawn up, planners were aware of the very urgent need to expand output in the Urals-Volga, in the face of stabilized production at Baku and Groznyy. If the country were to be provided with additional oil, this oil would have to come from the eastern regions. Consequently, the plan called for output in the Urals-Volga to increase to 7 million tons by 1942.

Geologist Gubkin believed that the best prospects for further growth in output were offered by exploration of the Devonian. Unfortunately, the inadequacy of equipment and the nature of the overburden, coupled with a "no-show" during the first test of the Devonian at Tuymazy in 1940, argued

TABLE 10

Regional Concentration of Oil Production in 1940

Area	Output (Million Metric Tons)	Percent of Total
Caucasus (Baku and Groznyy)	27.05	87.1
Urals-Volga	1.85	6.0
Kazakhstan and Central Asia	1.50	4.8
Others (Ukraine, North, Far East)	0.70	2.1
Total	31.10	100.0

Source: M.M. Brenner, Ekonomika neftyanoy promyshlennosti SSSR (Moscow: 1962), p. 51.

against the expenditure in any significant quantity of investment funds for this purpose. In the interim, production at Ishimbay began to decline, but minor output from Tuymazy and other new fields served to raise the total Urals-Volga output in 1940 to 6 per cent of the national supply. Thus, at the close of 1940, production of crude oil in the Soviet Union was still concentrated in the Caucasus (Baku and Groznyy), and this concentration was an extreme liability in time of war.

Reflecting the combination of enemy occupation and the transfer to the east of prospecting and drilling crews, coupled with the relative stagnation inherited from the late 1930's, output from the Baku fields fell sharply during the war years, to 11.5 million tons by 1945, or roughly 50 per cent of the prewar high. Despite accelerated activity in the Urals-Volga, output from this region had increased to only 2.8 million tons by the close of the war. The most dramatic find of oil in the Urals-Volga during this period--the tapping of the Devonian at Tuymazy in the fall of 1964--came too late to be regarded as a significant contribution to the war effort. Nevertheless, production of crude oil at Tuymazy increased four-fold in 1945 and represented about 20 per cent of total oil produced in the Urals-Volga in that year.

The effect of World War II on the crude oil producing industry is clearly illustrated in Table 11.

TABLE 11

Soviet Crude Oil Production During World War II
(in million metric tons)

Year	Output	Index
1941	33.0	100
1942	22.0	66.7
1943	18.0	54.5
1944	18.3	55.5
1945	19.4	58.8

Source: Robert E. Ebel, The Petroleum Industry of the Soviet Union (Washington, D.C.: 1961), p. 74.

Soviet publications offer little insight into foreign trade during World War II. The most recent, a statistical handbook covering the years 1918-66,[2] devotes just one table to the period 1941-45, and expresses imports and exports in ruble values only. The contribution of the United States to the Soviet war effort, through Lend-Lease, is generally ignored or treated very lightly. From United States primary source materials, however, the extent of aid in petroleum products can be ascertained. Exports of petroleum products from the United States to the Soviet Union during 1941-45 are shown in Table 12.

In sum, the United States provided the petroleum industry of the Soviet Union during World War II with, among other things, shipments of fuels and lubricants totaling almost 2.1 million tons. These shipments continued during 1946 and 1947 for an additional 89.5 thousand tons, all motor fuel.[3] Soviet trade statistics for these latter two years omit, purposely or otherwise, mention of these imports from the United States. It had been the policy of the United States to give full and complete

TABLE 12

Soviet Imports of Oil
from the United States, 1941-45

Year	Quantity (Metric Tons)
1941	301,342
1942	149,037
1943	362,067
1944	609,300
1945	538,608

Source: Jack B. Pfeiffer, Petroleum Industry of the U.S.S.R., a Preliminary Study (Washington, D.C.: Council for Economic and Industry Research, Inc., 1954) Table 14, p. 48.

support to the Soviet war effort. The USSR lacked inter alia facilities to produce aviation gasoline in amounts sufficient to meet their air requirements. A representative of the United States was sent to Moscow in 1941 for the purpose of developing data that would have assisted in expediting the refinery construction program, but he was not afforded an opportunity to do more than make a superficial survey.[4] And despite repeated requests during the ensuing years, adequate information was never available on Russian requirements, indigenous production and productive capacity, and refinery facilities and output. It can be emphasized that this stiuation still prevails.

CHAPTER 4   RECONSTRUCTION: 1946-55

Published statistics show that the Soviet Union was able to resume the export of oil in 1946, but only on a very limited scale. Exports that year totaled a reported 0.5 million tons, all petroleum products, which were more than offset by the import of 0.9 million tons of petroleum products and a minor quantity of crude oil.

The position of the Soviet Union in international oil markets in the early postwar period was tempered somewhat by the need to import considerable quantities of petroleum products from Eastern Europe--products of a quality higher than that obtainable from domestic refineries. Although these domestic refineries were capable of meeting indigenous demand volumetrically, the absence of sophisticated secondary refining processes meant that the Soviet Union had to look to other sources for certain grades of kerosene and diesel fuel, which were provided largely by Rumania and East Germany. Moreover, production of crude oil, at least from 1946 through the early 1950's (see Table 13) was adequate only to meet requirements of the oil refining industry. In fact, from 1946 through 1950, or during the whole of the fourth Five Year Plan, the Soviet Union was a net importer of crude oil.

The weak position in crude oil, combined with the continuing need to import high-quality products, kept the Soviet Union as a deficit trader in oil through 1953 (see Table 14), although this dependency extended only to other Communist countries. In the following year, the export of crude oil and petroleum products increased by 2.3 million tons, a growth of more than 50 per cent, whereas imports declined by about 15 per cent. The Soviet Union at last had regained some degree of respectability in oil trade; it now exported more oil than it imported.

TABLE 13

Production of Crude Oil
in the Soviet Union, 1946-55

Year	Production (Thousand Metric Tons)
1946	21,746
1947	26,022
1948	29,249
1949	33,444
1950	37,878
1951	42,253
1952	47,311
1953	52,777
1954	59,281
1955	70,793

Source: Narodnoye khozyaystvo SSSR v 1958 godu: Statisticheskiy yezhegodnik (Moscow, 1959), p. 185.

TABLE 14

Trade in Oil by the Soviet Union, 1946-55

(in thousand metric tons)

Quantity	1946	1947	1948	1949	1950	1951	1952	1953	1954	1955
Exports										
Crude Oil	0	0	0	100	300	900	1,300	1,500	2,100	2,900
Petroleum Products	500	800	700	800	800	1,600	1,800	2,700	4,400	5,100
Total	500	800	700	900	1,100	2,500	3,100	4,200	6,500	8,000
Imports										
Crude Oil	9.1	74.9	74.0	131.9	336.6	59.9	197.6	104.6	193.0	574.8
Petroleum Products	900	500	800	1,700	2,300	2,600	3,600	4,600	3,800	3,800
Total	909.1	574.9	874.0	1,831.9	2,636.6	2,659.0	3,797.6	4,704.6	3,993.0	4,374.8
Net Trade[a]										
Crude Oil	- 9.1	- 74.9	- 74.0	- 31.9	- 36.6	+ 841.0	+1,102.4	+1,395.4	+1,907.0	+2,325.2
Petroleum Products	-400	+300	-100	-900	-1,500	-1,000	-1,800	-1,900	+ 600	+1,300
Total	-409.1	+225.1	-174.0	-931.9	-1,536.6	- 159	- 697.6	- 504.6	+2,507.0	+3,625.2

[a] Plus (+) sign denotes net exports; minus (-) sign denotes net imports.

Source: Vneshnyaya torgovlya SSR, statisticheskiy sbornik, 1918-1966 (Moscow, 1967), pp. 76-77, 98-99.

Soviet trade in oil with non-Communist countries had also resumed in 1946, again in very minor amounts. But while most observers point to the year 1955 as the take-off point for the postwar resumption of oil sales to the West by the Soviet Union, official trade statistics show such resumption to have occurred somewhat earlier. Indeed, the turning point appears to have been made in 1951, the first year of the fifth Five Year Plan (1951-55). In that year oil sales to the West showed a net of 344.8 thousand tons, compared to the deficit of 99 thousand tons incurred in the previous year, reversing the trend of continually declining net exports which had begun in 1947. Thereafter, net trade in oil with non-Communist countries began to increase rapidly, buoyed by concomitant high rates of growth in production of crude oil.

The overwhelming portion of Soviet oil exports to non-Communist countries in the postwar period up through 1954 was directed to Italy, Sweden, and Finland, the latter country from 1949 through 1953 accounting for at least one-half of all such sales. Soviet oil trade with non-Communist countries for the years 1946-54 is depicted on a country-by-country basis in Table 16.

By 1955 the position of the USSR as a net exporter of oil had been fully established, as within a relatively short three years it had moved from net imports of about 0.5 million tons to net exports of more than 3.6 million tons. Although the economy was progressing rapidly in its postwar reconstruction, much remained to be done in providing heavy industry with the most modern equipment and technology. Soviet engineers, metallurgists, and scientists simply lacked the facilities and the know-how to perform on a level comparable with their Western colleagues, and if Soviet industry were to be equipped with the best, the only alternative was to find some means to finance large-scale imports. Most certainly this desired equipment and technology could have been developed over time, but this would mean further delays. Perhaps more important, however, Western science and engineering could not be expected to stand still in the interim and as a result the Soviet Union would lose ground in its effort to catch up with the West.

The majority of Soviet manufactured items could not compete in foreign markets with those of Western origin. Shoddy in appearance, poor in quality, overpriced, and of questionable

TABLE 15

Soviet Oil Trade with
Non-Communist Countries, 1946-55

(in thousand metric tons)

Year	Exports	Imports[a]	Net Trade
1946	58.6	0	58.6
1947	356.7	0	356.7
1948	200.5	0	200.5
1949	251.3	55.5	195.8
1950	164.4	263.4	- 99.0[b]
1951	344.8	0	344.8
1952	783.9	121.6	667.7
1953	1,304.7	23.9	1,280.8
1954	2,550.2	0	2,550.2
1955	4,039.2	467.1	3,572.1

[a] All crude oil from Austria.
[b] Minus (-) sign designates net imports.

Source: Vneshnyaya torgovlya SSSR, statisticheskiy sbornik, 1918-1966 (Moscow, 1967), pp. 162-229.

TABLE 16

Soviet Oil Trade with Non-Communist Countries,
by Country, 1946-54

(in thousand metric tons)

Quantity	1946	1947	1948	1949	1950	1951	1952	1953	1954
Exports									
Great Britain	0	57.1	55.1	9.6	5.5	7.9	0.7	1.4	11.2
Italy	0	55.2	6.4	6.8	5.7	110.4	382.4	138.2	183.2
W. Germany	0	0	0	0	0	0	0	59.5	12.1
Finland	18.4	49.5	45.1	197.1	147.3	221.6	380.7	716.6	897.7
France	0	0	0.9	0	0	0	0.6	28.4	123.4
Sweden	0.8	4.9	1.7	21.8	5.3	2.7	15.0	269.0	743.9
Yugoslavia	39.1	175.3	78.6	15.2	0	0	0	0	48.2
India	0	0	0	0	0	0	0	4.6	207.9
UAR	0	0	0.1	0	0	0	0	0	74.8
Greece	0.3	0	0.5	0	0.1	1.5	0.9	0.5	0.2
Denmark	0	14.7	12.1	0.8	0.5	0.1	1.1	20.3	231.9
Ireland	0	0	0	0	0	0	0	65.5	15.7
Turkey	0	0	0	0	0	0.6	2.5	0.7	
Total	58.6	356.7	200.5	251.3	164.4	344.8	783.9	1,304.7	2,550.2
Imports									
Austria	0	0	0	55.5	263.4	0	121.6	23.9	0
Total	0	0	0	55.5	263.4	0	121.6	23.9	0
Net Trade[a]	58.6	356.7	200.5	195.8	-99.0	344.8	667.7	1,280.8	2,550.2

[a] Minus (-) sign denotes net imports.

Source: Vneshnyaya torgovlya SSSR, statisticheskiy sbornik, 1918-1966 (Moscow, 1967), pp. 162-229.

35

durability, they could scarcely hope to attract the eye of the potential buyer. Agricultural commodities were needed at home; reserves of hard currency were far from adequate to support any prolonged buying campaign. Yet the Soviet Union unquestionably was rich in natural resources--raw materials easily comparable in quality to those being offered by other nations. With only minimum additional capital investment, production of selected commodities could be expanded to sufficient levels both to cover domestic requirements and to provide increasing quantities for use in international trade. And Europe, in the midst of unparalleled economic expansion, with an accompanying seemingly insatiable thirst for oil, represented a natural market. Furthermore, it possessed the technical know-how and proven equipment the USSR was searching for.

In order to develop exportable surpluses of oil, it was necessary to re-orient the primary energy balance of the country. In the postwar period, coal had had a predominant and rising share in the primary energy balance, while oil had a small and declining share. This "mineralization" of the fuel balance was the principal feature in the expansion of the fuels industry. Emphasis had been given to the development of local, low-quality fuels and to the development of regional self-sufficiency in terms of fuel. But in May, 1955 an article in the Party newspaper _Pravda_, signed by an obscure engineer, provided the justification and framework for a shift in the fuels and energy balance in favor of crude oil and natural gas.[1] More important, the Soviet government underscored its intent to bring such a shift about by a concomitant redirection of capital investment funds.

The next step was to put these funds to work, not necessarily in finding oil, for extremely rich deposits had already been discovered in the Urals-Volga in the late 1940's and early 1950's, but to further develop the production base, to expand refining capacity, and to enlarge the oil pipeline network. By 1955, even without this emphasis, the Urals-Volga had already become established as the key producing area in the country and in that year provided almost 59 per cent of the total national output of crude oil.[2]

There was no question of potential, but only of orderly development of this potential, to assure maximum recovery at lowest possible costs.

RECONSTRUCTION: 1946-55   37

The combination of governmental support, capital, and
waiting reserves enabled the crude oil extracting industry to
achieve unprecedented annual increments. In 1954 the production of crude oil in the USSR was only slightly more than
59 million tons; within four years production had been virtually
doubled, to 113.2 million tons. The means to finance the
much desired imports had been found.

But having the means--oil--to finance imports was not
sufficient in itself. A market had to be found for this oil, at
a time when Western trade with Communist nations was barely
perceptible, and when the tensions of the Cold War were still
much in evidence in East-West relations.

The approach of the Soviet Union in searching for markets
for its exportable oil was quite simple. Any newcomer hoping
to break into an established market not only has to be able to
offer goods of at least equal quality, but prices for these goods
have to be so adjusted as to present sufficient justification for
the buyer to shift from his established source of supply. So
it was with Soviet crude oil and petroleum products in the late
1950's and early 1960's. Prices were pegged at levels well
below the previously accepted standard in the international
market. Moreover, the Soviet Union not only offered oil at
bargain rates, but offered to barter this oil in exchange for
manufactured goods, agricultural commodities, and various
raw materials, or was willing to accept payment in local "soft"
currencies. This made Soviet oil most attractive to the developed and developing nations alike. Such maneuverings in
the marketplace were well beyond the possible range of operations of the international oil companies who, in order to stay
in business, had to demonstrate to their stockholders an ability
to produce a reasonable profit, expressed in convertible currencies.

This approach to marketing paid off handsomely and, so
far as is known, the Soviet Union was able to find buyers for
all of the oil it had for sale in any given period. A place for
Soviet oil was found in Italy in exchange for large-diameter
steel pipe and associated equipment, in Brazil in return for
coffee, in Cuba in return for sugar, and in Egypt in return
for cotton, to cite some of the more well-known instances of
barter arrangements concluded.

The appearance in the marketplace of oil originating in a country where every decision taken has overriding political motivations would in itself be sufficient reason for an expression of apprehension on the part of the international oil companies. Moreover, these companies found themselves in an almost untenable position in certain of their historic markets, in competition with a seller whose modus operandi placed it outside their capacity of effective normal commercial response.

Finally, the emergence of Communist oil as a force to be reckoned with came at a time when the international oil industry already was attempting to readjust to excess supplies[3] and was faced as well with the prospect that production would continue to exceed requirements for some time to come. Thus, with markets already saturated in Western Europe, a place for Communist oil had to come at the expense of Western, largely Middle East, oil and there was considerable uncertainty as to just how much Communist oil might become available for disposal in Western Europe in the coming years. And although the relative share which the Soviet bloc carved out for itself was but a small fraction of the total, the disruptive influence which such sales generated probably was well beyond the fondest hopes of Soviet officials; those sales had at the same time fulfilled their primary purpose of securing advanced equipment and technology.

To sum up: Outside the normal concern for another seller in a market already distinguished by oversupply, apprehension on the part of many of the international oil companies centered around (1) the fear that certain importing countries might find themselves dangerously dependent on a source of supply which might be cut off for political and/or economic reasons; (2) the very high percentage rates of growth attributable to these oil sales; (3) the initial low price offerings of the Communist oil; and (4) the professed inability to compete with a seller able to manipulate prices without regard for cost considerations.

CHAPTER 5 THE POSTWAR DEVELOPMENT OF OIL EXPORT FROM THE SOVIET BLOC

The oil industry of the Soviet Union has exhibited sustained high rates of growth since 1955 and now ranks, in terms of the amount of crude oil produced annually, second only to the United States. As output of crude oil has increased from 70.8 million tons in 1955 to 309 million tons in 1968--at an average annual growth of 12 per cent -- the Soviet Union has been able to meet growing domestic requirements for oil incurred by the concomitant expansion of all branches of industry, transport, agriculture, and the household use of various fuels. It has in addition been able to support a modern military force, and further to set aside increasingly larger amounts of oil for export to other Communist countries and to non-Communist countries.

Table 17, shows the steady inroads trade in oil has been making into indigenous supply. The growth in net exports from only 3.7 million tons in 1955 to 85.8 million tons in 1968, representing in that year 27.8 per cent of production, has meant that virtually one out of three incremental tons of crude oil production during these years has been directed to consumers outside the country.

The percentage allocation of exports of crude oil and petroleum products from the USSR among Communist and non-Communist buyers has held remarkably stable since the late 1950's, when exports to the West began to be expanded very rapidly. In the succeeding years, between 44 and 45 per cent of the total export of oil has been directed without fail to other Communist countries, the remaining 55 to 56 per cent to the Free World. That such stability in allocation of exports has been maintained for such an extended period strongly suggests the imposition of deliberate policy by Soviet

TABLE 17

Soviet Net Oil Exports as Percentage
of Crude Oil Production, 1955-68

Year	Production[b]	Trade[a] (Million Metric Tons) Exports	Imports	Net Exports	Net Exports as Percent of Production
1955	70.8	8.4	4.7	3.7	5.2
1956	83.8	10.6	5.6	5.0	6.0
1957	98.3	16.3	4.5	11.8	12.0
1958	113.2	18.9	4.6	14.3	12.6
1959	129.6	26.5	4.7	21.8	16.8
1960	147.8	34.5	4.7	29.8	20.2
1961	166.1	42.7	3.8	38.9	23.4
1962	186.2	47.1	3.0	44.0	23.6
1963	206.1	53.1	3.0	50.1	24.3
1964	223.6	58.3	2.3	56.0	25.0
1965	242.9	66.2	2.1	64.1	26.4
1966	265	75.6	1.8	73.8	27.8
1967	288	81.0	1.5	79.5	27.6
1968	309[c]	88.5[d]	2.7[e]	85.8	27.8

[a] Except for 1968, data have been derived from the various annual statistical trade handbooks issued by the Ministry of Foreign Trade of the USSR. Trade in petroleum products has been expressed in crude oil equivalents (1 ton of crude = 0.92 ton of petroleum products).

[b] Data for 1955-60 from Narodnoye khozyaystvo SSSR za 1960 (Moscow, 1961), p. 262; data for 1961-65 from Narodnoye khozyaystvo SSSR v 1965 g. (Moscow, 1966), p. 175; for 1966, from Pravda, January 29, 1967, p. 1; for 1967, from Izvestiya, January 25, 1968, p. 1.

[c] Izvestiya, January 26, 1969, p. 1.

[d] Vneshnyaya torgovlya, No. 7 (July, 1969), p. 53.

[e] Estimate.

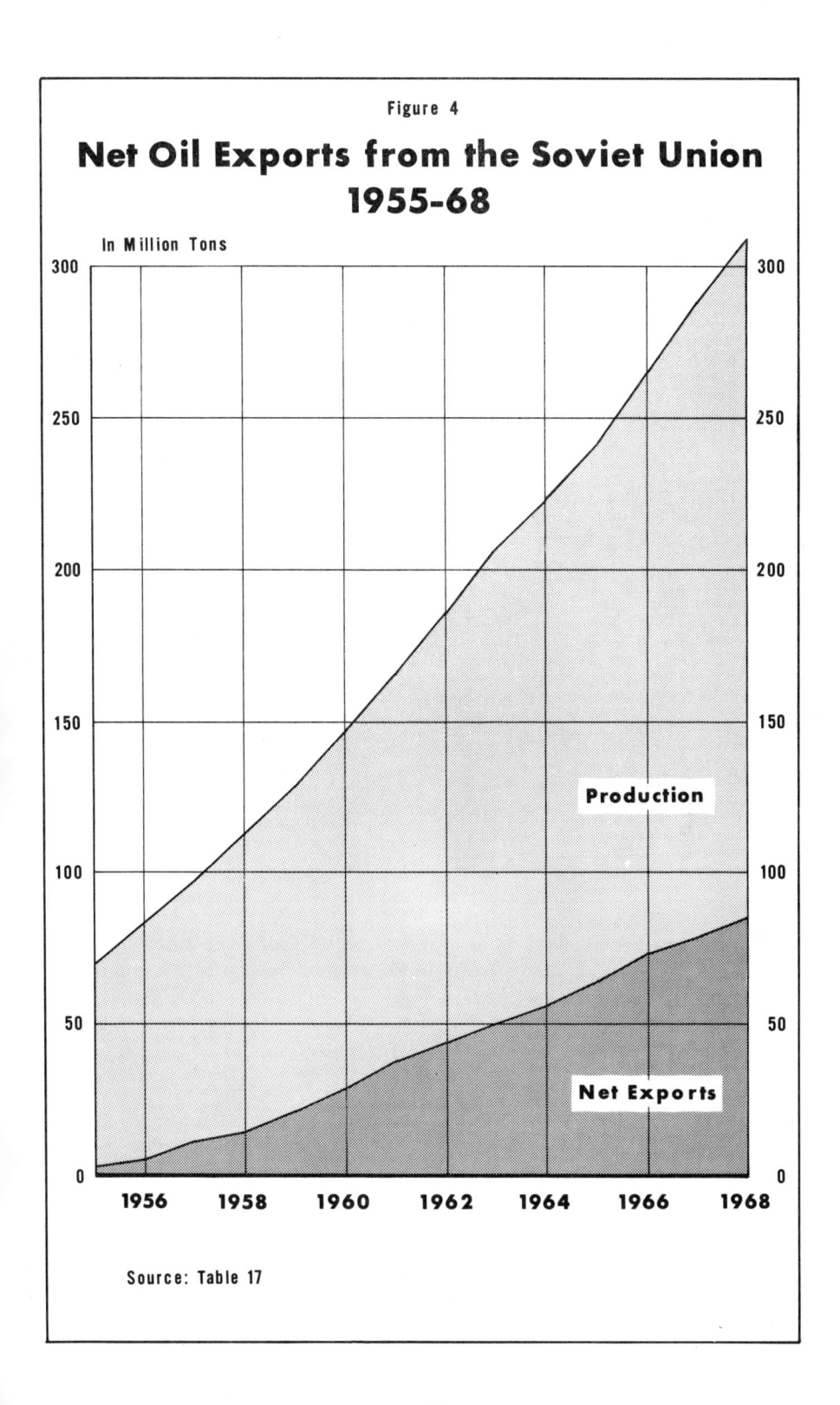

Figure 4
Net Oil Exports from the Soviet Union 1955-68

TABLE 18

Export of Oil from the Soviet Union, 1955-68[a]

(in million metric tons)

Year	Crude Oil	Petroleum Products	Total
1955	2.9	5.1	8.0
1956	3.9	6.2	10.1
1957	5.9	7.8	13.7
1958	9.1	9.0	18.1
1959	12.5	12.9	25.4
1960	17.8	15.4	33.2
1961	23.4	17.8	41.2
1962	26.3	19.1	45.4
1963	30.2	21.1	51.3
1964	36.7	19.9	56.6
1965	43.4	21.0	64.4
1966	50.3	23.3	73.6
1967	54.1	24.7	78.8
1968[b]	59.2	27	86.2

[a] Except for 1968, data have been derived from the various annual statistical trade handbooks issued by the Ministry of Foreign Trade of the USSR.

[b] Vneshnyaya torgovlya, No. 7 (July, 1969), p. 53.

TABLE 19

Import of Oil by the Soviet Union, 1955-68

(in million metric tons)

Year	Crude Oil	Petroleum Products	Total
1955	0.575	3.8	4.375
1956	1.510	3.8	5.310
1957	1.331	2.9	4.231
1958	1.078	3.2	4.278
1959	1.083	3.3	4.383
1960	1.166	3.2	4.366
1961	0.888	2.7	3.588
1962	0.496	2.3	2.796
1963	0.543	2.3	2.843
1964	- -	2.1	2.1
1965	- -	1.9	1.9
1966	- -	1.7	1.7
1967	0.061[b]	1.339	1.4
1968[b]	- -	2.5	2.5

[a] Except for 1968, data have been derived from the various annual statistical trade handbooks issued by the Ministry of Foreign Trade of the USSR.

[b] From Algeria.

[c] Estimate. About 1 million tons of the indicated imports in reality represent swaps of petroleum products with Western oil companies. Such imports are listed to balance out deliveries of oil to third-country customers on Soviet accounts by these Western firms.

TABLE 20

Distribution of Soviet Oil Exports Among
Communist and Non-Communist Buyers,[a] 1955-68

Year	Communist[b] Buyers Million Tons	Percent Total	Non-Communist Buyers Million Tons	Percent Total
1955	4.2	52.5	3.8	47.5
1956	5.2	51.5	4.9	48.5
1957	7.6	56.3	5.9	43.7
1958	8.9	49.1	9.2	50.9
1959	11.1	44.0	14.1	56.0
1960	15.2	45.6	18.1	54.4
1961	18.4	44.7	22.8	55.3
1962	20.7	45.6	24.7	54.4
1963	23.0	44.7	28.4	55.3
1964	25.3	44.7	31.3	55.3
1965	28.9	44.9	35.5	55.1
1966	32.3	43.8	41.4	56.2
1967	35.3	44.8	43.5	55.2
1968[c]	42.2	49	44	51

[a] Except for 1968, data have been derived from the various annual statistical trade handbooks issued by the Ministry of Foreign Trade of the USSR.

[b] Defined as the Communist countries of Eastern Europe (Bulgaria, Czechoslovakia, East Germany, Hungary, Poland, and Rumania) plus Cuba, Yugoslavia, North Korea, North Vietnam, Mongolia, Communist China, and Albania.

[c] Estimate.

Figure 5

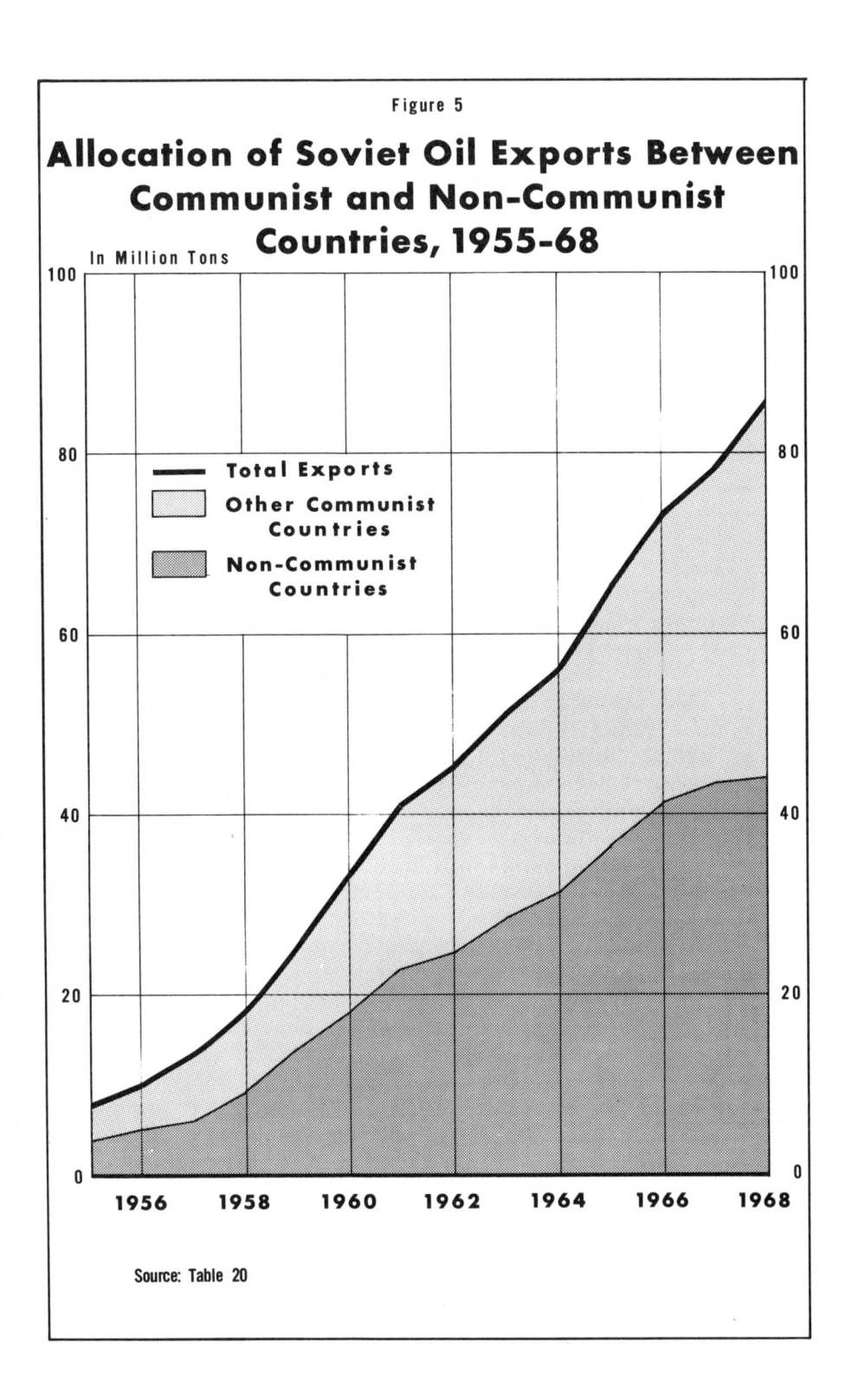

Allocation of Soviet Oil Exports Between Communist and Non-Communist Countries, 1955-68

In Million Tons

- Total Exports
- Other Communist Countries
- Non-Communist Countries

Source: Table 20

officials to maintain some form of reasonable balance in the distribution of its exports, a balance which if maintained over time would become a base upon which to forecast allocations of future exportable oil. This policy would then serve as a guide in determining to what extent the oil requirements of other Communist countries, particularly those of Eastern Europe, could be met by the Soviet Union and in turn would indicate to Eastern Europe what quantities of oil it might have to arrange to purchase from the West. In addition, the All-Union Oil Export Organization (Soyuznefteeksport) would have an acceptable approximation of how much oil it might have for sale to non-Communist buyers and, with certain assumptions, would be able to estimate foreign exchange earnings from these sales. This latter estimate would be of considerable value in determining the scale of available means for financing desired imports of goods and equipment. Deliveries to other Communist countries again jumped sharply in 1968, to an estimated 42.2 million tons, representing the highest demand on the Soviet exportable surplus--49 per cent--since 1958.

The preponderant share of exports of oil to other Communist countries is made up of crude oil, virtually all of which is directed to Eastern Europe and Cuba (see Table 21). Eastern Europe has been expanding its refining industry quite rapidly in recent years; this capacity is approaching self-sufficiency in terms of volume and quality of output, as is reflected in the generally declining trend in product imports since 1961. To meet the growing thirst of Eastern Europe for crude oil as well as supplying Cuban requirements has meant that crude oil deliveries to other Communist countries during the last several years have exceeded total crude oil sales to the Free World. But product sales to the West, in contrast to such sales to other Communist countries, have gradually increased since at least 1955, and the volumetric difference in these product sales represents essentially the volumetric difference between Soviet oil sales to the West and to other Communist countries. (See Table 22 for the export of crude oil and petroleum products from the Soviet Union to non-Communist countries, 1955-68.)

The emergence of the Soviet Union as a net exporter of oil to the Communist world, particularly since 1954, has enabled it to assume a position alongside the Western oil companies as a major factor in the international oil market.

TABLE 21

Soviet Exports of Crude Oil and Petroleum Products
to Other Communist Countries,[a] 1955-68[b]

(in million metric tons)

Year	Crude Oil	Petroleum Products	Total	Annual Gain
1955	2.2	2.0	4.2	- -
1956	2.9	2.3	5.2	1.0
1957	4.6	3.0	7.6	2.4
1958	5.2	3.7	8.9	1.3
1959	6.4	4.7	11.1	2.2
1960	8.8	6.4	15.2	4.1
1961	10.1	8.3	18.4	3.2
1962	12.7	8.0	20.7	2.3
1963	14.8	8.2	23.0	2.3
1964	17.9	7.4	25.3	2.3
1965	22.4	6.5	28.9	3.6
1966	25.5	6.8	32.3	3.4
1967	27.3	8.0	35.3	3.0
1968[c]	32.7	9.5	42.2	6.9

[a] Defined as the Communist countries of Eastern Europe (Bulgaria, Czechoslovakia, East Germany, Hungary, Poland, and Rumania) plus Cuba, Yugoslavia, North Korea, North Vietnam, Mongolia, Communist China, and Albania.

[b] Except for 1968, data have been derived from the various annual statistical trade handbooks issued by the Ministry of Foreign Trade of the USSR.

[c] Estimate.

TABLE 22

Soviet Exports of Crude Oil and Petroleum Products
to Non-Communist Countries, 1955-68[a]

(in million metric tons)

Year	Crude Oil[b]	Petroleum Products[c]	Total[d]	Annual Gain
1955	0.7	3.1	3.8	-
1956	1.0	3.9	4.9	4.1
1957	1.3	4.6	5.9	4.0
1958	3.9	5.3	9.2	3.3
1959	6.1	8.0	14.1	4.9
1960[e]	9.0	9.1	18.1	4.0
1961	13.3	9.5	22.8	4.7
1962	13.6	11.1	24.7	1.9
1963	15.4	13.0	28.4	3.7
1964	18.8	12.5	31.3	2.9
1965	21.0	14.5	35.5	4.2
1966	24.8	16.6	41.4	5.9
1967	26.9	16.7	43.5	2.1
1968[f]	26.5	17.5	44	0.5

[a] Except for 1968, data have been derived from the various annual statistical trade handbooks issued by the Ministry of Trade of the USSR. For a simplified methodology for estimating the probable levels of oil exports from the Soviet Union in a given year, see Table 23.

[b] Gross exports; imports of crude oil from the West are indicated in Table 24.

[c] Gross exports; imports of petroleum products from the West are indicated in Table 24.

[d] Totals derived from unrounded data.

[e] In 1960 the USSR assumed responsibility for meeting the crude oil requirements of Cuba, following the Communist revolution in that country which resulted in the withholding of historical Free World sources of oil supply. With no spare or unutilized productive capacity available in the Soviet oil industry, but faced with the political necessity of honoring this responsibility, domestic consumption in the USSR was reduced across the board to free oil for shipment to Cuba. It should not be considered, then, that the USSR was using oil which otherwise might have been disposed of in Free World markets.

[f] The Soviet Union reportedly developed an exportable surplus in crude oil of 59.2 million tons in 1968, and in petroleum products of 27 million tons. Deliveries of crude oil to other Communist countries probably totaled 32.7 million tons and deliveries of petroleum products about 9.5 million tons. Fulfillment of these obligations left the Soviet Union with 26.5 million tons of crude oil and 17.5 million tons of products for sale to non-Communist countries.

TABLE 23

Simplified Method for Estimating Crude Oil and
Petroleum Product Exports from the Soviet Union

Indicator	Amount (Million Metric Tons)
Crude Oil	
Production of crude oil	a/
Crude oil losses and increments to storage	b/
Charge to refining	c/
Available for export:	d/
Petroleum products	
Yield from refining:	e/
Imports:	f/
Natural gas liquids, synthetics	g/
Total product supply	h/
Consumption	i/
Available for export	j/

a/ Annual plan, plus probable overfulfillment, generally on the order of 1 million tons.

b/ It has seemed reasonable to assume that a total of 3 per cent of the crude oil (related to production) is lost in handling, in increments to storage, and in use as a fuel, although there is recent evidence which strongly suggests that an average of 5 per cent is so consumed.

c/ Estimate obtained from the current press, usually expressed as a percentage increase over the previous year, or given as a link relative in terms of a number of preceding years.

d/ Derived as a residual after deducting losses and charge to refining from production.

e/ Estimated at 92 per cent of the charge of crude oil to refining.

f/ Estimate based on past trends. Imports are minor in quantity, they are declining, and any error in estimation is not likely to be of any influence.

g/ Estimate composed largely of the output of natural gas liquids, data concerning which usually can be found in the technical press. Synthetics are only minor today, mostly oils obtained from the refining of shale, and probably can be ignored over the long run.

h/ Summation of e, f, and g.

i/ Estimate based on the extrapolation of past trends.

j/ Estimate derived as a residual.

TABLE 24

Soviet Imports of Oil from
Non-Communist Sources, 1955-67

(in thousand metric tons)

Year	Crude Oil[a]	Petroleum Products	Total
1955	467.1	0	467.1
1956	1,375.2	0	1,375.2
1957	1,124.1	0	1,124.1
1958	1,004.9	0	1,004.9
1959	1,012.2	0	1,012.2
1960	1,002.8	0	1,002.8
1961	772.6	0	772.6
1962	495.8	3.1	498.9
1963	543.2	4.4	547.6
1964	0	35.5	35.5
1965	0	46.0	46.0
1966	0	267.2	267.2
1967	61.3	187.0	248.3

[a] Except for 1967, from Austria. In 1967 crude imports originated with Algeria.

Source: Data have been derived from the various annual statistical trade handbooks issued by the Ministry of Foreign Trade of the USSR.

This place has been secured with virtually no investment in foreign marketing facilities and with a tanker fleet which, although greatly enlarged, ranks only eighth in the world, with 3.3 per cent of the ship tank fleet deadweight tonnage (dwt).[1]

To exports of oil from the Soviet Union have been added small volumes of petroleum products originating with the countries of East Europe. Each of the countries of East Europe has been able to make some quantities of oil available for sale to non-Communist countries, but with the exception of Rumania, these quantities are very small and are of virtually no significance in world oil trade. Again, with the exception of Rumania, such exports are in reality little more than a re-export of Soviet crude oil in the form of petroleum products.

Rumania is the only country in East Europe whose oil supply is completely independent of the Soviet Union. Indeed, its current production of 13.4 million tons (1968) is roughly double domestic demand; all of the surplus oil is exported as petroleum products, allocated almost equally between non-Communist and Communist buyers.

The development of East European oil sales to non-Communist countries, beginning in 1955, is documented in Table 25. It is estimated that in 1968 East Europe sold a total of 7 million tons of petroleum products to the West. When added to exports by the USSR, total sales by the Soviet bloc that year were 51 million tons, to which East Europe contributed only 23.7 per cent. For comparison, in 1955 the Soviet bloc marketed about 8 million tons of oil in the West, and the East European contribution at that time was 22 per cent.

During the period 1955-60 more than three-quarters of the export of oil from East Europe to the West originated with Rumania. In subsequent years, however, the easy availability of Soviet crude oil through the Friendship oil pipeline has enabled the other East European countries together to equal Rumanian exports. Since 1960, then, most of the growth in exports from East Europe have been provided not by Rumania, but collectively by Bulgaria, Czechoslovakia, East Germany, and Poland. Those increases achieved by Rumania since 1955 may for the large part be attributed to a decline in exports to the USSR. From 1955 to 1966 this decline can be measured at 2.4 million tons.

TABLE 25

Exports of Oil[a] from Eastern Europe
to Non-Communist Countries, 1955-68

Year	Exports (Thousand Metric Tons)[b]
1955	1,779
1956	1,776
1957	1,903
1958	2,191
1959	3,072
1960	3,582
1961	4,500
1962	4,700
1963	5,500
1964	6,000
1965	6,700
1966	6,800
1967	6,500
1968	7,000

[a] All petroleum products.

[b] Figures for 1955 through 1961 from National Petroleum Council, Impact of Oil Exports From the Soviet Bloc, Volume II (Washington, 1962), p. 447-449; figures for 1962 and 1963 from National Petroleum Council, Impact of Oil Exports From the Soviet Bloc, Supplement to the 1962 Report (Washington, 1964), p. A-12; figures for 1964-68, estimated.

POSTWAR DEVELOPMENT 53

Since 1965 East European oil sales to the West have leveled off, as domestic consumption has caught up with supply. Because of the probability that indigenous supplies of crude oil in Eastern Europe are not likely to be expanded significantly, future increases in exports will have to be based on the import of crude oil, either from the Soviet Union or the Middle East, in volumes in excess of requirements.

While the international oil industry was evaluating Soviet oil with some cause for alarm, it is prudent to observe that the regard of the USSR for its competitors was at best unflattering.

The position of Communist officials on oil and the measure of influence which they believe the international oil industry wields is perhaps best reflected in the following excerpt from a leading mass propaganda Soviet journal:

> It should be borne in mind that oil concessions represent, as it were, the foundations of the entire edifice of Western political influence in the (less developed) world, of all military bases and aggressive Blocs. If this foundation cracks, the entire edifice may begin to totter and then come tumbling down.[2]

Although direct confrontations have been avoided and today there are even semblances of commercial cooperation between East and West in the marketing of oil (see Chapter 8, section titled "Impact of 1967 Middle East Crisis"), the campaign to malign the international oil companies continues unabated.

This campaign scarcely conceals the scornful regard in which the international oil companies are held by the Soviet Union. Every opportunity is taken by the Communist press to present the activities of the oil companies in the worst possible light. To the Communist mind, nothing better represents the very worst of the capitalist system than the Western oil companies, whose mountainous profits have been wrested from the developing nations through exploitation and plunder. Typical of the propaganda directed against the oil companies is the following item from *Pravda*, the official organ of the Communist Party.

"Here they come - the magnificent six from ESSO"--so are advertised the productions of the American oil octopus in the newspaper Hindustan Times. But in light of such dirty bandistic activities, which ESSO undertakes in developing countries, this innocent advertisement suddenly acquires a double meaning: yes, here they are, bandits, plunderers from ESSO! Our correspondent from Madras tells about one of the aspects of the activities of this American company. (The full range of the Communist press barrage against the international oil industry is covered in Part II). This item was accompanied by the advertisement referred to, strangely reminiscent of the advertisement for the motion picture "The Magnificent Seven"; the caption reads "Plunderers from ESSO."

American oil monopolies, striving to preserve the Indian market for themselves, for a long time have inhibited the development of a national oil industry in India, and have endeavored to prove that the country is lacking in prospects for oil. But these efforts have proved futile. And in the face of the successful development of the Indian national oil extracting industry the monopolies of the US have become more compliant.

Some days ago one of the leaders of the American oil company ESSO (the oil monopoly which has in direct investments in India alone of 700 million rupees) appeared before journalists in Madras. He confirmed the readiness of the company to cooperate in the development of an oil industry belonging to the government.

During the Fourth Five Year Plan India, for the further development of the petrochemical industry, has assigned more than 4 billion rupees. And the American monopoly has resolved to put a portion of this sum in its own pocket. This is why they are willing to cooperate.[3]

Yet on record the ingenuity and inventiveness of the American oil industry is openly admired by those who are involved in the day-to-day operations of the Soviet oil industry and who are less concerned about the official party line. Confrontations between the Communist oil man and the capitalist oil man, made possible through the technical exchange program established between the two governments of the Soviet Union and the United States, undoubtedly have contributed to an awareness of the strengths (and weaknesses) and advantages (and disadvantages) of Communism and capitalism.

Under this exchange program, Soviet (and American) experts can move about the country on predetermined itineraries responsive to predetermined areas of interest. For the Soviet experts, it is a unique opportunity to inspect and contemplate that which they have already read about in our technical publications. For the Americans, it is a unique opportunity to make a first-hand appraisal of Communist industry and often it means travel in areas and visits to industrial facilities rarely open to Westerners. [4]

Although the Communist oil man is always prepared to defend his system of government and the superiority of his oil industry, at the expense of the free enterprise way of getting things done, the functional beauty of American design, the high degree of automation, and the simplicity of operation of American equipment is a source of envy. Yet the leader of the Soviet oil delegation which toured American oil facilities during October-November 1960, in sizing up one of the continuing headaches of U.S. oil, was quick to balance this envy with a good-humored warning: "Don't drill too many wells. You can't afford it." [5]

The Communist oil man is an avid reader of U.S. oil trade publications. Some are sufficiently schooled in the English language to be able to utilize these materials directly. Others are dependent upon a broad abstracting service which reportedly has 22,000 translator-specialists on call. [6] All of the fields of science and engineering, except those covered by specialized abstracting services (i.e., medicine, agriculture, and construction technology) are handled by the All-Union Institute for Scientific and Technological Information. This institute, under the Academy of Sciences of the USSR, coordinates a network of more than sixty information centers which collect and abstract scientific and technical information from Soviet and foreign-language publications, and issues 25 volumes of abstracts monthly; these volumes, known as Referativnyy zhurnal (Reference Journal), are divided into a total of 160 subdivisions. In addition, for each field of science, there is a weekly publication entitled Ekspress informatsiy (Express Information) which, as its title implies, is designed to overcome the time lag so characteristic of attempts to cope with today's almost overwhelming flow of new information. By limiting its coverage to just the most important articles, abstracts can be in the hands of interested researchers within a month of receipt of the original material. The institute also issues the annual publications Itogi nauki (Results of Science) and Itogi nauki i tekhniki (Results of Science and Technology) which summarize the main achievements of world science and technology during the year. [7]

Coverage in particular of U.S. oil trade publications is admirable and it embraces the complete spectrum from a single-line abstract in Referativnyy zhurnal to a full Russian-language version of the monthly publication Petroleum Engineer which is then made available for subscription at a nominal fee. In addition, important reference works are translated cover-to-cover. Because the Soviet Union is not a subscriber to the International Copyright Convention, it pays no fees for the privilege of producing such translated versions, except in those isolated instances in which special arrangements are made with the author. On the other hand, the absence of a copyright from all Soviet publications means that translations of this material can be freely made.

CHAPTER 6 — THE OIL PRICING POLICY OF THE SOVIET UNION

The prices that the Soviet Union receives for the crude oil (and petroleum products [1]) which it markets outside the country have become a matter of major dispute. The Soviet pricing policy has a particular emotional appeal to all critics, whether capitalist or socialist, for official statistics can be, and have been, applied to demonstrate that the USSR is taking advantage of the captive markets in the other Communist countries to charge a much higher price for crude oil compared to prices obtained in the West. Similarly, low price offerings in Western Europe for Soviet crude oil have been of particular concern to the international oil companies, a reaction reflecting fear that if attempts to meet these artificially lowered prices were made, buyers in other markets around the world would insist on similar treatment.

Of the higher prices paid by the other Communist countries, it is argued that the price discrimination is really a form of subsidy of cheap oil sales to non-Communist countries. The magnitude of this discrimination is illustrated in Tables 26, 27, and 28.

Although one might suppose that at least part of the differential in the prices paid for crude oil by various buyers may reflect varying costs of transportation within the Soviet Union, in at least one glaring example this is not the case: Crude oil has been continually sold to East Germany at prices which in earlier years were as much as double those paid by its neighbor to the west, West Germany. The average price to East Germany is about 46 per cent higher than that to West Germany; the decline in prices charged to East Germany since 1964 reflects the benefit of use of the Friendship crude oil pipeline for delivery.

TABLE 26

Average Prices[a] Paid for Soviet Crude Oil
by Other Communist Countries, 1955-67

Year	Price (U.S.$ Per Barrel)[b]
1955	$ 3.38
1956	3.30
1957	3.28
1958	2.97
1959	3.01
1960	3.01
1961	2.54
1962	2.52
1963	2.55
1964	2.57
1965	2.42
1966	2.18
1967	2.10

[a] F.o.b. the Soviet border.

[b] Conversion based on 7.3 barrels per metric ton and 1 ruble = $1.11.

Sources: The various annual statistical trade handbooks issued by the Ministry of Foreign Trade of the USSR, the Communist countries of Eastern Europe (Bulgaria, Czechoslovakia, East Germany, Hungary and Poland), Cuba, Communist China, North Korea, North Vietnam, and Yugoslavia.

TABLE 27

Average Prices[a] Paid for Soviet Crude Oil
by Non-Communist Countries, 1955-67

Year	Price (U.S.$ Per Barrel)[b]
1955	$2.16
1956	2.17
1957	2.55
1958	2.08
1959	1.88
1960	1.57
1961	1.26
1962	1.26
1963	1.43
1964	1.41
1965	1.40
1966	1.39
1967	1.50

[a] F.o.b. the Soviet border.

[b] Conversions based on 7.3 barrels per metric ton and 1 ruble = $1.11.

Sources: The various annual statistical trade handbooks issued by the Ministry of Foreign Trade of the USSR.

Figure 6

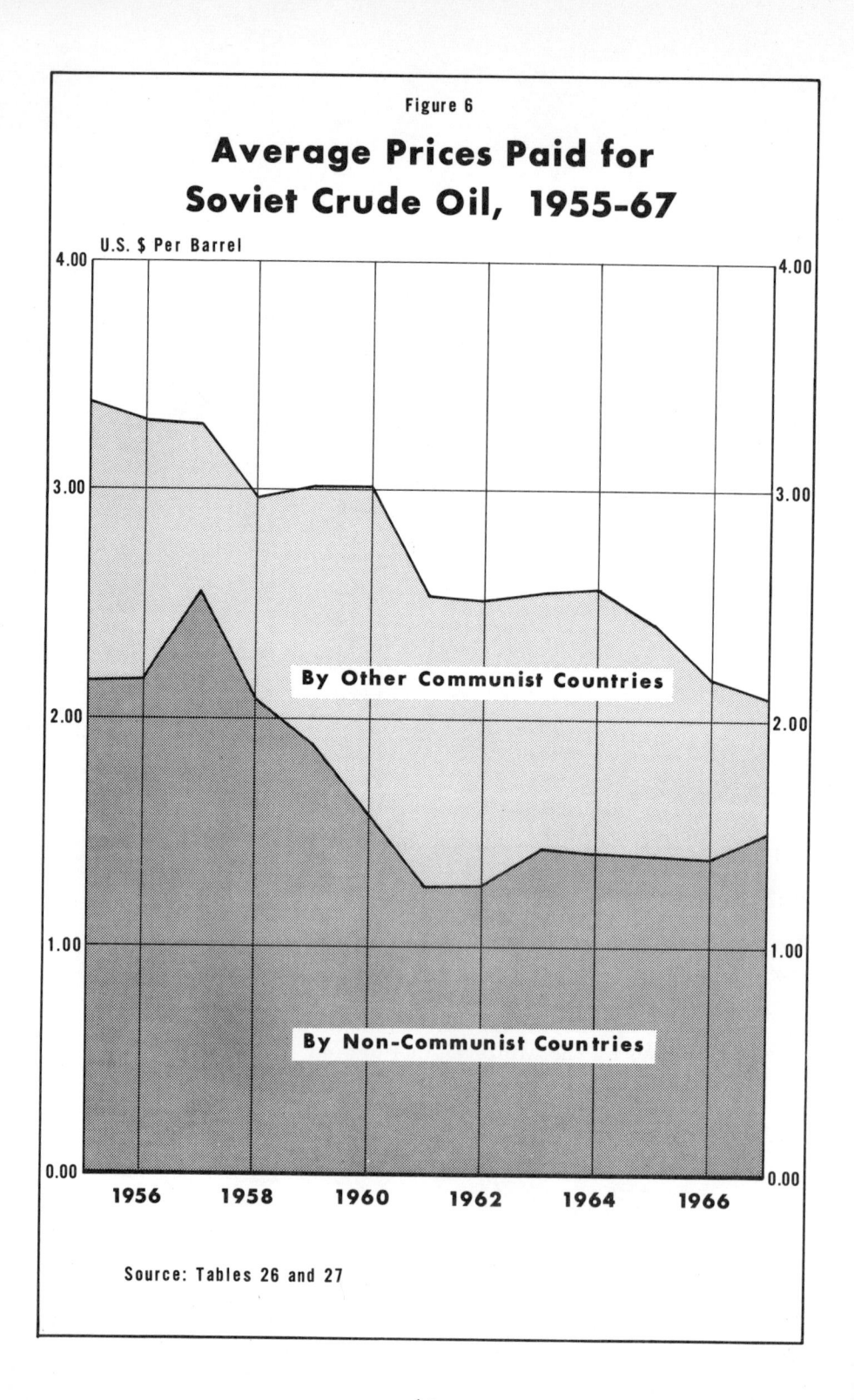

Average Prices Paid for Soviet Crude Oil, 1955-67

U.S. $ Per Barrel

By Other Communist Countries

By Non-Communist Countries

Source: Tables 26 and 27

TABLE 28

Apparent Annual Average Prices[a] Paid
for Soviet Crude Oil, 1955-1967

(in U.S. dollars per barrel)[b]

Country	1955	1956	1957	1958	1959	1960	1961
Latin America							
Argentina	$2.12	$ --	$ --	$1.61	$1.52	$ --	$ --
Brazil	--	--	--	--	2.14	1.73	1.65
Uruguay	--	--	--	2.35	2.16	2.16	2.12
West Europe							
Austria	--	--	--	--	2.57	2.57	2.15
Belgium	--	--	--	--	--	--	1.32
Finland	--	--	--	--	1.75	1.72	1.50
France	2.00	2.08	2.48	2.24	1.76	1.61	1.52
Germany, West	--	--	--	--	1.69	1.39	1.27
Greece	--	--	--	--	2.05	1.94	1.85
Italy	1.95	2.18	2.68	2.00	1.75	1.42	1.31
Portugal	--	--	--	2.33	--	--	--
Spain	--	--	--	--	--	--	--
Sweden	--	--	--	--	--	--	--
Switzerland	--	--	--	--	--	--	--
Other Eastern Hemisphere Countries							
Burma	--	--	--	--	--	--	--
Egypt	1.95	1.93	2.41	2.23	1.80	1.63	1.43
Ghana	--	--	--	--	--	--	--
Japan	--	--	--	1.78	1.68	1.34	1.26
Morocco	--	--	--	2.06	1.94	1.83	1.65
Average to Free World	$2.16	$2.17	$2.55	$2.08	$1.88	$1.57	$1.26
Other Communist Countries							
Bulgaria	--	--	--	--	--	--	--
China	5.12	5.12	5.12	3.05	2.96	2.92	--
Cuba	--	--	--	--	--	1.53	1.49
Czechoslovakia	3.58	3.04	3.08	3.11	3.18	3.14	3.17
Germany, East	2.99	3.13	3.15	2.72	2.70	2.69	2.61
Hungary	2.39	2.77	2.97	3.02	3.08	3.06	3.04
Mongolia	--	--	--	--	3.48	3.50	3.68
Poland	2.70	2.82	3.26	2.94	3.26	3.28	3.17
Yugoslavia	2.38	2.38	2.75	2.55	2.36	2.27	1.92
Average to Other Communist Countries	$3.38	$3.30	$3.28	$2.97	$3.01	$3.01	$2.54

(Continued)

[a] F.o.b. the Soviet border.

[b] Conversions based on 1 metric ton = 7.3 barrels and 1 ruble = $1.11.

Sources: The various annual statistical trade handbooks issued by the Ministry of Foreign Trade of the USSR.

TABLE 28 (Continued)

Country	1962	1963	1964	1965	1966	1967
Latin America						
Argentina	$ --	$ --	$ --	$1.31	$ --	$ --
Brazil	1.61	1.60	1.59	1.53	1.45	1.35
Uruguay	--	--	--	--	1.25	--
Europe						
Austria	1.67	1.67	1.67	1.67	1.67	1.67
Belgium	--	--	--	--	--	--
Finland	1.52	1.81	1.86	1.90	1.84	1.76
France	1.51	1.58	1.42	1.26	1.25	1.29
Germany, West	1.30	1.30	1.39	1.40	1.39	1.40
Greece	1.85	1.84	1.83	1.75	1.74	1.75
Italy	1.30	1.24	1.22	1.21	1.22	1.45
Portugal	--	--	--	--	--	--
Spain	--	--	1.55	1.54	1.53	1.79
Sweden	--	1.54	1.49	--	--	--
Switzerland	--	--	1.24	1.26	1.34	--
Other Eastern Hemisphere Countries						
Burma	--	--	1.53	1.52	1.40	1.27
Egypt	1.45	1.68	1.73	1.76	1.78	1.80
Ghana	--	--	--	1.52	1.28	1.20
Japan	1.26	1.23	1.18	1.21	1.22	1.42
Morocco	1.61	1.62	1.58	1.52	1.53	1.57
Average to Free World	$1.26	$1.43	$1.41	$1.40	$1.39	$1.50
Other Communist Countries						
Bulgaria	--	2.49	2.45	2.45	2.47	1.97
China	--	--	--	--	--	--
Cuba	1.49	1.58	1.56	1.60	1.58	1.58
Czechoslovakia	3.10	3.09	3.09	2.75	2.35	2.34
Germany, East	2.66	2.64	2.61	2.33	2.05	2.04
Hungary	3.03	3.03	2.96	3.03	2.69	2.32
Mongolia	3.49	3.32	3.49	3.49	3.49	3.49
Poland	3.10	3.05	3.01	2.60	2.23	2.22
Yugoslavia	1.81	1.68	1.63	1.62	1.61	1.64
Average to Other Communist Countries	$2.52	$2.55	$2.57	$2.42	$2.18	$2.10

OIL PRICING POLICY

TABLE 29

Soviet Oil Price Differentials,
East and West Germany, 1959-67

Year	Price (U.S.$ Per Barrel) West Germany	East Germany
1959	$1.69	$2.70
1960	1.38	2.69
1961	1.27	2.61
1962	1.30	2.66
1963		
1964	1.39	2.61
1965	1.40	2.33
1966	1.39	2.05
1967	1.40	2.04

Source: Table 28.

Since at least 1955, however, average prices paid for soviet crude oil by other Communist countries have been steadily declining, from a peak of $3.38 per barrel in 1955 to $2.10 per barrel in 1967. In comparison, average prices paid by non-Communist countries increased through 1957, then began a steady decline through 1962 as market penetration in Western Europe was accomplished by means of the time-honored method of reduced prices. Once a place in the market has been secured, effort has been made to secure the greatest return for the oil sold, which means price increases if at all feasible. Beginning in 1964, however, and extending through 1966, prices held relatively stable around a base of $1.40 per barrel; prices increased in 1967 to an average of $1.50 per barrel. The increase in 1967 mirrors the opportunism of the USSR, which was quick to take advantage of the closure of the Suez Canal and the resultant disruption in normal flows of oil out of the Middle East to raise its prices. Large importers of Arab oil (for example, Italy and Japan) bore the brunt of the price increase. Yet the monetary gain derived from higher prices charged to non-Communist buyers was virtually offset by the drop in per-barrel income from purchases made by other Communist countries. Nevertheless, despite the fact that other Communist countries as a unit pay much less per barrel for Soviet oil today compared with earlier years, they are still confronted with a comparative overcharge of 40 per cent. And this price discrimination has been too large to escape the attention of those being discriminated against.

The decline in prices charged East Europe for Soviet crude oil has been the result of two separate and distinct actions. First, member countries of the Council for Economic Mutual Assistance (CEMA)[2] are now utilizing a new world price base period for determining intra-CEMA prices. A shift in the price base from 1957-58 to 1960-64 has been undertaken, with the result that the price difference between the 1960-64 price and the current world price has narrowed substantially in comparison with the differences between the 1957-58 price and the prevailing world price during the early 1960's. In other words, this shift in base periods has served to bring intra-CEMA prices much closer to prevailing world prices. Second, a portion of the decline in prices charged may be attributed to improvements in the oil distribution system within the Soviet Union and, so far as the countries of East Europe are concerned, to the construction of the Friendship crude oil pipeline network (depicted in Figure 7) which provides a direct link between the crude-oil producing fields in the Urals-Volga with refineries in East Europe. As confirmation, it can be noted that, with the exception of 1967, average prices paid by Bulgaria per barrel of Soviet crude have held relatively steady, whereas those paid its neighbors have declined. Bulgaria is not linked to the USSR on the Friendship pipeline but continues to receive oil by tanker.

The standard rebuttal of the Soviet Union to intimations on the part of other Communist countries that they are being overcharged for crude oil has been this: For the goods it buys from East Europe the USSR pays in excess of the going market price and, perhaps more important, offers a market for machinery and equipment which, because of poor quality and/or technical obsolesence, would be difficult to dispose of in the West. The development of this argument is illustrated in the following panel discussion broadcast by Radio Prague.

(Question:) Could you speak briefly about the fundamental problems and difficulties which hinder the development of cooperation within CEMA? Another question very much under discussion is the role of

Figure 7

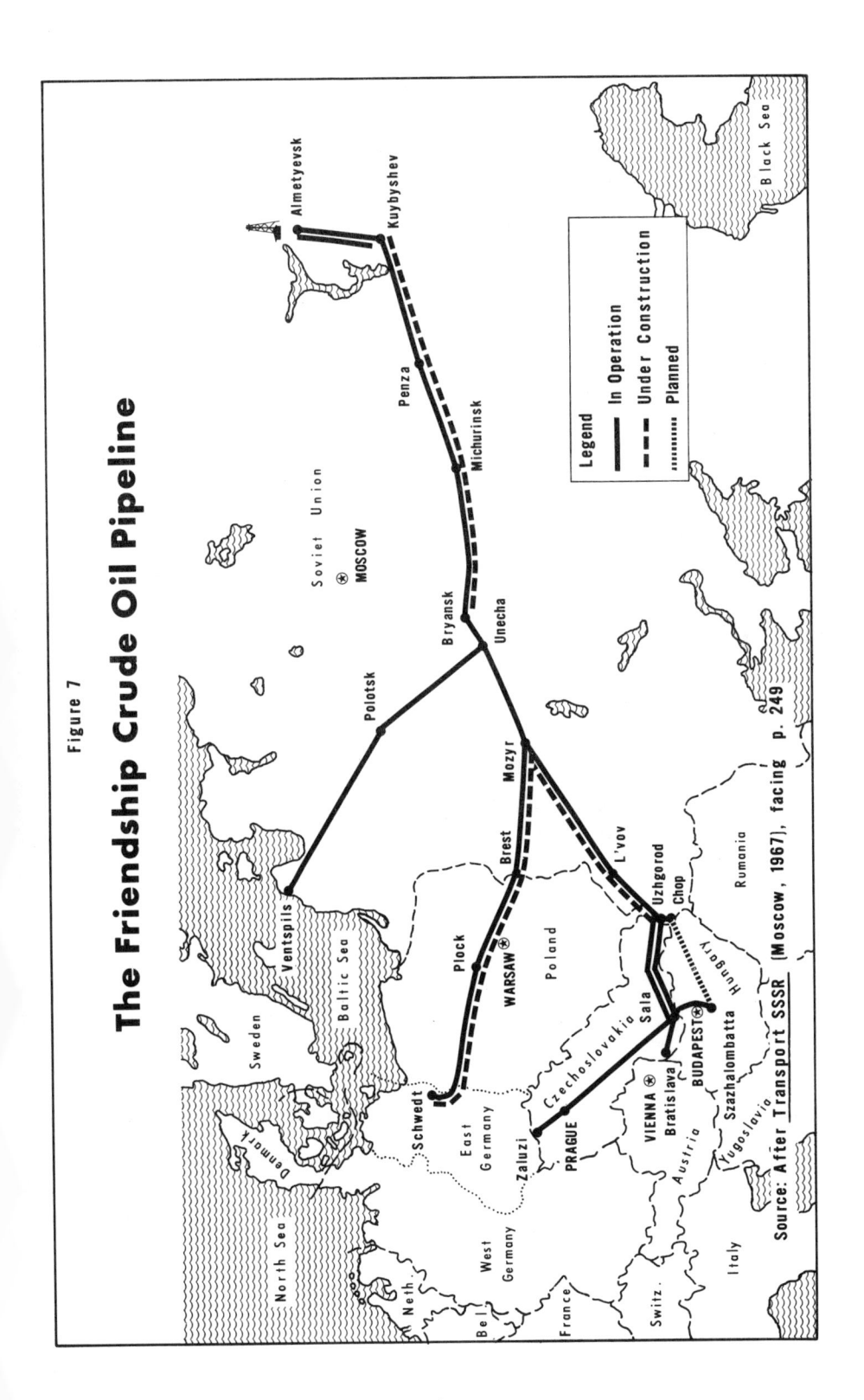

The Friendship Crude Oil Pipeline

Source: After Transport SSSR (Moscow, 1967), facing p. 249

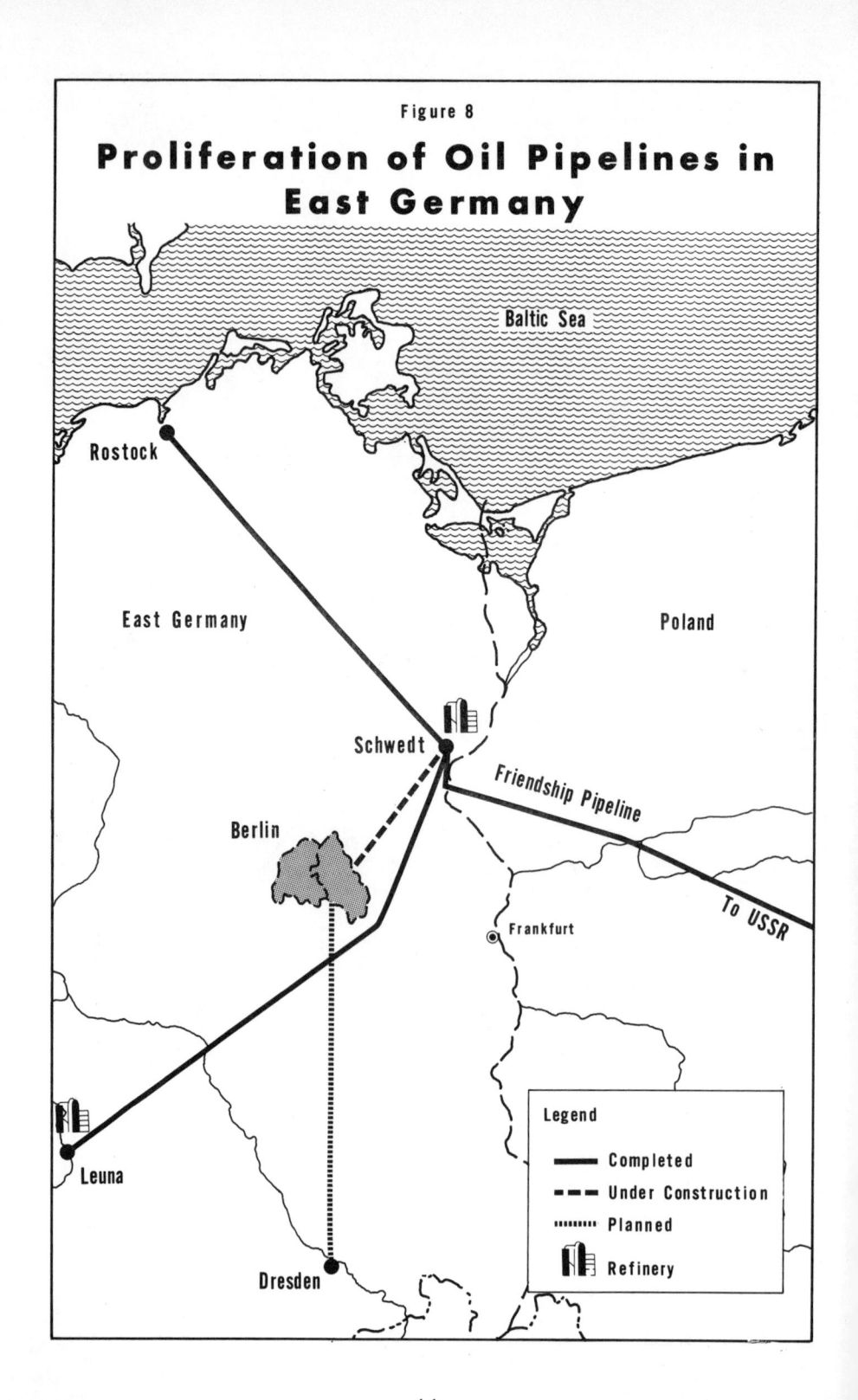

Figure 8
Proliferation of Oil Pipelines in East Germany

OIL PRICING POLICY 67

the USSR in CEMA. Often questions are asked
whether the USSR, as the strongest member of
that organization, does not draw from it the biggest
benefit, whether it does not develop to the
detriment of its partners. The main argument
in this is usually the question of imports and
prices of Soviet crude oil.

(Answer:) That is a fairly involved question.
The USSR basically plays a double role in its
relations with the socialist countries. In the
case of the less advanced socialist countries it
is a supplier of raw materials and equipment.
In the case of the advanced socialist countries
it mostly plays the role only of a supplier of raw
materials.

(Question:) I think that it is a very important
role because the majority of the advanced socialist
countries have built industries needing a lot
of raw materials. This leads, I think, to a certain
economic dependence. We, for instance, badly
need crude oil but do not have the currency to buy
it in the West. The question can be asked why we
do not export to the West and buy crude oil at lower
prices? But this does not come into consideration
as long as there are great difficulties with the sales
of our goods on the world markets.

(Answer:) The USSR, because it has become the
supplier to socialist countries of raw materials,
means that the USSR has constantly to expand its
raw material base. At the same time the demands
of the Soviet industry are growing.

As far as we are concerned: There are two opposing
views--the Czechoslovak and the Soviet. Investments
for the development of Tyumen oil are higher
than those in other areas.[3] The question is whether
that credit is to be granted to cover the particular
costs of Tyumen oil or to cover production costs
usual on the world markets. As far as I know Czechoslovakia
holds the latter view, the USSR, understandably,
the other. These are precisely the problems

where the interests of the individual Socialist countries clash.

(Question:) I think that the new system of management will support these tendencies; of course, and not only our new system, but also the economic reforms which are being carried out in other socialist countries, because everywhere there is a stress on a greater weight of economic criteria. That means that every country will think much more carefully before it concludes a trade agreement whether it is economically beneficial to it.

(Question:) To go back to the crude oil question and, in particular, to the prices of crude oil, because listeners frequently write in to say that we are paying two or even three times more for Soviet oil than the USSR charges Italy or the German Federal Republic. I know that it is much more complicated because, on our part, we supply industrial goods which cost more than the price we would obtain in the West.

(Answer:) It is not two or three times more than the prices paid by Italy, but it is a fact that we pay more for Soviet crude than Italy. That, however, is balanced by our trade agreements. The USSR offers us better prices for a whole range of industrial equipment [4] supplied by us to the USSR. [5]

Several months later the Czechoslovakian Permanent Representative to CEMA attempted to argue that the price of 14 rubles per ton for Soviet oil at the Czechoslovakian border was advantageous.[6] He noted that oil on other markets including transportation was 4 to 5 rubles more expensive, and if sometimes allegations were made that the Soviet Union sold oil to the Italians for 8 rubles, this was a misunderstanding of statistical data, since for this price the oil was delivered to Soviet ports and transport costs were paid by the buyer. However, he failed to make the distinction that while both the 14-ruble price per ton to Czechoslovakia and the 8-ruble price to Italy were f.o.b., it most certainly would not cost the difference of 6 rubles ($6.66) to move one ton of oil from a Soviet Black Sea port to Italy, a distance of only some 1,700

OIL PRICING POLICY

miles. Most probably, transportation costs would be no more than one-third the difference.

There have been lingering doubts that the USSR in fact is granting East Europe favored prices for its exports and there has been implied criticism that Czechoslovakia, for one, has been underselling uranium ore to the USSR. To this accusation the Minister of Foreign Trade of Czechoslovakia responded by stating: "It can be openly said that we have not sold and do not sell uranium ore to the USSR under worse conditions in comparison with those conditions which existed in the past in the world and which exist today. The price per kilogram of uranium ore reaches 15 to 17 American dollars. We sell [uranium ore] to the USSR at a higher price. Moreover, as a result of surpluses in the world market, prices for uranium ore fluctuate, but we have a stable selling price to the USSR up to 1970."[7]

The latter point, stable selling prices, has become particularly important in the current Soviet defense of its pricing policies. The closure of the Suez Canal in June, 1967 following the outbreak of hostilities between Israel and the Arab nations, the subsequent diversion of oil traffic around the Cape of Good Hope and the resultant increases in the delivered cost of oil to consumers in England and on the continent have all been used to good advantage by the Soviet press. The newspaper *Izvestiya* carefully outlined for its readers how the stable prices in the world socialist market work to the benefit of East Europe in its purchases of oil from the Soviet Union.

> In sharp contrast to the pricing nightmare that shakes the capitalist oil market and enriches a handful of oil magnates, there is price stability and soundness in the oil market of the socialist countries. No type of price increase is threatened this year, when such CEMA member countries as Bulgaria, Hungary, East Germany, Poland and Czechoslovakia will be importing from the USSR about 30 million tons of crude oil and petroleum products. If the socialist camp did not have its own stable oil market and if prices here would increase by the same degree, as they did for England, that is, on the average of 4 rubles per ton, then these countries would have to pay an additional minimum of 120 million rubles a year.

TABLE 30

Foreign Exchange Earnings from the Sale of
Soviet Oil to Non-Communist Countries, 1955-67

	Earnings		Volume Marketed	
Year	(Thousand U.S. $)[a]	Annual Growth (Percent)	(Million Metric Tons)	Annual Growth (Percent)
1955	$ 81,600	--	3.8	--
1956	112,800	38.2	4.9	28.9
1957	157,400	39.5	5.9	20.4
1958	178,700	13.5	9.2	55.9
1959	244,200	36.7	14.1	53.3
1960	263,200	7.8	18.1	28.4
1961	284,400	8.1	22.8	26.0
1962	313,700	10.3	24.7	8.3
1963	365,000	16.4	28.4	15.0
1964	383,000	4.9	31.3	10.2
1965	421,000	9.9	35.5	13.4
1966	485,100	15.2	41.4	16.6
1967	539,000	11.1	43.5	5.1

[a] Conversion based on the official exchange rate of 1 ruble = U.S. $1.11.

Source: The various annual statistical trade handbooks issued by the Ministry of Foreign Trade of the USSR.

TABLE 31

Shifts in Earnings from the Sale of Soviet Oil
to Non-Communist Countries, 1961-67

Year	Volume Exported[a] (Million Metric Tons)	Annual Return (Million Dollars)[b]	Average Price (Dollars Per Barrel)[c]
1961			
Crude oil	13.3	122	$1.26
Petroleum products	9.5	134	1.93
1962			
Crude oil	13.6	126	1.26
Petroleum products	11.1	157	1.94
1963			
Crude oil	15.4	157	1.43
Petroleum products	13.0	208	2.19
1964			
Crude oil	18.8	193	1.41
Petroleum products	12.5	190	2.08
1965			
Crude oil	21.0	214	1.40
Petroleum products	14.5	207	1.96
1966			
Crude oil	24.8	252	1.39
Petroleum products	16.6	233	1.88
1967			
Crude oil	26.9	295	1.50
Petroleum products	16.7	244	1.94

[a] Data have been derived from the various annual statistical trade handbooks issued by the Ministry of Foreign Trade of the USSR.

[b] Rounded to the nearest million. Rubles converted to U.S. dollars at the official exchange rate of 1 ruble = U.S. $1.11.

[c] Conversions on basis of 1 metric ton = 7.3 barrels for crude oil, and 7.5 barrels for petroleum products. Values based on 1 ruble = U.S. $1.11.

Figure 9

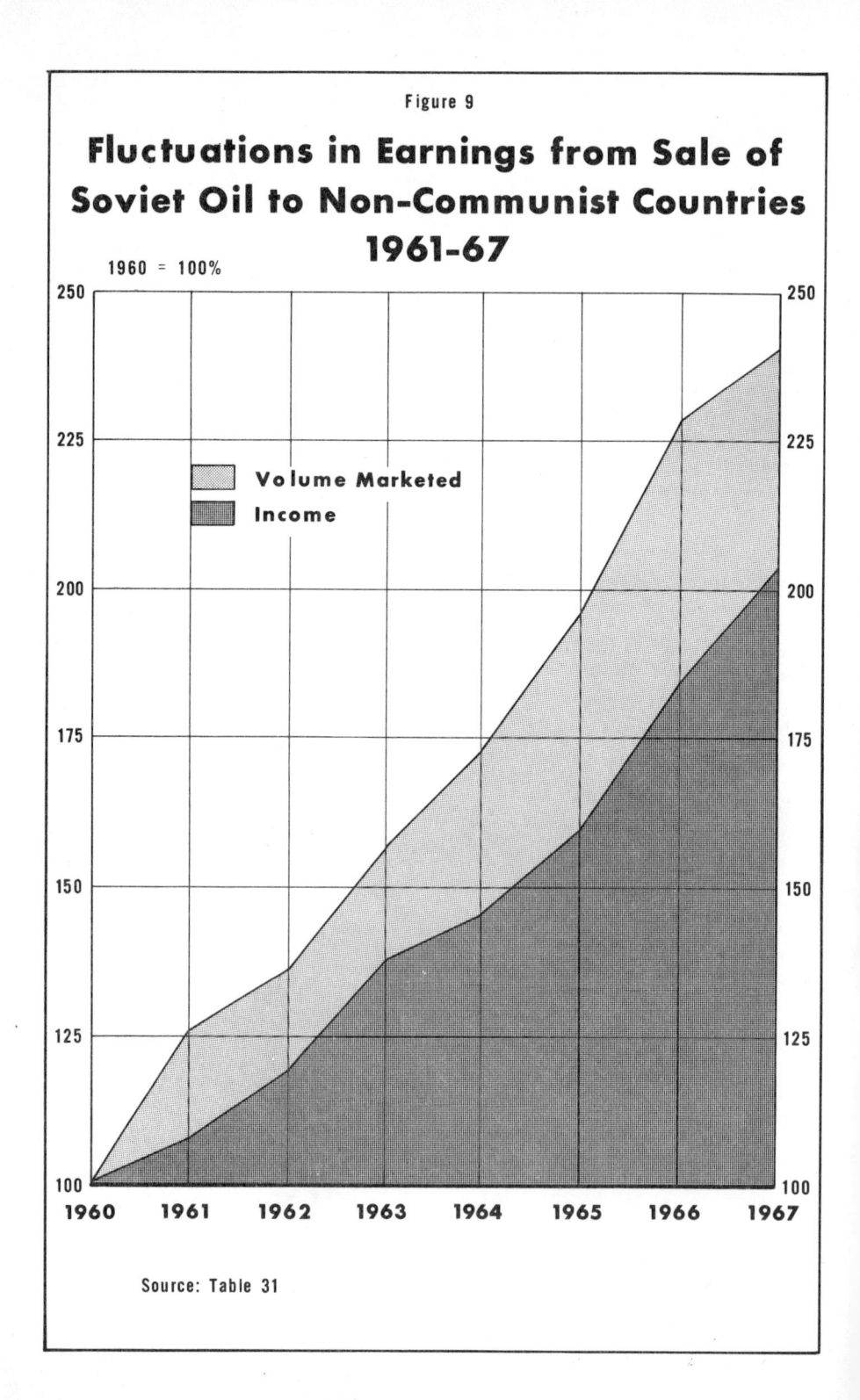

Fluctuations in Earnings from Sale of Soviet Oil to Non-Communist Countries 1961-67

1960 = 100%

Volume Marketed
Income

Source: Table 31

OIL PRICING POLICY

In order to secure such an additional sum, these countries would have to additionally produce and export, for example, more than 6 thousand diesel locomotives, manufactured by Czechoslovakia and East Germany, or more than 2 thousand passenger (railroad) cars, which are exported by Poland, East Germany, and Hungary, or more than 700 thousand tons of fruits and vegetables, grown in Bulgaria and Hungary, or 10 thousand Hungarian motor buses, or about 9 million tons of Polish hard coal. [8]

At the same time that the Soviet Union has been describing the advantages of stable prices to its colleagues in East Europe, it has issued a call for at least continued stable prices in Western Europe. The reasoning behind this becomes quite clear upon an examination of trends in foreign exchange earnings from the sale of Soviet oil to non-Communist countries, as indicated in Tables 30 and 31. For the most part, the growth in foreign exchange earnings simply has failed to keep pace with the growth in oil exports in recent years. Average returns per barrel of crude oil exported have been barely stable, declining by 4 cents per barrel in 1966, compared to 1963. This comparatively small decline cost the USSR the equivalent of almost $8 million in foreign exchange earnings that year. Average returns for petroleum products sold to non-Communist countries have also declined since 1963; a good portion of this decline, however, can be traced to a shift in the pattern of product sales away from the higher-priced distillates to the lower-priced residual fuel oils.

There have been recent accumulating instances of a reluctance on the part of Soviet trading representatives to adjust their tenders downward solely for the purpose of making a sale. The growing tendency is not to give away more than is necessary to secure the foreign exchange, equipment, or technology which they are seeking. With limited amounts of oil for sale, and directed to secure the greatest return for this oil, these representatives have no alternative other than to withdraw. In dealing with countries where the possible sale of Soviet oil is motivated largely by political considerations, the reluctance to engage in discounting disappears to the extent warranted by the prospect of enlargement of the Soviet sphere of influence.

CHAPTER 7 CONFLICT WITH THE ARAB WORLD

The appearance of Communist oil in Western Europe has not been without its political drawbacks for Soviet foreign policy from the very beginning. In view of the then-prevailing surpluses of crude oil in the world, Soviet oil sales could only lead to a depression in prices and an actual loss in oil sales by established sellers. Because most of the Arab oil-producing countries derive the major part of their revenue from the sale of oil in markets sought by the USSR, the Soviet export drive was cooly received in the Middle East.

It is perhaps fitting that political events in the Middle East were in part responsible for the generous reception accorded Soviet oil by certain buyers in Europe. It is suggested that the way for enlarged imports of Soviet oil by West Europe in the late 1950's was considerably eased by the Suez Crisis of 1956-57. The closure of the Suez Canal and the attendant danger of a petroleum shortage in Europe outlined very clearly the need for at least some diversification in origin of oil supplies for these consuming countries. Moreover, there were emerging indications that oil could be purchased from the Soviet Union at attractive savings compared with the delivered cost of Middle East oil, if only because the Black Sea ports were considerably closer to western and northern Europe than were ports of export in the Persian Gulf. And of course the possibility of bartering for this oil was sufficient incentive in itself. The intent was not to make a complete shift, for this would be physically impossible--the Soviet bloc fell far short of matching the Persian Gulf in the volumes of oil that could be made available for export. But increased imports from the Soviet bloc might serve to reduce consumer vulnerability somewhat should another oil crisis develop in the Middle East.

As its oil trade was rapidly expanding in the late 1950's and early 1960's, the Soviet Union on several occasions argued that it was just as much entitled to a share in the export market as any of the Western countries, that its "historic" share of European oil trade was 14 per cent, and that intentions were to recover this position. [1]

As explained by E. P. Gurov, head of Soyuznefteeksport, who attempted to justify the heightened Soviet oil sales campaign when questioned at the second Arab Oil Congress, held in Beirut in 1960, the Soviet Union was "only renewing our legal place among the exporting countries, which belonged to us in the prewar years and which was forfeited by us owing to conditions of wartime as well as due to the necessity of reestablishing home industry immediately after the war." [2]

This argument failed to placate certain spokesmen for the Arab crude-oil producing nations for, although they attempted to play down the significance of Soviet oil exports relative to the decline in oil prices in recent years, the comparatively low prices at which Soviet oil was being offered had not escaped their attention. At the third Arab Petroleum Congress in Alexandria in 1961, Emile Bustani of Lebanon, in a direct confrontation with the Soviet delegates, posed the following:

> We have observed lately that Russian oil has been dumped in the Free World markets at much lower prices than Arab oil. [You say] that this is a free market, free competition. It is not free competition. . . You can sell at any price you like, for I have statistics here to show. . . that you sell your oil at very low prices in the Free World market and you sell at high prices to satellite countries. . . .
> If it were a free market, would our oil be able to go to these free markets there and compete with your oil? [3]

In the months that followed, the policy of the Soviet Union toward the Arab nations was one of concomitant deprecation of the international oil companies, the presentation of the socialist countries as the only ones really interested in their well-being, and the advocacy of national oil companies to take the place of the oil "monopolies." Technical assistance and economic aid programs were greatly enlarged, followed by an increasing presence of military advisers.

Yet, within a relatively few years--in 1967--Soviet and Arab oil interests were again to clash, as it appeared that the Soviet Union was taking advantage of the disruption in normal flows of oil from the Middle East to Europe brought about by the Arab-Israeli war of June, 1967. The USSR, cut off from its customers in Southeast Asia and the Far East because of the closure of the Suez Canal, concentrated its oil sales in Europe, and with admirable results, as exports to that marketing area increased by 17 per cent in 1967, while total oil exports rose by only 7.1 per cent. The Western oil trade press with considerable relish drew the attention of their readers to Soviet marketing activities, which seemingly were running contrary to the traditional Soviet posture of support of the Arab World.[4] Soviet officals were particularly sensitive to any adverse publicity which might damage USSR-Arab relationships. When they were publicly accused of supplying oil to Israel after President Gamal Nasser of the United Arab Republic had cut off Israel's normal supply of oil on May 23, 1967 by blockading the Strait of Tiran,[5] Soviet officialdom was beside itself. The chairman of Soyuznefteeksport took the unusual step of calling a press conference with Western journalists to deliver a scathing denial of such accusations:

> Question. In recent days in the western press there have been stories to the effect that the Soviet Union is attempting to take the place of the Arab countries in the oil markets in connection with the decision, taken by the Arab countries, not to deliver oil to those Western countries which supported the Israeli aggression. In part, these stories state that the Soviet Union has offered crude oil and petroleum products to England. Would you comment on these articles?
>
> Answer. First of all, I would like to say that all of these stories uniformly are fictitious, the purpose of which, as well as many other insinuations in this same plan, is to bring doubt among the Arab peoples in relation to their true friend--the Soviet Union. There have been no proposals regarding the delivery of crude oil and petroleum products to England.

CONFLICT WITH THE ARAB WORLD 77

> Question. The American press has published information that an Israeli tanker is making a regular run between ports of the Soviet Union and Israel. What can you say about this?
>
> Answer. This is also a clumsy lie from beginning to end and carries clearly a provocative nature. The Soviet Union is no way sells crude oil and petroleum products to Israel. [6]

While ostensibly lending support to the policies and programs of the Arab nations, the Soviet Union, through action or inaction, has embarked on a course which conceivably could lead to a collision with the ultimate interests of the Arab oil-producing nations. This possibility stems from the importance of oil exports to both the Arabs and the USSR. Most experts very early discounted the probability that the USSR could come up with any significant quantities of "extra" oil to take advantage of the temporary denial in 1967 of Arab oil to European markets. Their investment philosophy simply does not permit the development of any shut-in capacity to produce. But whatever flexibility was available was put to good use, as attention was concentrated in those relatively small markets where political and economic gain would far outweigh the amounts of oil involved. The larger markets were left alone.

The oil supply emergency of 1967 provided the Soviet Union with ample arguments to support future oil sales campaigns in Western Europe; namely, that the Middle East as a supplier of oil must be considered politically unreliable, that Europe would do well to re-examine the feasibility of continued high dependence on the Middle East and, as an alternative, look to the Soviet Union. With only minor exceptions, the Soviet Union can point to an unblemished decade as a responsible trader in oil. Whether such arguments will be advanced publicly is highly unlikely, but they can be expected to be obliquely put into play during seller-buyer negotiations.

Most observers regard the Soviet Union as the major winner in the Arab-Israeli conflict, despite initial evaluation to the contrary. Although replacement of arms and equipment lost by the Arabs during the six-day war has been costly, the expansion of Soviet influence throughout the whole Middle East

Figure 10
The Israeli Python and U.S. Oil Monopolies

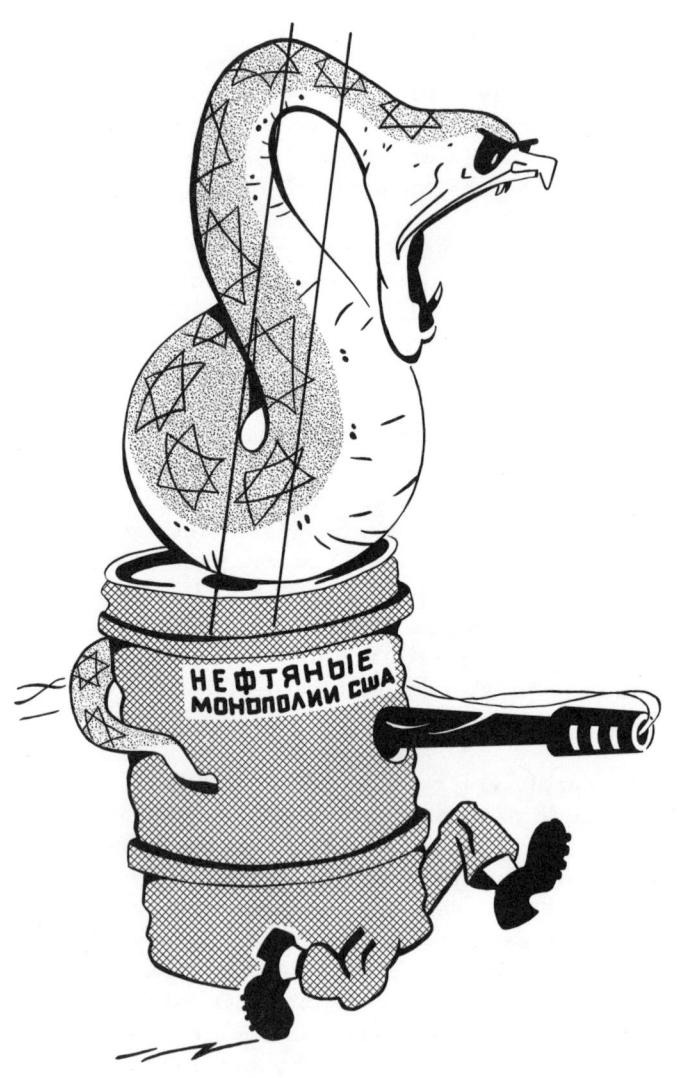

An Israeli python in the shape of the dollar-sign rides on a barrel of oil labeled "Oil Monopolies of the U.S.," which support military intervention in the Middle East, as implied by the smoking barrel.

Source: <u>Pravda vostoka</u>, February 11, 1968, p. 3

CONFLICT WITH THE ARAB WORLD

has been more than ample compensation. Much of this has been in the form of a greatly enlarged military presence. But a large part of their success has been gained simply through the show of backing for the Arab cause, through the use of American support for Israel to widen further the gap between the Arab world and the United States, and through the vigorous pursuit of expanded technical aid and economic assistance programs in the Middle East.

It is difficult to assess correctly the Soviet attitude toward Middle East oil. Discounting Soviet interest in Middle East oil per se had been the accepted attitude of the past, but the very heavy demands for oil being placed on the Soviet Union by East Europe, plus rapidly increasing domestic requirements, may have altered this attitude in recent months. Certainly the recent contracts negotiated by Saudi Arabia and Iran to supply oil to Rumania and Czechoslovakia, respectively, remove some of the Soviet burden for meeting the growing thirst for oil in East Europe; such contracts most probably could not have been negotiated without at least the tacit approval of the Soviet Union. Moreover, in recent months the USSR itself has concluded negotiations which will make certain quantities of Middle East oil available for import and/or re-export. Nevertheless, it is probable that the major Soviet interest in the Middle East with respect to oil lies in taking full advantage of available opportunities to reduce the influence of the international oil companies, lending at least verbal support to the emerging national oil companies. Those purchases of Middle East crude oil which the Soviet Union may make probably can be regarded as politically motivated, and less as evidence of a real need for outside oil to augment domestic supplies. American prestige and influence in the Arab world have suffered as a consequence of the aftermath of the Arab-Israeli war, while, as noted, that of the Soviet Union has shown concomitant gain. Although most Arab nations belatedly accepted the fact that the United States did not provide military support to Israel during the war--an accusation that was employed by the Arabs as sufficient reason for imposing an oil embargo on deliveries to the United States and to the United Kingdom and West Germany as well-- the absence of American support, or understanding, of their cause has been very difficult for the Arab nations to accept.

Many of the Arabs still distrust the oil pricing policy of the Soviet Union. In view of the growing number of contracts to export oil from the Middle East to East Europe and to the USSR, the Saudi Arabian Minister of Oil and Mineral Resources, Sheikh Ahmed Zaki Yamani, felt it prudent to warn his colleagues that "the Soviet Union exports oil at prices below the world level, thus shaking international prices."[7] By this, Yamani implied that imports of Middle East oil by the Soviet bloc would in effect enhance the latter's capability to export to the West, that these exports would be offered at below prevailing world prices, that other suppliers would be forced to follow suit, and that as a consequence the oil-derived income of the Middle East countries would suffer.

This distrust is not likely to abate abruptly in the coming months. The Arabs are increasingly conscious of developments in the marketplace, particularly those which might impact upon income levels. Soviet moves undoubtedly will be watched just as closely as those of the Western oil companies, and Arab spokesmen are not apt to confer any immunity to criticism upon the USSR. Should these spokesmen come to recognize that sales of Middle East oil to Eastern Europe and the Soviet Union in effect will extend the life of the Soviet bloc oil export campaign, in view of the announced reduction in exportable surpluses originating with the USSR in the coming years, (see Chapter 8, section on "Soviet Oil Supplies, 1970"), the bloc will have to exert great care in the disposal of its Middle East purchases.

CHAPTER **8** THE FUTURE OF
COMMUNIST TRADE IN OIL

RECAPITULATION

Western interest in the development of the oil and gas industries of the Soviet bloc has not stemmed from any idle academic curiosity. Rather, this interest has been largely commercial and those who are active in one way or another in international oil and gas markets have considered it to their advantage to be cognizant of the export potential of the bloc. If competition is to be met, and met successfully -- and today most international oil companies have come to recognize the Soviet bloc as just another competitor to be reckoned with-- prior assessment of the bloc's short-term and long-run capability becomes a meaningful asset.

The re-emergence of the Soviet Union as an oil exporter of international standing, although following in unmistakable parallels those lines of development set down in the late 1920's and early 1930's, triggered a response in the West of unusual proportions.

In retrospect, the growth in bloc oil sales to a level exceeding one million barrels daily by 1967 was accompanied by waves of publicity often well out of proportion to the quantities being marketed. The apprehension of being swept under by a "sea of Red oil" never found basis in fact, although in the days when bloc oil sales were exhibiting extremely high relative rates of growth, that is, in the late 1950's and early 1960's, the apprehension was indeed genuine. The mounting fear on the part of governments and industry of not being able to compete or cope with a commodity which could, and probably would, be offered for sale in the marketplace at prices well below cost, simply for the purpose of bringing the so-called international oil cartel to its knees, also never found basis in fact. This concern derived from the fear that Soviet bloc oil sales in every instance would be used for political gain, that oil in the most strict sense would be nothing more than a convenient extension of the political aims and aspirations of the Soviet Union, and that the rules of the game, as understood by the established international companies, would be completely ignored.

On the contrary, most observers have come to recognize that bloc oil sales to the West for the most part have been economically motivated, and that this oil has been used as barter in payment for advanced equipment and technology. It might be generalized that the oil has purchased "time," time which otherwise would have been spent in the development of processes and in the accumulation of know-how to produce the advanced equipment and technology the bloc was now gaining in barter for its oil.

That trade in oil has been primarily economically motivated does not contradict the posture of the Soviet Union toward trade in general, as expressed in the oft-quoted assertion by Khruschev: "We value trade least for economic reasons and most for political purposes." (This comment, made by Khruschev to a visiting delegation of U.S. Congressmen in September, 1955, is presented in its usual form--somewhat out of context. Krushchev said in full: "We value trade least for economic reasons and most for political purposes as a means of promoting better relations between our two countries.")[1] It simply became a matter of fact that the Soviet Union valued its trade in oil with the West as a partial solution to its desire to industrialize at the most rapid rate possible. The handling of oil, considering there was little else to offer, dictated getting maximum return for quantities sold abroad and placing these quantities in markets where this maximum return, hopefully in the form of advanced equipment and technology, could be assured.

In retrospect the Soviet Union (and East Europe as well) has conducted itself in a manner not generally distinguishable from that of its Western competitors. This does not mean that the Soviet Union has never used oil as a political weapon, for it has,[2] but for the most part its conduct in international

THE FUTURE OF COMMUNIST TRADE IN OIL

trade may be termed "proper," for it most certainly recognized that failure to conduct itself properly in the marketing of oil could well jeopardize its trading position in all commodities.

The "historic" 14 per cent share of European markets to which Soviet officials laid early claim has yet to be regained. Although the amount of Communist oil imported by Western Europe has increased from year to year, the percentage participation of Communist oil in the Western European supply has held relatively constant, reflecting growth in indigenous demand at a pace in keeping with these increased imports. In 1960, Communist oil represented an estimated 8 per cent of the demand for oil in Western Europe and 4.5 per cent of the demand in the Free World outside the United States. In the succeeding seven years, no important changes in these shares have taken place. The dependence on Communist oil varies greatly by country; generally, however, for all West European countries, with the exception of Finland, if the dependence on Communist oil is relatively high the quantities involved are relatively small and the development of substitute supply, should the need ever arise, would not be particularly difficult.

It is unlikely that the Soviet Union, and the Soviet bloc as a whole, will abruptly change its approach to the marketing of fuels in the coming years. Lacking, at least for the foreseeable future, a wide variety of goods and materials for which an export demand could be found in the West, the Communist countries in all probability will continue to emphasize the exporting of raw materials and other items considered attractive in terms of international demand and which can compete in quality with those offered by Western suppliers.

At the same time, the bloc may be expected to continue, and even to enlarge, its import of plant and equipment from the West. If past patterns are indicative, payment for these imports will be sought in mutually advantageous barter arrangements.

This means that the sales of energy will be largely concentrated in Western Europe. Trade with developing countries should continue along those lines and within those limits exhibited in the past. It is believed that the activity in 1965, when the growth in oil sales to developing countries outstripped the growth in deliveries to the industrialized countries, was unusual and not the making of a precedent.

The bloc is expected to continue to conduct itself as just another competitor in the marketplace. Indeed, the profit motive seems to have crept into oil sales somewhat earlier than it was accepted by the domestic economy. Interest in gaining maximum return should not wane. Over the long run, the bloc probably will attempt to solidify its position in those markets which it considers to be the most lucrative, and to move cautiously when approaching new markets.

PROBLEMS OF FORECASTING

Any attempt to forecast the possible levels of export of oil and natural gas from the Soviet bloc through 1975 must reflect the acceptance of a number of basic assumptions, including that (a) the energy economies of the Soviet bloc will continue to develop in an orderly fashion, (b) world political conditions will remain essentially unchanged, (c) there will be no marked shifts in governmental attitudes regarding East-West trade, and (d) there will be no change in evaluation by the bloc of energy as a marketable commodity in international trade. The denial of any of these assumptions, for whatever the reason, probably would have a recognizable impact on energy export levels.

The Soviet Union possesses a tremendous primary energy potential, to the extent that resources should pose no limitation on levels of production in the coming years. It can be expected that the energy sector of the Soviet economy can develop, within reason of course, along whatever guidelines are deemed appropriate by Soviet planners. Similarly, the quantities of oil and gas that become available for export will not be by accident but will be a careful recognition of the economic and political programs of the government.

East Europe, on the other hand, is poor in energy. This area became a net importer of fuel in the early 1960's and indigenous production of energy should lag even further behind demand in the coming years. Just how this deficiency in supply is to be met will bear heavily on the amounts of energy the bloc will have for sale to non-Communist countries. The Soviet Union for some time had been reluctant to commit itself fully to meeting East European energy requirements, particularly after 1970, and it was thought that East Europe might look to the West as a major supplier of crude oil. However, recently concluded trade agreements between several of the East European countries and the Soviet Union, covering 1971-75, although sketchy in detail, have reduced somewhat the potential for Western crude oil sales in this area. The intent of the Soviet Union to cover at least a significant portion of the growing requirements for crude oil in East Europe is also underlined by the initiation of construction of a second crude oil pipeline parallel to the Friendship pipeline. Such action does not erase completely the prospects for the marketing of Western crude oil in East Europe, but these prospects have been reduced.

Unquestionably the events of August, 1968--the invasion of Czechoslovakia by Warsaw Pact forces--have in their own way weakened the prospects of large-scale exports of oil to Eastern Europe from Iran and other oil producers in the Middle East. Although a number of tentative barter agreements (between the individual East European countries and Iran, for example) have been signed, outwardly involving considerable amounts of oil, the participating countries have failed to follow through with specific arrangements on prices and delivery terms. Clearly the Soviet Union, in attempting to mend its political fences with East Europe, somewhat damaged of late, has been reassuring its East European neighbors that Soviet oil and gas will continue to be made available in increasing amounts. Undoubtedly some room will remain for imports of oil from the West, in particular the Middle East and North Africa, and Venezuela to a lesser extent, if for no other reason than the economic benefits to be derived from such trade, which the USSR would be among the first to recognize. Equally important would be the closer political ties between the Middle East and the Communist world that this trade would bring about. But the greater political rigidity which inevitably follows such severe action as the "invited" invasion of one Communist country by its political allies may give rise to heightened rigidity in the relationships of the so-called socialist commonwealth with outsiders.

It should be made clear that difficulties in forecasting are not necessarily limited to a distant year, and the imprecision inherent in attempts to develop forward-looking estimates in advance of announcement of official Soviet plans for the development of the economy in a given year can be abundantly illustrated.

For example: Based on trends in the expansion of crude oil production in the USSR during the first three years of the Five Year Plan (1966-70), wherein increments to output averaged 22 million tons per year, it had been considered reasonable to project 1969 output to about 332 million tons, with output in 1970 reaching perhaps 356 million tons. Despite all of this educated reasoning, the 1969 plan for crude oil production was announced as 326.5 million tons.[3] Even after allowing for what has come to be regular annual plan overfulfillment, extraction in 1969 most probably will not exceed 328 million tons. This of course meant a concomitant reduction in the 1970 estimate which, reflecting other pertinent information culled from the technical press,[4] was now reduced to 348 million tons.

While in retrospect it is easy to find correlative evidence to justify the sharp dip in crude oil growth rates, clear-cut indicators of an immediate slowdown were absent. It is apparent that after enjoying a succession of years of prosperity and continued growth in increments to production, the oil extracting industry of the Soviet Union has been forced to pause, temporarily at least, to allow other sectors of the oil industry which had begun to lag badly, to catch up. This rather abrupt slowdown reflects admitted failures in pipeline construction and in the development of new oil finding and producing equipment responsive to the more demanding conditions of greater depths, higher formation temperatures and pressures, and exacting climate and terrain now being encountered in the expanding search for oil and gas in the Soviet Union.

Having accepted reduced estimates of crude oil availability, it then became necessary to re-examine the overall supply and demand for oil in the Soviet Union, with reappraisal of exports to other Communist countries and to non-Communist countries and of the domestic demand for oil in the USSR as well, in an attempt to determine where the impact of reduced supply might fall.

THE FUTURE OF COMMUNIST TRADE IN OIL

Over the years fluctuations in increments to crude oil production in the Soviet Union have been extreme at times, and these fluctuations have been partnered with uneven increments to apparent consumption of petroleum, although not always in parallel. The extent of these divergencies is depicted in Figure 11, which provides data for 1960 through 1968 and the production plan for 1969. It is beyond the scope of this study to investigate those factors underlying the uneven annual growth in consumption of petroleum. Similarly, analysis in detail of the causes of uneven annual growth in crude oil extraction is left to other scholars. But the fact that annual increases in consumption have been so erratic over the years, for reasons not readily explained, greatly compounds the difficulty of forecasting on a year-to-year basis. Fortunately, these fluctuations can be smoothed out somewhat when looking into the future.

Annual deliveries of crude oil and petroleum products to other Communist countries are held relatively rigid, having been fixed through previously negotiated trade contracts, and are normally subject to comparatively small upward or downward deviation. This eases somewhat the burden of forecasting.

Thus, declines in oil supply have to be accounted for in adjusted allocation to the domestic market and/or in those surpluses available for export to non-Communist countries. The absence of severe deviations in the growth of oil exports to non-Communist countries, when compared with the fluctuations in local consumption, seems to indicate that Soviet planners have found it more expedient to make whatever adjustments are necessary in that sector--local consumption--where such adjustments could, if only because of the much larger base, be assimilated much easier.

OUT OF OIL?

In recent months published analyses of prospects for the export of oil from the USSR have been discolored by reasoning to the effect that the USSR would not have oil in quantities sufficient to satisfy East European demand by 1980, which in

Figure 11

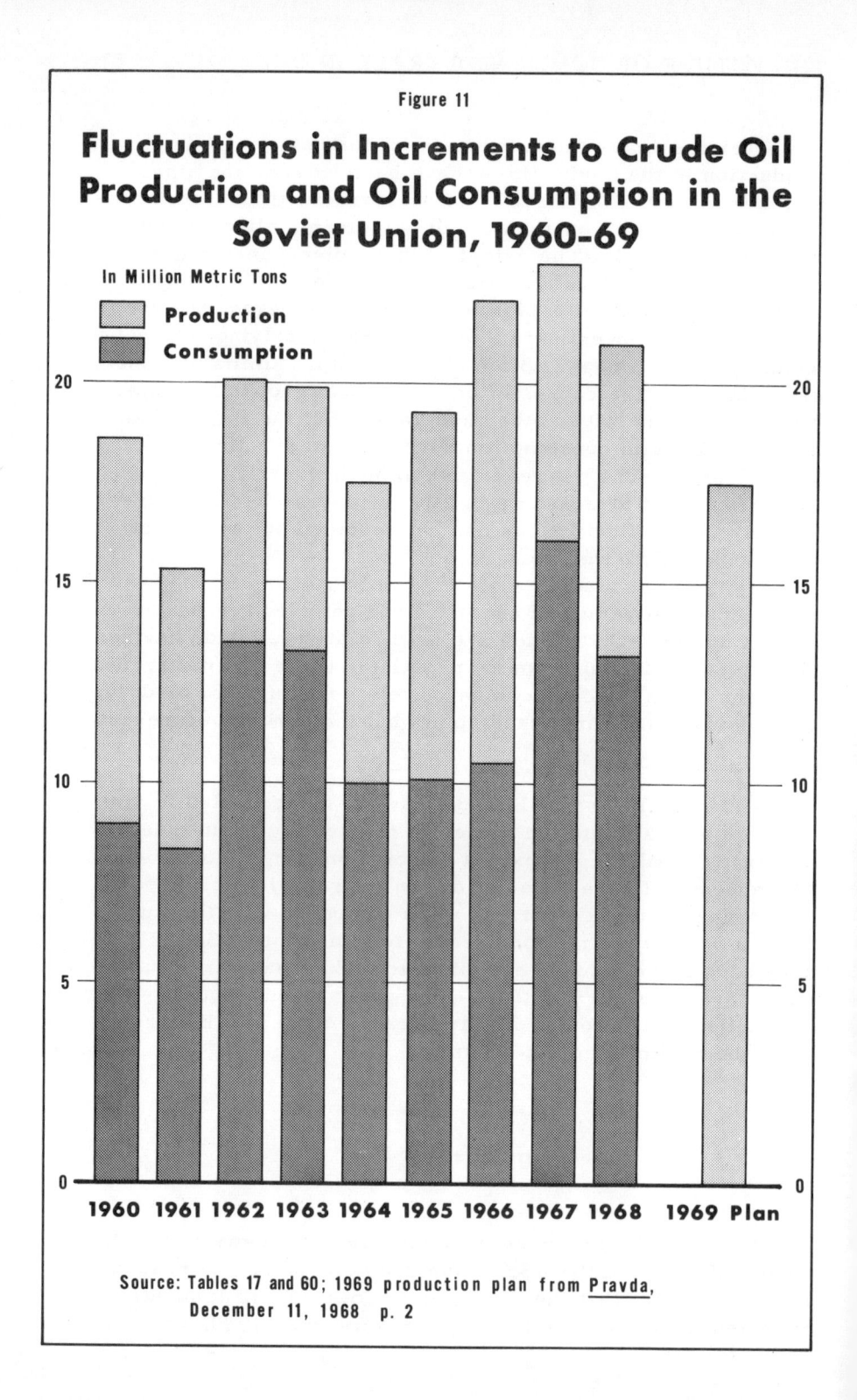

Fluctuations in Increments to Crude Oil Production and Oil Consumption in the Soviet Union, 1960-69

In Million Metric Tons
- Production
- Consumption

Source: Tables 17 and 60; 1969 production plan from Pravda, December 11, 1968 p. 2

essence would mean the disappearance of Soviet oil from world markets.

The presumption that the Soviet Union would be unable to meet East European crude oil requirements and that major imports from the West would be required by East Europe has derived largely from a 1966 article titled "Where Is the Oil To Come From?" which appeared in the Polish newspaper Polityka. Its author, Stanislaw Albinowski, a respected Polish journalist who specializes in foreign trade, argued that in view of limited oil supplies on part of the Soviet Union--limited because of rapidly expanding domestic demand--opportunities for importing crude oil from selected countries of the Third World (that is, the Middle East) would have to be considered. His calculations indicate the possibility of imports from the Third World reaching 90 million tons by 1980. Albinowski had laid the background for his proposal in an earlier, less definitive article, in which he pointed to the economic desirability of importing fuels from the Third World over purchases from the Soviet Union or attempts to develop indigenous production to the levels desired. (The full text of both articles appears in Part II, in section titled "Soviet Trade in Oil.")

Responsible journalists have erred by accepting without question all of Albinowski's findings and in particular his major conclusion of an impending oil shortage within the Soviet bloc, brought about by growth in demand outstripping growth in supply.[5] While this makes eye-catching headlines, it simply is not true and creates false illusions about the future of Communist oil and gas, illusions which could become dangerous if allowed to enter into the formulation of Western foreign policy posture toward the Middle East.

Albinowski is reasonably correct in his evaluation of energy supply in the Soviet Union, but he far overstates the future domestic demand of that country for crude oil. He observes that per capita consumption of crude oil in the USSR in 1965 was 870 kilograms, whereas calculations based on official government data show it to have been about 771 kilograms, derived as shown in Table 32. Proceeding from this base, then, for 1965, and accepting his estimates of the percentage growth in per capita crude oil consumption through 1980 (about 5.7 per cent per year) and of crude oil production in the USSR, per capita consumption reaches 1.77 tons in 1980,

TABLE 32

Computation of Per Capita Consumption
of Crude Oil in Soviet Union, 1965[a]

Item	Quantity			
Crude oil production	243.0	million metric tons		
Imports of petroleum products	2.07[a]	"	"	"
Total supply	245.07	"	"	"
Export of crude oil	43.4	"	"	"
Subtotal	201.67	"	"	"
Export of petroleum products	22.8[a]	"	"	"
Subtotal, available to domestic economy	178.87	"	"	"
Per Capita Consumption	0.771 tons[b]			

[a] In terms of crude oil equivalents. It is estimated that one ton of crude oil is the equivalent of 0.92 ton of petroleum products.

[b] The population of the Soviet Union in 1965 was reported at 232 million. Thus: 178.9/232 yields a per capital consumption in 1965 of about 0.771 tons, or 771 kilograms.

Source: Tables 17, 18, and 19. Population data from Narodnoye khozyaystvo SSSR v 1965, statisticheskiy yezhegodnik (Moscow, 1966); p. 7.

or a total of 495 million tons for the country as a whole, which implies an exportable surplus of 135 million tons of crude oil equivalent--roughly double his estimate.

The demand for crude oil in the other CEMA countries as developed by Albinowski is also a gross exaggeration. Demand for petroleum products in East Europe in 1965 was on the order of only 27 million tons and projections suggest that demand will likely increase to perhaps 47 to 48 million tons by 1970 and further to 75 million tons daily by 1975, for an average annual rate of growth of roughly 10.8 per cent. Most probably this growth rate will begin to decline by the mid-1970's, and consumption by 1980 may be on the order of 110 million tons, roughly two-thirds of Albinowski's estimate of 170 million tons. After allowing for local output, this means that East Europe in 1980 will be faced with an import requirement on the order of 80 million tons which, if met fully by the USSR, would require less than 50 per cent of that country's oil surplus (expressed in crude oil equivalents) for the year, as derived above.

It might be argued that Albinowski deliberately overstated his case in an apparently unsuccessful attempt to shake the hold of the coal industry on the energy sector of the East European economy. But his forceful handling of the questionable undue reliance for oil on a non-Communist area may have accomplished what he was hoping for. It may have completed the rationale that diversification of trade, to include the developing countries of the world, not only was economically desirable, in that these countries offer an attractive market for East European goods and equipment, but was also politically desirable, and tenable, as well.

COMMUNIST OIL THROUGH 1970

East European Search For Oil

East Europe has long recognized the dangers inherent in having become so overwhelmingly dependent upon the Soviet Union for oil to supply the civilian and military sectors of its economy. Until recently this apprehension was not openly discussed nor admitted. But one highly placed Communist source in Czechoslovakia, commenting on the liberalization process then taking place in his country, observed that the Soviet Union could bring pressure to bear, for example, on oil supplies or on iron ore supplies, virtually all of which were derived from that country.[6] Yet he volunteered that his country was going to go ahead and do what it must, Soviet reaction notwithstanding.

Some three months later, the invasion of Czechoslovakia by Warsaw Pact forces presented the Soviet Union with ample opportunity to impose economic sanctions, but none were taken. Indeed, the Soviet press and radio took great pains to point out to its home audience and to the rest of the world that the invasion in no way disrupted the orderly flow of goods and equipment into Czechoslovakia from the USSR.[7]

Nevertheless, the dependence of East Europe on the Soviet Union is expected to continue, for there is no alternative, and the inroads that other Communist countries will make on whatever amounts of oil the Soviet Union has for export cannot be lightly dismissed as falling within the easy capability of the USSR. On the contrary, to meet the energy requirements of East Europe and of other Communist countries, such as Cuba, will place a heavy burden on the Soviet Union. Fortunately the Soviet Union has made a number of major discoveries of oil and gas in recent years, the development of which will permit substantial increases in production. During 1966-70 the extraction of crude oil in the Soviet Union should increase by an estimated 105 million tons, to about 348 million tons. Some 68 per cent of this increment in new supply will be required to meet the anticipated growth in domestic demand, a somewhat greater imposition than during the preceding five-year period, when 60 per cent of the increment in new supply was needed to cover growth in demand. But because additions to supply in absolute terms will continue to outstrip increases in demand, the Soviet Union can be expected to make considerable increments to its exportable surplus of oil in the five-year period ending in 1970.

It is believed, but with less authority, that the Soviet Union will continue to add to its exportable surplus of petroleum during 1971-75. Based to a large degree on acceptance of

Soviet claims[8] regarding the potential of the newly discovered oil and gas fields in West Siberia and of those on the Mangyshlak Peninsula of Kazakhstan and in Belorussia, production of crude oil is scheduled to increase to not less than 460 million tons by 1975,[9] thus permitting the USSR to challenge the United States for the position as the leading producer of crude oil in the world. For these years, claims on the new supply by the growth in domestic demand may reach 71 per cent, leaving a relatively smaller percentage available for export, but a larger share in absolute terms.

It has been noted that obligations to meet the crude oil and petroleum product needs of other Communist countries will make sharp inroads into whatever exportable surpluses the USSR may develop. How much so is summarized in Table 35.

Deliveries of crude oil to East Europe alone during 1966-70 are to total well in excess of 1 billion barrels. In addition, East Europe can be expected to import some oil from the Middle East and North Africa. Such imports might reach as much as 6 million tons by 1970, given the proper political and economic circumstances, and, together with Soviet supplies, should permit the development of product availability in these countries to levels in excess of local requirements. As a result the capability to export oil from East Europe should increase during these years, if only marginally in terms of international oil movements, reaching perhaps 8 million tons by 1970. (See Table 36). Again, at least one-half of these exports probably will originate with Rumania.

The closure of the Suez Canal in early June, 1967 as an adjunct to the Arab-Israeli conflict effectively denied Middle East oil to East Europe. To illustrate, in 1966 the Oil Consortium in Iran agreed to make crude oil available to the National Iranian Oil Company (NIOC) in amounts of up to 2 million tons in 1967 and of up to 1 million tons additional each year over the previous year, through 1971, for a total approaching 20 million tons.[10] This crude oil was to be sold only to East Europe (excluding Yugoslavia), which would avert NIOC oil coming into direct competition with Consortium oil in European markets. But the closing of the canal and the subsequent high costs of shipment around the Cape of Good Hope priced large amounts of Middle East oil out of the East

Figure 12
Oil and Gas Deposits in Western Siberia

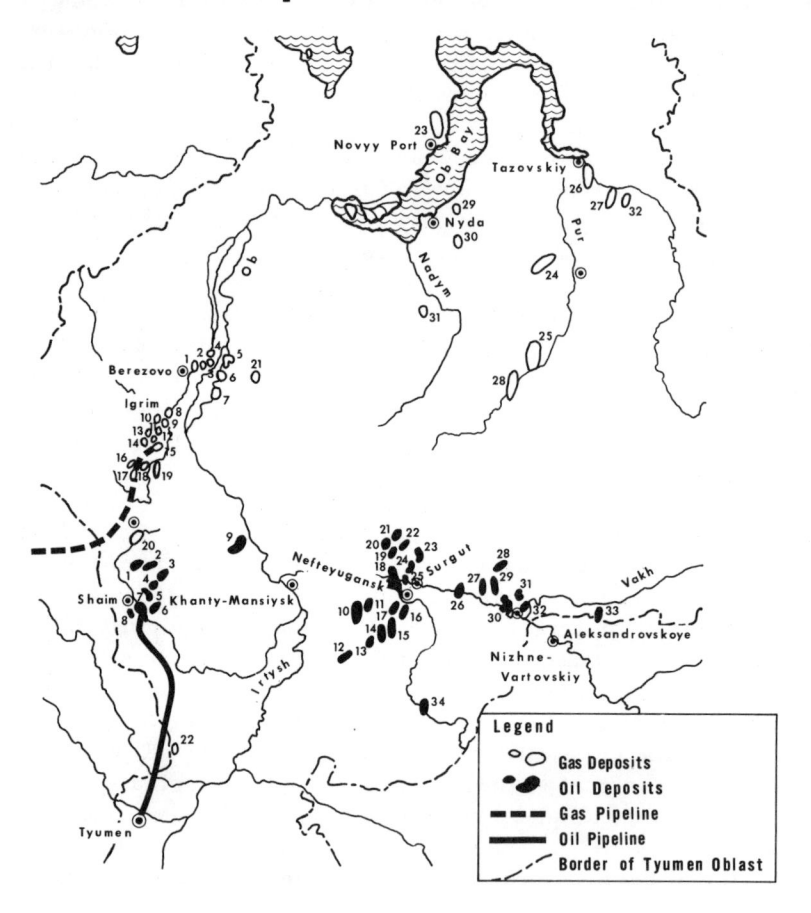

Legend
- Gas Deposits
- Oil Deposits
- Gas Pipeline
- Oil Pipeline
- Border of Tyumen Oblast

Gas Deposits:
1-Berezovo
2-Deminsk
3-Northern Alyasov
4-Southern Alyasov
5-Pokhromo
6-Chuel
7-Tugiyan
8-North Igrim
9-South Igrim
10-Paul-Tur
11-Nuliy-Tur
12-Eastern Syskonsyninsk
13-Western Syskonsyninsk
14-Southern Syskonsyninsk
15-Punga
16-Gornyy
17-Ozero
18-Shukhtungort
19-Sote
20-Upper Konda
21-Northern Kazym
22-Lenin
23-Novyy Port
24-Urengoy
25-Gubkin
26-Tazov
27-Zapolyarnyy
28-Komsomol

Promising areas:
29-Nydin
30-Medvezhe
31-Nadym
32-Russkiy

Oil Deposits:
1-Lemin
2-Danilov
3-Ubin
4-Central Mylymin
5-Mortymin
6-Teterev
7-Trekhozera
8-Mulymin
9-Kamen
10-Salym
11-Pravda
12-Upper Salym
13-Malo-Balyk
14-Southern Balyk
15-Middle Balyk
16-Mamontov
17-Ochimkin
18-Ust-Balyk
19-Bystryy
20-Vingin
21-Minchimkin
22-Vershina
23-Northern Surgut
24-Western Surgut
25-Saygotin
26-Lokosov
27-Northern PoKura
28-Agan
29-Vatin
30-Megion
31-Lower Vartovsk
32-Samotlor
33-Vakh
34-Taylakov

Source: A.K. Kortunov *Gazovaya promyshlennost SSSR* (Moscow: 1967), p. 38

Figure 13
Oil and Gas Deposits in Western Kazakhstan and Western Turkmen

Legend ◯ Gas
　　　　 ◐ Gas/Oil

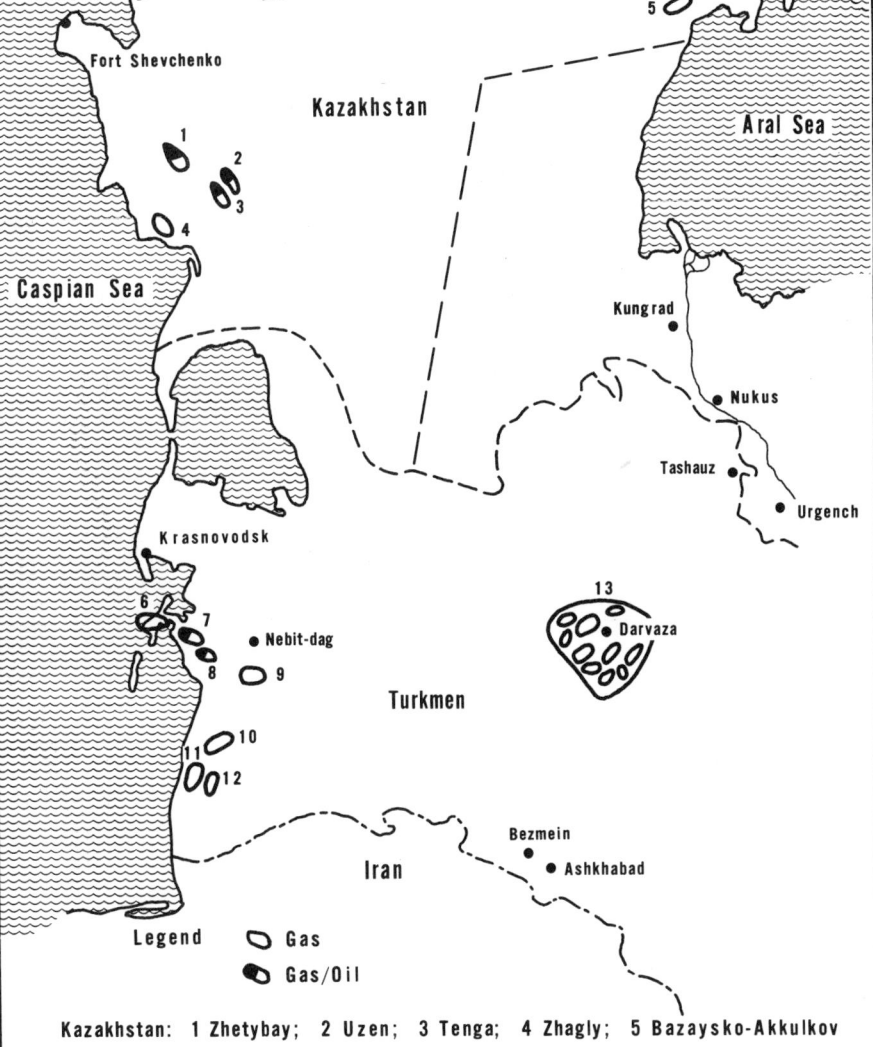

Kazakhstan: 1 Zhetybay; 2 Uzen; 3 Tenga; 4 Zhagly; 5 Bazaysko-Akkulkov
Turkmen: 6 Cheleken; 7 Kotur-Tepe; 8 Barsa-Gelmes; 9 Kyzylkum;
　　　　　10 Kamyshldza; 11 Okarem; 12 Miasser; 13 Zeagli-Darvaza Group

Source: A.K. Kortunov, Gazovaya promyshlennost SSSR (Moscow: 1967) p. 40

TABLE 33

Fluctuations in Short-Term and Long-Term
Crude Oil Production Planning in the Soviet Union

(in million metric tons)

Year	Original Plan	Revised	Actual
1959	128	128	129.6
1960	144	144	147.9
1961	161	164	166.1
1962	181	185	186.2
1963	200	205	206.1
1964	220	222	223.6
1965	230 to 240[a]	242	242.9
1970	390[b]	345 to 355[c]	--
1975	545[b]	450 to 470[d]	--
		460 to 470[e]	
		not less than 460[f]	
		450[g]	
1980	690 to 710[b]	600 to 620[d]	--

[a] This goal, and those for 1959-64, were a part of the Seven Year Plan (1959-65).

[b] As indicated in the Twenty Year Plan (1961-80), released in October 1961.

[c] A goal of the Five Year Plan (1966-70).

[d] Neftyanoye khozyaystvo, 10 (October, 1967), 1-10.

[e] Vyshka, October 12, 1967, pp. 2-3.

[f] The New York Times, January 11, 1969, p. C-39.

[g] Bakinskiy rabochiy, March 13, 1969, p. 2. Note that this most recent statement of anticipated crude oil production in 1975 is also the least optimistic.

TABLE 34

Revisions in Soviet Coal and Gas Production Plans,
1965 and 1970

Year	Commodity	Unit of Measure	Original Plan	Actual or Revised
1965	coal	million metric tons	612	577.7
	gas	billion cubic meters	150	129.4
1970	coal	million metric tons	665 to 675	a
	gas	billion cubic meters	225 to 240	215[b]

[a] The goal for 1969 for coal was set at only 595 million tons, unchanged from actual output in the preceding year. That coal output has fallen so far behind original intentions has been ignored in the Soviet press.

[b] An early revision reflecting difficulties encountered in the extraction sector during the first two years of the Five Year Plan. The plan for 1969 called for the extraction and production of only 185.8 billion cubic meters, and it seemed unlikely that by 1970 output will exceed 200 billion cubic meters, if that.

TABLE 35

Soviet Crude Oil and Petroleum Product Exports
to Other Communist Countries, 1965, 1970, and 1975

(in million metric tons)

Importer	1965[a]	1970[b]	1975[b]
East Europe[c]	22.436	39.5	57
Yugoslavia	1.011	2	4
Cuba	4.727	5	6
Communist China	neg.[d]	neg.[d]	neg.[d]
North Korea	0.392	0.5	1[e]
North Vietnam	0.121		
Total	28.7	47.0	68.0

[a] Vneshnyaya torgovlya SSSR za 1965 god (Moscow, 1966).

[b] Estimated.

[c] Bulgaria, Czechoslovakia, East Germany, Hungary, and Poland. Rumania is not an importer of oil from the USSR.

[d] Specialty products only.

[e] Order of magnitude only.

TABLE 36

Probable Supply and Demand
for Oil in East Europe, 1970

	Amount (Million Metric Tons)
Supply	
Crude oil	
Domestic production	17.5
Imports	
From USSR	36
From non-Communist sources	6[a]
Total crude oil	59.5
Petroleum products	
Yield from refining	53.7[b]
Imports	3.5[c]
Natural gas liquids	neg.
Total petroleum products	57.2
Demand	
Crude oil	
Charge to refining	57.7
Losses, increments to storage, direct use	1.8[d]
Available for export	0
Petroleum products	
Domestic demand	47.7
Available for export	9.5
To other Communist countries	1.5[e]
To non-Communist countries	8

[a] The Middle East, North Africa, and Venezuela.
[b] Estimated at 93 per cent of the refining charge.
[c] A small portion obtained from Free World sources.
[d] Estimated at 3 per cent of the crude oil supply.
[e] The Soviet Union.

European market. Continued closure of the canal means that the prospects for Middle East oil moving to East Europe will be limited to the Iraqi pipelines, through the projected UAR pipeline paralleling the canal, through the existing and/or projected pipeline(s) across Israel, or perhaps via the newly completed Syrian pipeline which terminates at Tartus on the Mediterranean. More easily perhaps, Arab oil from North Africa might become available.

In April, 1968 The General Petroleum and Mineral Organization (Petromin), the Saudi Arabian state-owned oil company, announced it had signed a four-year agreement covering the shipment of 9 million tons of crude oil to Rumania. These exports were to start in June, 1968, with repayment in the form of about $100 million of goods and industrial equipment. The crude oil would be supplied to Petromin by the Arabian-American Oil Company, the major producer in Saudi Arabia. In October, however, Saudi Arabia abruptly halted shipments, indicating displeasure with tardiness of Rumanian repayment.

Impact of 1967 Middle East Crisis

It is not yet clear just how the events of August, 1968 will impact upon the possible future movement of oil from the Middle East into Eastern Europe, if indeed there is any impact at all. Perhaps the major deterrent to such movement is the lack of hard currency on the part of the East European countries to pay for this oil. Those agreements which so far have been negotiated generally have involved the barter of East European goods and equipment for Middle East oil. Any disenchantment with the quality of these goods and equipment most likely would be reflected in a reluctance to participate in additional barter arrangements, with future sales, if any, to be conducted on a hard currency basis.

However, there is evidence that East Europe may be willing to look outside the Middle East for additional supplies of oil. Rumania, for one, signed an agreement in May, 1968 covering the import of crude oil from Venezuela for a period of ten years. Deliveries reportedly are to total 11 million tons, for an average of 1.1 million tons annually.[11]

THE FUTURE OF COMMUNIST TRADE IN OIL

Small, almost token, amounts of oil from North Africa moved into East Europe in 1968. Rumania tested Libyan crude oil in its refineries,[12] but as far as is known, no long-term contract with Libya has yet been signed. Bulgaria purchased small amounts of crude from Algeria and Syria; East Germany imported an estimated 100,000 tons from Egypt and minor quantities from Syria.

If East Europe had been counting on Middle East oil in 1967 and then was deprived of it, this may serve as partial explanation of the increase in imports of Soviet oil products that year, contrary to the trend of declining imports of products as crude imports increase. For undoubtedly any unforeseen shortfall in oil supplies in East Europe would have to be made up by the USSR if industrial activity were to keep pace with plan.

The Middle East crisis of 1967 brought about other difficulties for the Soviet Union, among the most pressing of which was finding some means of supplying oil to its customers in Southeast Asia and the Far East, now cut off from the Black Sea ports of the USSR because of the closure of the Suez Canal.[13] The solution it came upon was unique for the Soviet Union, although certainly not uncommon to Western oil companies.

In its search for alternatives, the Soviet Union has entered into a number of agreements which in essence involve the swapping of Soviet crude oil and/or products for Middle East oil. This Middle East oil will be delivered to Soviet Asian customers and similar quantities of Soviet oil will find their way into Western Europe on international oil company account. The first such swap was engineered during the last half of 1967 when Naftamondial, a Swiss-based oil broker. arranged a (cost-insurance-freight) deal whereby Shell[14] and British Petroleum provided 100,000 tons of gasoline and gas oil from the Persian Gulf and Aden to the USSR for delivery to Ceylon. In return these two companies obtained a comparable amount of Soviet gasoline and gas oil in the Black Sea for their markets in Western Europe.

Some six months later the Soviet Union worked out a three-way swap under which Soviet crude oil, originally destined for Japan, will instead be delivered to British Petroleum for disposal in Western Europe. In turn, British Petroleum will undertake to supply an equivalent amount (an estimated 15,000 to 20,000 barrels daily for eight to twelve months) of Abu Dhabi (Persian Gulf) crude to Japan. Deliveries were to begin in July, 1968.

This was followed shortly by a second East-West crude oil exchange, with France's Compagnie Française des Petroles (CFP) as the beneficiary at the Western end of the contract. [15] The new arrangement covers 3.5 million barrels of crude, with deliveries to begin in June, 1968 and to continue through 1968. The USSR was to be provided with Murban supplies f.o.b. Abu Dhabi from independent Naftamondial for Soviet customers in Japan and Burma. A similar amount of Soviet crude will be picked up in the Black Sea for transfer c.i.f. to CFP in Italy.

Such arrangements can provide little more than temporary relief, as apparently recognized by Soviet officials, who have announced plans to begin construction in 1971 of a 4,000-mile, presumably 40-inch crude oil pipeline linking prolific new oil fields in West Siberia with the Pacific Ocean. [16] The terminus for this pipeline on the Pacific Ocean is likely to be the port of Nakhodka, located about 100 kilometers east of Vladivostok. This pipeline has long been on the drawing board and has been the subject of sporadic negotiations with Japan during the past several years, but until now has lacked any acceptable rationale for construction.

Under one proposal, made by the USSR at a joint economic committee meeting in Tokyo in 1965, the Japanese would finance the construction of this oil pipeline and the USSR would repay with crude deliveries of between 200,000 and 330,000 barrels daily (the equivalent of 10 million to 16.5 million tons annually) for a twenty-year period starting in 1974. [17] But differences over the repayment period derailed negotiations at a second meeting of the committee held in Moscow in June, 1967. Japanese interest in Soviet oil, as a means of reducing dependence upon the Middle East--about 90 per cent of Japanese oil needs are met by imports from the Middle East--undoubtedly has heightened as a result of the Middle East oil crisis of 1967. At the same time, the Japanese government has been under considerable pressure by the steel industry to enter into a long-term purchase arrangement with the Soviet

THE FUTURE OF COMMUNIST TRADE IN OIL

Union, for the steel manufacturers foresee a tremendous market for steel line pipe if a pipeline to the Pacific Ocean were built.[18]

A crude oil pipeline extending from West Siberia to the Pacific Ocean would serve a multitude of purposes. It would bring oil into an oil-deficit area and thus promote industrial development in heretofore neglected East Siberia and the Far East. It would provide the Soviet Union with an opportunity to obtain for itself a larger share of the growing Japanese market. But perhaps more important, particularly so in light of the interruption to normal flow of oil out of the Middle East in 1967, completion of this pipeline, which is geographically and politically isolated from the Middle East, will greatly strengthen the hand of the Soviet Union in Southeast Asia and the Far East in the event of another denial of Middle East oil.

Indeed, preparations already are well advanced to transform Nakhodka into a major petroleum export base. An office of Soyuznefteeksport has been established at Nakhodka in view of the growing export of oil through Far Eastern ports to Japan, North Korea, and North Vietnam, and in view of anticipated exports to Ceylon and Burma.[19] And in early January, 1968 the first stage of an oil terminal was opened at Nakhodka to handle oil originating in Siberia and the Urals-Volga.[20]

Soviet Oil Supplies, 1970

In early January, 1969, at a press conference sponsored by the Foreign Ministry, the Minister of the Oil Extracting Industry of the Soviet Union, Valentin D. Shashin, predicted that Soviet oil exports to non-Communist areas would not continue to rise significantly in the future because of growing requirements at home and in Eastern Europe.[21] As usual, the magnitude of domestic requirements was not discussed, nor was the scale of future deliveries of crude oil to Eastern Europe indicated. Shashin did note, however, that the capacity of the Friendship crude oil pipeline, which supplies oil to all of Eastern Europe except Bulgaria and Rumania, was being expanded from its present 40 million tons per year to 100 million tons.

Although Shashin's remarks concerning the future of Soviet oil trade with the West are subject to varying interpretation, it is abundantly clear that the "Soviet oil offensive" which had its beginnings in the late 1950's has largely run its course. This does not mean that within the next few years Soviet oil will disappear from the marketplace. On the contrary, some form of growth can be expected in Communist oil sales in the coming years, but in relative terms this growth will be quite small compared to those very high rates previously exhibited and, moreover, the absolute increments in sales are likely to decline on the average as well.

Shashin reasoned that increasing domestic demand, together with rising requirements in Eastern Europe, were responsible for the forthcoming slowdown in exports. Without a doubt these factors will bear heavily on unallocated surpluses. But he failed to mention other developments in the energy sector which in turn may be of equal importance. These include the current stagnation in the coal extracting industry and the failure of the gas industry to live up to expectations. Underachievements in expanding the supplies of coal and gas implies that a larger-than-planned share of the growth in energy requirements has to be assumed by oil and, although the amounts involved may be comparatively small, the additional above-plan consumption of oil could easily mean the difference between a relatively significant growth in sales to the West or no growth at all.

The probable supply and demand for crude oil in the USSR in 1970 is developed in Table 37. It can be seen that in 1970 the Soviet Union will have to allocate the larger share of its anticipated exportable surplus of crude oil (41 million tons out of an indicated exportable surplus of 70 million tons) to other Communist countries. Nevertheless, such large deliveries of crude oil obviate the need for increased deliveries of petroleum products to these importers, to the extent that exports of both crude oil and products to Communist countries that year may total 47 million tons, leaving 46 million tons of crude and products for sale to the non-Communist world.

Thus the gross export of petroleum to the West by the Soviet bloc in 1970 may approach 54 million tons, with net exports on the order of 48 million tons, after deducting probable imports of crude oil from North Africa, Venezuela, and

TABLE 37

Probable Supply and Demand
for Oil in the Soviet Union, 1970

	Amount (Million Metric Tons)
Supply	
Crude oil	
Domestic production	348
Imports	0.5[a]
Total crude oil	348.5
Petroleum products	
Yield from refining	247.0[b]
Imports	1.5[c]
Natural gas liquids	6
Total petroleum products	254.5
Demand	
Crude oil	348.5
Charge to refining	268.5
Losses, increments to storage, direct use	10[d]
Available for export	70
To other Communist countries	41
To non-Communist countries	29
Petroleum products	254[e]
Domestic demand	231[f]
Available for export	23
To other Communist countries	6
To non-Communist countries	17

[a] The USSR is to purchase 0.5 million tons of crude oil from Algeria during 1969-70 and a total of 4 million tons through 1975. Iraq is to pay for Soviet technical assistance in the development of its oil industry through the export of crude oil (The New York Times, May 5, 1968, Sec. 3, p. 1).
[b] Estimated at 92 per cent of the refinery charge.
[c] From East Europe.
[d] Estimated at 3 per cent of crude oil production.
[e] Rounded to the nearest million tons.
[f] Based on extrapolation of past trends in consumption.

Figure 14

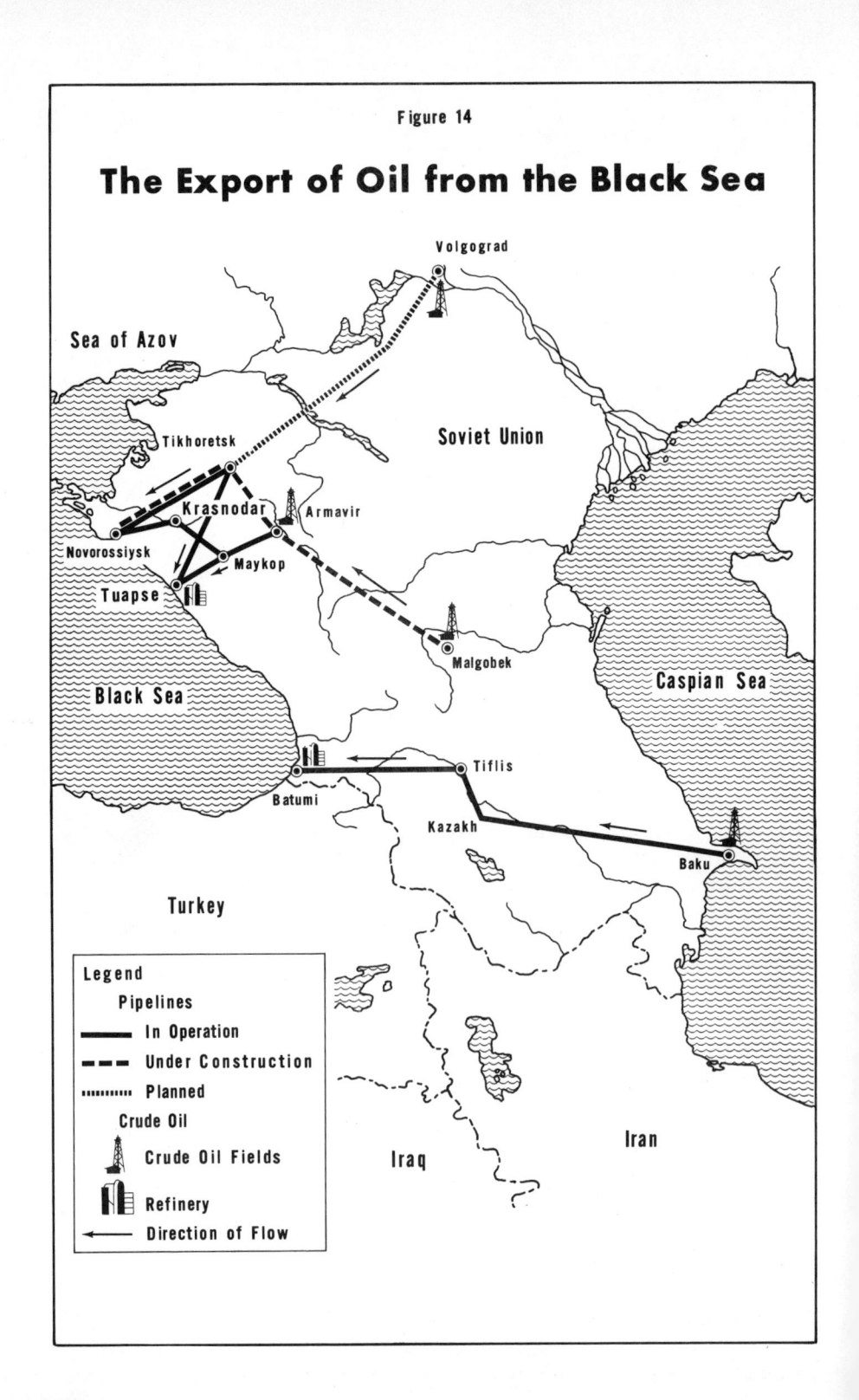

The Export of Oil from the Black Sea

the Middle East. There is no reason to believe that the bloc will not be successful in finding a place for this oil. Worldwide demand for oil continues to exhibit substantial annual gains, particularly in Western Europe where the bulk of the Communist oil most likely will be marketed, and it is anticipated that bloc exports can be accommodated with minimum disruptive impact on the position of other sellers in the market.

That the Soviet Union may have only 46 million tons of crude oil and petroleum products for sale to the West by 1970, for a gain in the 1966-70 period equal to exporting just 1 out of every 10 tons added to production, is clear testimony to the growing thirst for oil both at home and in Eastern Europe. Allocation of the growth in crude oil production during 1965-70 is shown in Table 38.

Most Western observers would agree that the primary motivation behind the expansion in Soviet oil exports has been the need to increase earnings of foreign exchange, and that the future of such exports is perhaps directly linked with the amount of imports--equipment and technology--they need to finance and by the absence of other items of equal attraction in international markets. Ultimate oil resources are sufficient both to support continued growth in oil exports and to meet the needs of expanding home use. But the willingness to continue these exports on an expanding scale has to recognize that increasing costs incurred in producing and transporting this oil, particularly in West Siberia and on the Mangyshlak Peninsula, as well as across-the-board increases for the oil extracting industry as a whole, has removed some of the attraction heretofore ascribed to oil as an item of export.

Campbell believes that the Soviet Union has a comparative advantage in oil and that this comparative advantage is a relatively long-term condition.[22] Although it can be demonstrated that the USSR in reality cannot afford to export oil--it would be more expedient to consume the oil domestically in place of high-cost coals--it can just as easily be demonstrated that the USSR can ill afford not to export oil, if this oil is freed for export through the production of an equivalent amount of natural gas, whose comparative costs are much lower than those for crude oil.

TABLE 38

Allocation of Growth in Soviet
Crude Oil Production, 1965-70

(in million tons)

Indicator	1965	1970	Increment	Allocation of Production Increment (Percent)[a]
Production of crude oil	243	348	105	100
Oil deliveries to other Communist countries[b]	28.9	47	18.1	17.2
Domestic demand	160.1	231	70.9	67.5
Oil deliveries to non-Communist countries	35.5	46	10.5	10.0

[a] The unallocated 5.3 per cent is accounted for in crude oil losses in handling, in processing, increments to storage, etc.

[b] Both crude oil and petroleum products, the latter expressed in crude oil equivalents.

Source: Tables 20, 37, and 42.

Most certainly the success--or lack thereof--in expanding the output of natural gas in the coming years is apt to impact upon the availability of oil for export. Gas instead of oil may increasingly be used to replace high-cost solid fuels which in turn would free more oil for export; the USSR might make more of its gas supplies available for export to Eastern Europe, which would tend to reduce their long-term oil requirements and temper demands upon Soviet oil supplies. In any event, the rapidity with which the new gas finds in West Siberia and in Central Asia can be developed and the success in laying down the ultra-large gas pipelines (diameters of 48, 56, and even 100 inches) which Soviet designers have on the drawing-board will bear heavily upon the levels of future Soviet oil sales.

Similarly, the future role of solid fuels, primarily coal, in the energy balance of the Soviet Union will have its own impact on oil and natural gas exports. Any bold effort to sharply cut back on coal extraction would almost immediately indicate less oil and gas for export, as available supplies of these latter fuels would be called upon to fill the coal gap. The attitude of Soviet experts--planners included--has undergone many changes over the past decade, varying from faith in the supreme value of coal for several hundred years to come, to neglect of it as being a fuel whose twilight has dragged on far too long. This latter view apparently prevailed during at least the first half of the Seven Year Plan (1959-65) as little conscious effort was made in those years to raise the level of coal extraction. Some improvement was discernable in the last two years of the plan, but not enough to enable the Seven Year Plan goal of 600 to 612 million tons of coal in 1965 to be met. Indeed, the plan for coal extraction in 1969 of some 595 million tons represents no increase over the amount produced in 1968, and it is likely that output in 1970 will do little more than equal the lower limit originally intended for 1965, if that.

Nevertheless, recent Soviet statements seem to bar any significant absolute lessening of the role of coal, although its relative share in the primary energy balance will continue to decline. The view that coal is in its twilight is currently regarded as premature by Soviet officials [23] and it is expected that coal will remain as the major source of fuel and raw material for the iron and steel industries and in the generation of electric power. If this view continues to prevail, and if

TABLE 39

Planned and Actual Soviet Coal Extraction, 1959-65

(in million tons)

Year	Annual goals under Seven Year Plan[a]	Actual[b]
1959	512.5	503.3
1960	536.3	509.6
1961	554.6	5υ0.4
1962	571.5	517.4
1963	586.4	531.7
1964	599.8	554.0
1965	612[c]	577.7

[a] A.F. Zasyadko, Osnovy tekhnicheskogo progressa ugolnoy promyshlennosti SSSR (Moscow, 1959), p. 34. These annual plans subsequently were revised as the plan unfolded and goals for a forthcoming year generally were a realistic appreciation of capability and achievements in the previous year.

[b] Narodnoye khozyaystvo SSSR v 1965 g. (Moscow, 1966), p. 178.

[c] Upper limit of 600-612 million ton range established by the plan.

annual plans and output levels are realistic in terms of requirements and capability, then a gradual replacement of coal by liquid and gaseous fuels will be possible, with minimum disruptive impact on export programs.

A LOOK AHEAD TO 1975

It is anticipated that crude oil production in the Soviet Union during the Five Year Plan 1971-75 will increase roughly by 110 million tons, to 460 million tons. This absolute increment will virtually repeat the growth in output estimated for the preceding five-year period, from 243 million tons in 1965 to an estimated 348 million tons in 1970. That the oil extracting industry of the Soviet Union will not be able to enlarge on its absolute growth is in reality clear recognition of the difficulties which face the industry in its forthcoming shift away from the Urals-Volga, which has been the center of oil production since the early 1950's, to the new fields in West Siberia and the Mangyshlak Peninsula.

The shift into Siberia will call for the development and installation of a completely new line of fully automated oil field equipment designed to operate under wide temperature fluctuations, from the heat of the short summer to the prolonged winters when temperatures can drop to -70°C. In addition, the challenge of an inhospitable terrain will have to be met, a frozen, featureless land which becomes impassable when the summer sun melts the permafrost, turning it into an oozing mass, a challenge only slightly less demanding than the cold of the Arctic winter. Moreover, the logistics of moving heavy oil drilling, producing, and pipeline equipment into such an area from distant supply points are frightening. Finally, certain amenities of life have to be provided to attract the drillers, pipeliners, and operators needed to find, produce, and move the oil. In all, for the Soviet Union to undertake the search for oil in its Siberian wastelands undoubtedly will turn out to be the most expensive program in its history.

Yet this is an expense which cannot be avoided if the oil extracting industry is to satisfy those growing requirements for fuel alluded to by Shashin, as well as the additional need to maintain an exportable surplus for trade with non-Communist countries. The importance of Siberian oil and gas is underlined in Table 40, which illustrates the rapid emergence of this area as a major contributor to energy supplies.

Adding to the Siberian increment the growth in output scheduled to be provided by the Mangyshlak Peninsula brings the total growth in these two regions during 1971-75 to 75 million tons [24] or more than two-thirds of the projected national growth.

Given the necessary priority claims on investment funds and equipment, the challenge of Siberia can be met. But to overcome the equally important task of moving labor into Siberia and getting it to stay will pose a much different problem for Soviet planners. Apparently the attraction of higher wages is not enough. In earlier years forced labor did much of the construction in Siberia. Then the call went out to the young people to volunteer for the exciting adventure of extending the frontier of Communism eastward. Young Ivan went, but he was not impressed by what he saw and even less by what he didn't see, and hurried back to the creature comforts of the large city. As a result labor turnover has run extremely high--reportedly 69 per cent of the labor force had left after less than a year's work and 54 per cent had not completed six months.[25] Under the free enterprise system, wage differentials as an inducement are normally accompanied by the best housing and services possible under the circumstances. In the Soviet Union, the rush to develop natural resources, such as oil and gas in Siberia, requires full attention to production processes, with the need for accompanying investment in worker facilities set aside.

Some of this oil and gas will be developed for the ultimate benefit of East Europe, and the Soviet Union, quite naturally because of the high costs involved, is looking to these countries to share in some way in the high costs through investment and through equipment deliveries. For the purpose of increasing deliveries of Soviet crude oil to Czechoslovakia after 1970 an agreement was signed between the two countries regarding co-operation in the development of crude oil production in the Soviet Union.[26] In accordance with this agreement Czechoslovakia will deliver to the Soviet Union on credit the necessary machinery and equipment for oil resource development. A similar arrangement has been negotiated between the USSR and East Germany.[27]

TABLE 40

Siberian Output as Percent of Total, 1965-1975

Year	Crude Oil USSR Output[a] (Million Tons)	From Siberia	Siberia as Percent of Total	Natural Gas USSR Output[b] (Billion Cubic Meters)	From Siberia	Siberia as Percent of Total
1965	243	1.0[c]	0.4	127.7	neg.	neg.
1968	309	12[d]	3.9	168.3	8[e]	5.1
1970(est.)	348	28 to 30[d]	8.3	200	12[f]	6
1975	460	70 to 80[g]	16.3	300 to 340	120[h]	37.5

[a] Data from Table 17 and Table 33.

[b] Data from Table 45.

[c] Ekonomika promyshlennosti, 12 (December, 1968), 12B3.

[d] Pravda, January 13, 1969, p. 2.

[e] Vyshka, December 23, 1968, p. 2. Tyumen Oblast only.

[f] Estimated, assuming at least partial operation of the Messoyakha-Norilsk gas pipeline.

[g] Geologiya nefti i gaza, 8 (August, 1968), 1-5.

[h] Stroitelstvo truboprovodov, 12 (December, 1968), 2-5. From northern regions of Tyumen Oblast alone.

Despite the obvious difficulties facing the oil extracting industry in the coming years, the goal of 460 million tons of crude oil by 1975 can be accepted as a reasonable statement, by a responsible Soviet official, of the capability of the industry. What, then, does the goal of 460 million tons of crude oil in 1975 portend for exports?

In essence, the growth in the export of crude oil by the USSR to other Communist countries is not expected to slacken after 1970, despite continued expansion in domestic demand and the offsetting influence of probable imports of oil by Eastern Europe from the Persian Gulf. It is believed that East Europe alone will require--and obtain--at least 57 million tons of crude oil from the Soviet Union by 1975, or roughly 58 per cent of the volume of crude oil the Soviet Union is expected to have available for export that year (Table 41). After satisfying the requirements in crude oil by the remaining Communist countries--an estimated 120,000 barrels daily or 6 million tons annually--more than 64 per cent of the exportable crude oil will have been spoken for before markets in the West can be considered.

Little change is anticipated in the allocation of petroleum product exports, with sales to non-Communist countries continuing to account for the predominant portion. It is estimated that by 1975 the Soviet Union may be able to offer as much as 55 million tons of oil to Free World buyers--made up of 35 million tons of crude oil and 20 million tons of petroleum products. Retention of emphasis on crude oil sales recognizes that the predominant market in Western Europe today is for crude oil and will remain so; it also underscores the reluctance of Soviet officials to permit the construction of refinery capacity in excess of domestic needs.

After 1970, the supply of oil in East Europe, with the notable exception of Rumania, should more nearly equate with demand, so that exports from this area are not likely to increase substantially (Table 43). Indigenous supply of crude oil in Rumania, expected to continue to be in excess of local needs, will be augmented by imports from the West, and it seems reasonable to assume that the major share of oil exported by East Europe will still originate with that country. On the other hand, imports of crude oil from the Middle East and North Africa may reach such proportions

TABLE 41

Probable Supply and Demand
for Oil in the Soviet Union, 1975

	Amount (Million Metric Tons)
Supply	
Crude oil	
Domestic production	460[a]
Imports	5[b]
Total crude oil	465
Petroleum products	
Yield from refining	325[c]
Imports	0
Natural gas liquids	10
Total products	335
Demand	
Crude oil	465
Charge to refining	353
Losses, increments to storage, direct use	14[d]
Available for export	98
To other Communist countries	63
To non-Communist countries	35
Petroleum products	335
Domestic demand	310[e]
Available for export	25
To other Communist countries	5
To non-Communist countries	20

[a] The New York Times, January 11, 1969, p. C-39. The feasibility of the Soviet Union reaching this level of output will become more apparent as the five year plan for 1971-75 unfolds.
[b] From the Middle East, through exchanges, direct purchase, and/or repayment for technical aid and assistance.
[c] Estimated at 92 per cent of the refinery charge.
[d] Estimated at 3 per cent of crude oil production.
[e] Based on extrapolation of past trends. For a comparison of growth in demand for oil in the Soviet Union and Eastern Europe during 1955-75, see Table 42.

TABLE 42

Comparative Growth in Demand for Oil
in the Soviet Union and in Eastern Europe, 1955-75

Year	Demand in Eastern Europe Million Metric Tons[a]	Percent Growth[b]	Demand in Soviet Union Million Metric Tons[c]	Percent Growth[b]
1955	12	--	60.4	--
1960	18	8.4	104.9	11.7
1965	28	9.2	160.1	8.8
1970	47.7	11.2	231	7.6
1975	75	9.5	310	6.0

[a] 1955 and 1965 are estimated; 1960 from National Petroleum Council, Impact of Oil Exports From the Soviet Bloc (Washington, 1962), p. 291; 1970 from Table 35; 1975 from Table 43.

[b] Average for 5-year period.

[c] Data from Table 60.

TABLE 43

Probable Supply and Demand
for Oil in East Europe, 1975

	Amount (Million Metric Tons)
Supply	
Crude oil	
Domestic production	22
Imports	
From USSR	57
From non-Communist sources	15
Total crude oil	94
Petroleum products	
Yield from refining	86[a]
Imports	neg.
Natural gas liquids	6
Total petroleum products	86
Demand	
Crude oil	
Charge to refining	91
Losses, increments to storage, direct use	3[b]
Available for export	0
Petroleum products	86
Domestic demand	75
Available for export	11[c]

[a] Estimated at 94 per cent of the refining charge.

[b] Estimated at 3 per cent of the crude oil supply.

[c] Probably all to non-Communist countries.

Figure 15

Probable Allocation of Oil Supply in the Soviet Union, 1970 and 1975

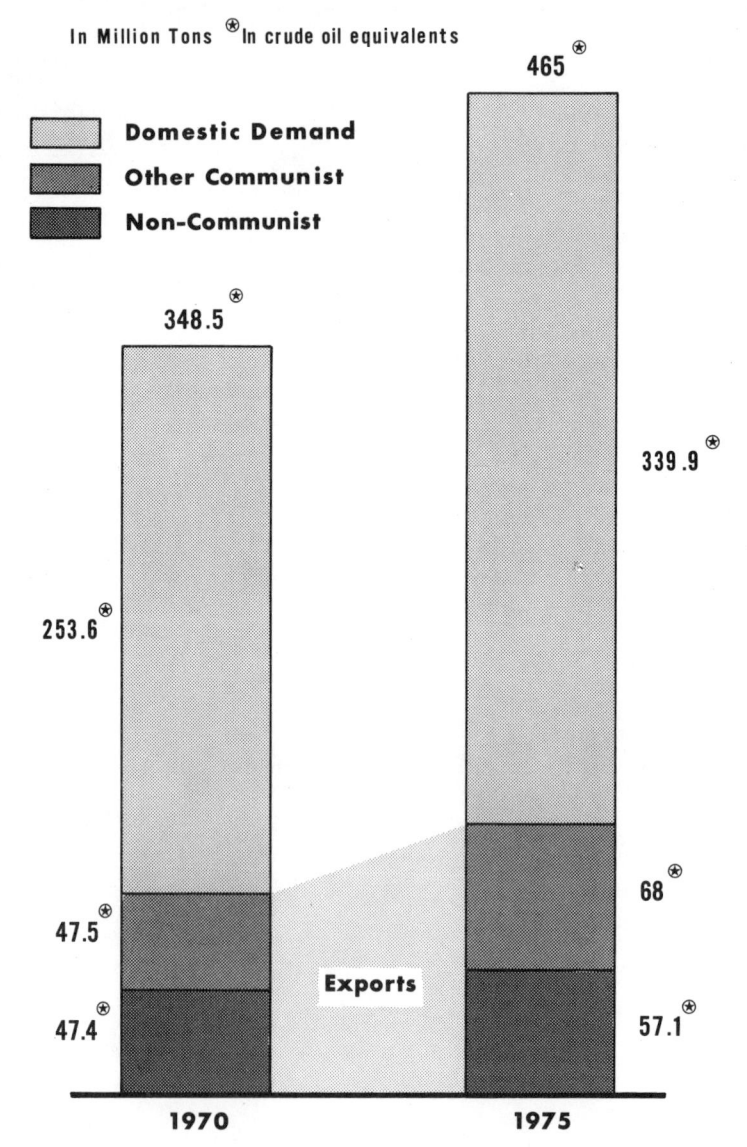

In Million Tons ⊛In crude oil equivalents

- Domestic Demand
- Other Communist
- Non-Communist

465 ⊛

348.5 ⊛

339.9 ⊛

253.6 ⊛

68 ⊛

47.5 ⊛

Exports

47.4 ⊛

57.1 ⊛

1970　　　1975

Source: Tables 37 and 41

as to wholly offset East European oil sales to the West, thus placing this trade in balance. Moreover, should the appropriate political and economic conditions develop, East Europe may even become a net importer of oil from the West. The scope of recently negotiated agreements and a reading of the general attitude toward additional agreements supports the belief that imports of oil by East Europe from the Third World in 1975 could well be triple those volumes forecast for 1970. It is on the basis of such imports that the forecast of an export of 11 million tons, all petroleum products, in 1975 has been made.

Conversely, there is at present no basis upon which to assume that the Soviet Union in 1970 and 1975 will be importing crude oil and petroleum products in relatively important quantities. But the possibility that the USSR, for political and/or economic reasons, may import Middle East oil in the coming years should not be overlooked.[28] Such oil might become available as payment for aid or technical assistance extended, or in return for services rendered under contract arrangements to explore for oil. Imports could be used to add to the capability to export; such imports would allow a reduction in domestic production and the diversion of freed investment funds to other sectors of the fuels economy; or some appropriate combination of these alternatives could be employed.

Any entry of the Soviet Union into the marketplace to secure crude oil, by whatever means, carries immediate political overtones. But care must be taken to differentiate between purely political arrangements and those made with geographic expediency in mind. For example, the United States imports crude oil and natural gas overland from Canada not because of a general national inability to meet demand but, among other reasons, because Canadian oil and gas is much closer to markets, say in the Pacific Northwest and the Chicago area, than are alternative indigenous sources of supply. Imports from Canada offer a savings in transport costs and in addition present Canada with an opportunity for larger-scale output, an opportunity which otherwise might not be available to that degree if only local markets were considered.

The countries of the Middle East and North Africa contain in themselves certain conditions which, taken together, to a certain extent lend favor to Communist economic and political penetration. These factors are:

    a.   Their economies are in an early state of development;

    b.   Expansion of these economies generally depends on the expansion of exports (and this means oil exports), and on the availability of foreign aid;

    c.   Independence has been theirs for only a relatively short period of time, and stronger political independence and broader economic independence are eagerly sought;

    d.   They seek to avoid any undue dependence or reliance on the Western capitalist or colonial powers;

    e.   They wish to develop and maintain broad political and economic ties.

It is reasonable to expect that the expansion of trade between the Communist world, largely the countries of East Europe, and the so-called Third World, will revolve around mutually advantageous barter arrangements. This would mean the exchange of Middle East oil for East European goods, equipment, and technology. By developing markets for its goods and equipment, whose quality standards are not yet on the level of West European requirements, East Europe could thereby promote its industrial development at a more rapid pace than what might be possible otherwise. This development, in turn, might add to the attractiveness of East Europe as a trader with its neighbors to the west.[29]

While East European trade with the Third World is expected to increase relatively sharply in the coming years, with the flow of Middle East oil into East Europe reaching perhaps 15 million tons by 1975, this by no means implies that the Middle East will abandon the West as its major economic partner. On the contrary, economic reality dictates that for the foreseeable future, Middle East oil will flow into Western Europe in increasing quantities. Western Europe today meets roughly 55 per cent of its oil requirements through imports from the Middle East, and although a decline in this relative dependence can be safely anticipated as larger quantities of oil from north and west Africa become available and as natural gas assumes increasing importance in primary energy supplies, in absolute terms this dependence can only increase. For there is no substitute for Middle East oil today;[30] and barring any technological break-through on the energy front relative to the recovery of oil from shale, tar sands, and/or coal in competitive quantities offered at competitive prices, or finds of European offshore oil or gas in comparable quantities, no substitute is likely to become available within the next fifteen years.

In summation, gross exports of petroleum by the Soviet bloc in 1975 may reach 66 million tons but net exports, with East Europe then expected to be a net importer from the West, would reflect a capability on the order of only 46 million tons, or less than projected net exports of 48 million tons in 1970.

Trends in trade in oil between the West and the Soviet bloc for the period 1965-75 are aggregated in Table 44. Clearly, two measurements of this trade are emerging--a measurement of the growth in gross oil exports from the bloc, and a measurement of the growth in net oil exports from the bloc. For the years 1966-70 both gross and net oil exports are to expand, but at varying rates. But for the subsequent five-year period 1971-75, as crude oil imports by East Europe and to a lesser degree by the USSR from the West assume more meaningful levels, relative growth in net bloc oil sales is to be negative.

In essence, an important moderation in the growth rate of Soviet bloc gross oil exports is anticipated for the ten-year period 1966-75, with much of this moderation to take place after 1967. The implied rate of growth for these ten years-- 4.6 per cent per year--should lag well behind the expansion in demand in Western Europe, its major market. This in turn implies a reduction in the relative role to be played by Soviet bloc oil in European energy markets compared to the participation of earlier years.

Inherent in the moderation in bloc oil sales, in addition to the factors already discussed as impacting on growth relative to Soviet oil exports, is the influence of changes in economic thinking currently taking place in the Soviet Union, particularly the acceptance by certain Soviet planners of the concept of opportunity cost in evaluating oil exports. Under this concept

TABLE 44

Actual and Estimated Soviet Bloc Trade in Oil
with Non-Communist Countries, 1965, 1970, and 1975

(in million metric tons)

Year	Gross Exports	Imports From Free World	Net Exports
1965			
Soviet Union[a]	35.5	0	35.5
East Europe[b]	6.7	0	6.7
Total	42.2	0	42.2
1970			
Soviet Union[c]	46	0.5	45.5
East Europe[d]	8	6	2
Total	54	6.5	47.5
1975			
Soviet Union[e]	55	5	50
East Europe[f]	11	15	- 4[g]
Total	66	20	46

[a] Data from Table 22.

[b] Data from Table 25.

[c] Data from Table 37.

[d] Data from Table 36.

[e] Data from Table 41.

[f] Data from Table 43.

[g] Minus (-) sign designates net imports.

the cost of exporting an incremental barrel of Urals-Volga crude oil (in terms of average costs, the cheapest in the country) would be evaluated not by its direct cost but by the cost of a marginal barrel of more expensive crude oil or the cost of the most expensive form of solid fuels which the Soviet Union otherwise would not have to produce for domestic use. Although not sophisticated economic reasoning by Western standards, it does represent a major departure in Soviet thinking.

Moreover, new wholesale prices for crude oil were introduced in 1967, at levels roughly double those previously in effect. Although internally this has accomplished little more than transferring profits from the refinery to the producer, some of the attraction of crude oil as an export commodity has been erased.

Certain Soviet writers continue to suggest that the monetary gain derived from oil exports could be enlarged, without any major increment in volume sales, if the export of products were to be emphasized. It is unlikely to expect, however, that major investment funds, which in themselves are limited, would be allocated for the construction of refineries solely to produce for the export market. For the mass disposal of petroleum products in Western Europe, quality standards of Soviet products would have to be upgraded considerably if they were to compete profitably and this would require the construction on a broad scale of expensive secondary refining processes now only in limited application in the Soviet Union.

Similarly, any broad-scale attempt by the Soviet Union to establish product marketing facilities in the West is not anticipated. To do so would require a considerable outlay in capital and the prospects for a reasonable return on this capital are minimal, particularly when marketing is attempted against a background of inexperience and in an area long characterized by its high degree of competition.

This does not mean, however, the complete absence of the USSR from the retail market in Europe. On the contrary, just prior to the outbreak of the Arab-Israeli war in June, 1967 the Soviet Union had taken several steps toward enlarging its presence in the United Kingdom and in Belgium. In May, 1967 Nafta (G.B.) Ltd., a wholly-owned Russian marketing company

Figure 16

Actual and Estimated Increments in Soviet Oil Sales to Non-Communist Areas 1955-75

In Million Metric Tons

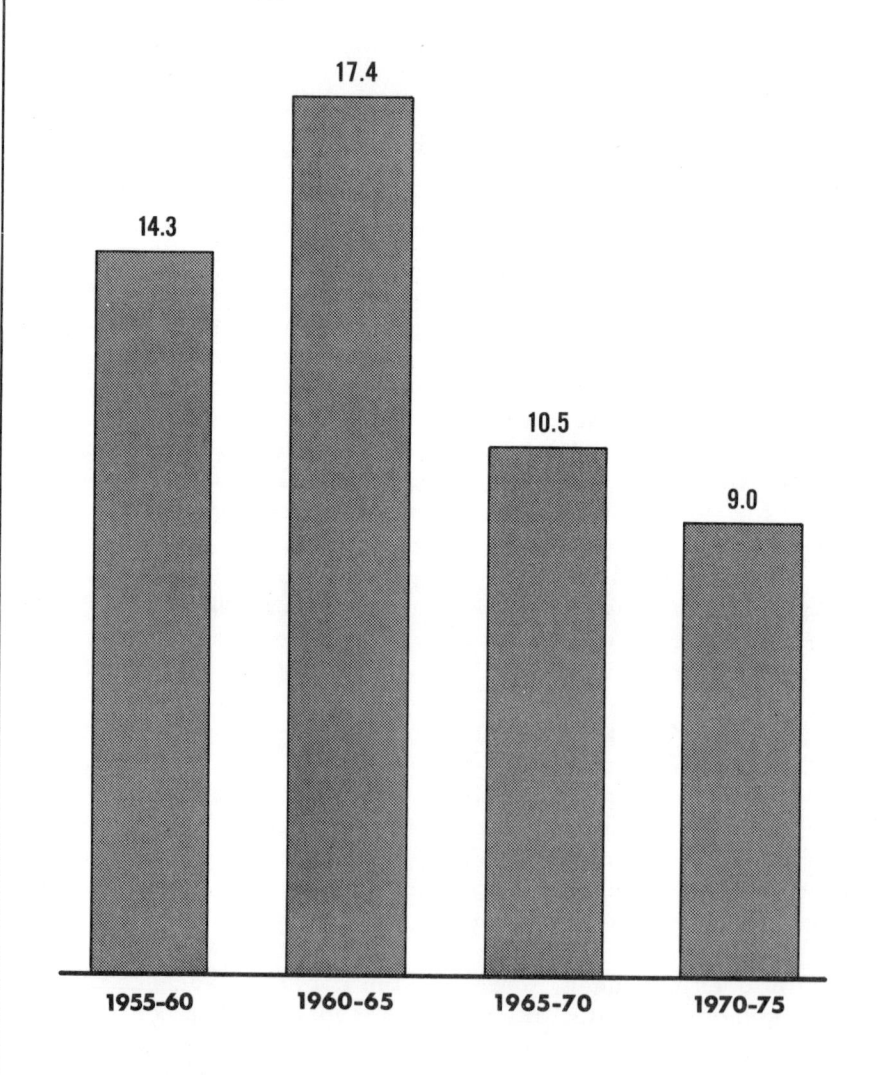

Period	Million Metric Tons
1955-60	14.3
1960-65	17.4
1965-70	10.5
1970-75	9.0

Source: Tables 20, 37 and 41

which advertises itself as importer, exporter, and distributor of crude oils and petroleum products, embarked on a promotional campaign to sell more motor gasoline in Britain. This campaign employed the most common of the Western promotional and marketing tactics--a new image, hard-sell advertising, and most important, prices 3 to 4 cents a gallon below the major brands. But Nafta's estimated annual sales of about 10 million gallons in Britain reportedly give it less than 1 per cent of the gasoline market and the indicated desire to increase sales by about 5 to 7 per cent a year is not likely to be particularly disturbing to the major oil companies operating in Britain. Rather, observers conclude that the USSR is attempting in a limited and inexpensive way to accumulate experience in marketing which can be applied both at home and abroad.

The specter of increased Soviet competition in the British gasoline market is not disturbing because Nafta (G.B.) is precluded from importing Soviet crude oil or oil products into Britain.[31] Thus Nafta (G.B.) is forced to buy its gasoline from other oil companies who at least until now have had no compunction about selling to a Communist competitor.

What the Soviet Union apparently would like to do is build or acquire a refinery in Belgium which would refine Soviet crude; the products could then be legally imported into Britain. At virtually the same time Nafta (G.B.) announced its sales campaign, Nafta (B) S.A., a joint Soviet-Belgian oil company [32] (owned 60 per cent by Soyuznefteeksport, the All-Union Oil Export Organization, and 40 per cent by the Belgian Bunkering and Stevedoring Co. and La Societé Commerciale Antone Vloeberghs and M. Stoop), obtained a sixty-year renewable lease on 62.5 acres in the Antwerp harbor for oil storage and berthing facilities. Belgian authorities gave Nafta (B) permission to import 500,000 tons of crude and 750,000 tons of refined products a year. But to import the crude oil Nafta (B) will need an import license and to obtain the license it must have refining capacity, either in a facility which it owns or available through a processing agreement. Early efforts to purchase the small Albatross refinery in Antwerp proved unsuccessful. If a processing agreement cannot be arranged, the remaining alternative would be for Nafta (B) to build its own refinery.

The import quotas assigned Nafta (B) are not sufficient in size to allow any meaningful manipulations in the marketplace, and it is probable that Nafta (B) has no intention of disrupting markets in Belgium or any country in Western Europe where it might attempt to sell its products. More realistically, the operations of Nafta (B) over the coming years, in keeping with the operations of Nafta (G.B.), probably can be loosely described as "live-and-learn," and their managers can be expected to make the most of the opportunity to study European business methods.

Although the Soviet Union seeks to enlarge its marketing activities in the West modestly, at least one international oil company has been partially successful in what may be described as a "reverse penetration" of Soviet bloc markets. Shell International announced in January, 1969 that it had reached agreement for establishing ten more gasoline stations in Hungary to sell gasoline and lubricants.[33] Just two and a half years ago Shell broke into Hungary with a service station in Budapest but sales were restricted to Western currencies, with tourists and diplomats the obvious customers. The new stations will be operated by Interag, a state agency, and Hungary will provide the gasoline, produced to meet Shell standards.

Other than whatever prestige goes with being the first to establish a beachhead in oil marketing in Eastern Europe, little short-term gain can be expected from this minute network. However, it will offer Shell International an opportunity to gain first-hand knowledge of working with the Hungarians, to build up goodwill and foster harmonious business relationships which could prove beneficial in the long run.

CHAPTER **9** SOVIET TRADE IN GAS

Only since 1955 has the Soviet Union undertaken the development of a natural gas producing and transmission industry to the point where this industry has been able to make a sizable contribution to primary energy supplies. In 1955 natural gas accounted for only 2.4 per cent of all primary energy produced in the Soviet Union, with extraction on the order of 9 billion cubic meters. By 1968 this share had been raised to about 18 per cent, an appreciable transformation, as the extraction of gas that year totaled 169.3 billion cubic meters.

Yet the gas industry has been a source of major disappointment to Soviet officials, as it has failed to live up to the hopes held out for it. Performance during the Five Year Plan has been lackluster and gas output in 1970 most probably will fall well short of the original goal for that year.

And judging from the extent of complaints in the Soviet press and judging from the inability of the industry to measure up to annual output plans, the Soviet Union has not been able to establish a natural gas industry whose various sectors--reserves, drilling, production, transmission, storage, and means to consume--are in reasonable balance with one another. In the end these factors would seem to indicate that the current potential for gas consumption in the Soviet Union is far from being realized.

Nevertheless, Soviet officials have attracted considerable international publicity in recent months by indications of a desire to market natural gas in Western Europe, and in Italy and West Germany in particular. That such offers would be made in the face of apparently unsatisfied local demand superficially would appear to underscore the intent to enlarge earnings of foreign exchange or to secure selected Western equipment,

TABLE 45

Trends in the Extraction of Natural Gas[a] in the Soviet Union[b]

(in billion cubic meters)

Year	Amount
1940	3.219
1945	3.278
1950	5.761
1951	6.252
1953	6.384
1954	7.512
1955	8.981
1956	12.067
1957	18.583
1958	28.085
1959	35.391
1960	45.303
1961	58.981
1962	75.525
1963	89.832
1964	108.567
1965	127.665
1966	142.965
1967	157.445[c]
1968	169.3[d]
1969 Plan	184.1[e]
1970 Plan (Original)	225 to 240[f]
(Revised)	215[g]
Estimate	200
1975 Plan	380 to 400[h]
	300 to 340[i]
	280 to 300[j]
1980 Plan (Original)	680 to 720[k]
(Revised)	640 to 650[l]

128

TABLE 45 (Continued)

[a] Includes both associated and non-associated natural gases, but excludes gas produced from shale and the underground gasification of coal. Manufactured gas is rapidly losing its relative importance in the Soviet Union, as Table 46 shows.

[b] Except as noted, from A.K. Kortunov, Gazovaya promyshlennost SSSR (Moscow, 1967), p. 62.

[c] Narodnoye khozyaystvo SSSR v 1967 g., statisticheskiy yezhegodnik (Moscow, 1968), pp. 236-242.

[d] Estimate, based on data given in Izvestiya, January 26, 1969, p. 1.

[e] Estimate, based on data given in Pravda, December 11, 1968, pp. 1-5.

[f] Ekonomicheskaya gazeta, 22 (June, 1966), 16-17.

[g] Geologiya nefti i gaza, 7 (July, 1968), 1-9.

[h] Stroitelstvo truboprovodov, 5 (May, 1968), 1-4.

[i] Stroitelnaya gazeta, February 5, 1969, p. 1.

[j] Bakinskiy rabochiy, March 13, 1969, p. 2. Note the gradual downward revision in the 1975 goal for extraction of natural gas as planners are able to make a more realistic appraisal of capabilities.

[k] As indicated in the Twenty Year Plan (1961-80), released in October, 1961.

[l] Geologiya nefti i gaza, 12 (December, 1967), 1-8.

TABLE 46

Trends in Manufactured Gas in the Soviet Union

Year	Gas produced from Shale and Coal (Million Cubic Meters)
1950	420
1958	1,807
1960	1,911
1965	1,683
1966	1,705
1967	1,735

Sources: A.K. Kortunov, Gazovaya promyshlennost SSSR (Moscow: 1967), p. 62 for the years through 1966; data for 1967 from Narodnoye khozyaystvo SSSR v 1967 godu, statisticheskiy yezhegodnik (Moscow: 1967), p. 237.

TABLE 47

Failure of the Soviet Natural Gas Industry
During the Five Year Plan 1966-70

(in billion cubic meters)

Indicator	1966	1967	1968	1969	1970
Five Year Plan Annual Goals[a]	142.0	158.3	170.3	191.3	225[b]
Revised Annual Goals	--	--	171.3[c]	184[d]	215[c]
Actual Production	143[c]	157.4[c]	169.3[c]	--	--
Estimated Production	--	--	--	185	200

[a] A.K. Kortunov, Gazovaya promyshlennost SSSR (Moscow, 1967), p. 70.

[b] The lower limit of the goal for 1970, which had been established at 225 to 240 billion cubic meters, including 1.7 billion cubic meters of manufactured gases.

[c] Data from Table 45.

[d] The 1969 goal for total gas was set at 185.8 billion cubic meters (Pravda, December 11, 1968, p. 1). It is estimated that output of gas produced from shale and the underground gasification of coal will again be on the order of 1.7 billion cubic meters, leaving 184 billion as the goal for natural gas.

Figure 17

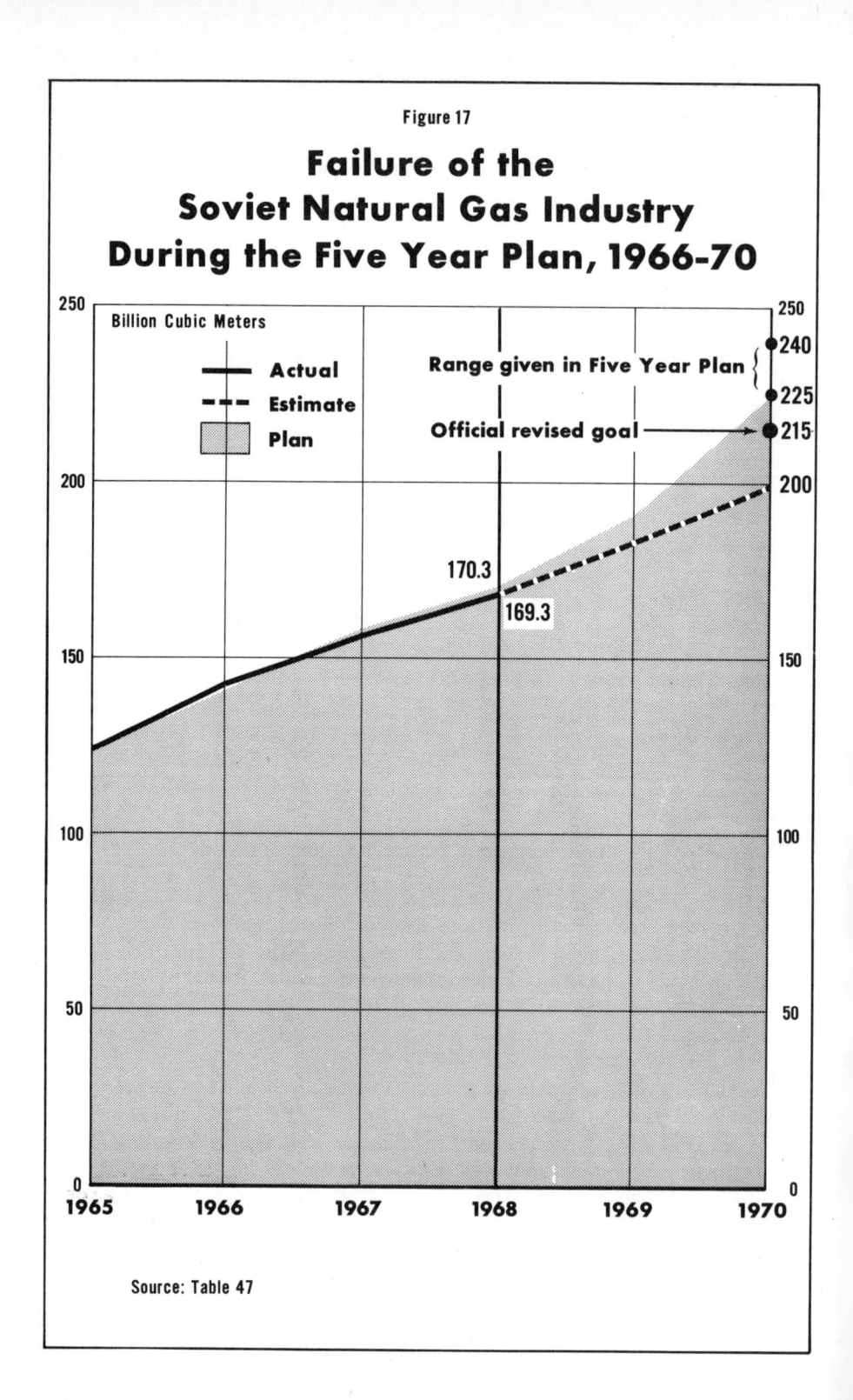

Failure of the Soviet Natural Gas Industry During the Five Year Plan, 1966-70

Billion Cubic Meters

— Actual
--- Estimate
▓ Plan

Range given in Five Year Plan { ●240 / ●225 }
Official revised goal ──→ ●215

170.3
169.3

1965 1966 1967 1968 1969 1970

Source: Table 47

SOVIET TRADE IN GAS 133

goods, or technology in exchange for this gas, whatever the gain might be, at whatever price necessary.

At the same time Soviet officials were indicating a desire to export gas, arrangements were being made to import gas into the Soviet Union. Such imports, from Iran and Afghanistan, are unprecedented in Soviet history and bring forth the intriguing prospect of selected regional dependence upon a non-Communist source of energy.

One widely accepted conclusion, drawn from these early maneuverings in gas trade, has been that such imports of gas in reality were undertaken to make possible the export of gas; that the USSR, should a major sale to the West materialize, in essence would be exporting Iranian gas. This is easy reasoning, particularly when viewed against the continuing shortfalls in natural gas production, and may have been valid at one time, prior to the assessment of the gas discoveries in West Siberia. But other evidence now points to long-term export gas to be provided by the new fields in West Siberia, located around the Ob River Bay in the northern part of Tyumen Oblast, with short-term export gas originating from the gas fields in the western and eastern portions of the Ukraine.

Those volumes of natural gas which the Soviet Union may export to Western Europe, and to Eastern Europe for that matter, should be regarded as more a product of the natural gas pipeline network of the country as a whole. This network, concentrated largely in the European part of the Soviet Union, and fed by gas deposits located on the peripheries of the system in the North Caucasus, Central Asia, the eastern and western Ukraine, and now in West Siberia, contains in itself a certain flexibility with respect to maneuverability of supplies. Therefore, any change in the input into the system can affect the input (or output) at the other extreme of the system.

To illustrate, gas arriving in the Moscow area from the new fields in West Siberia and in Central Asia can, and apparently will, permit the diversion elsewhere of gas in the eastern Ukraine, which otherwise would have been directed to consumers in Moscow. Based on recently released pipeline construction plans, it would appear that gas from new fields in the eastern Ukraine will be diverted to the Soviet-Czechoslovakian border for export. Such diversion is a logical move, if viewed only in terms of the geographic proximity of these Ukrainian gas fields to markets outside the country, whereas the gas fields in West Siberia lie thousands of kilometers distant. That Ukrainian gas is much closer to the export market offers considerable savings in pipeline construction time and costs and in transport expenditures.

More important, such diversification will enable the Soviet Union to respond almost immediately to the prospects for sales of gas outside its borders, thus to gain entry into new markets, to solidify its position in these markets, and then to attempt to enlarge this position when additional supplies of gas, from West Siberia, become available for export.

EXPORTS

The anticipated slow-down in Soviet oil sales to the West, discussed above, has important implications for the economy as a whole. Oil has been a major earner of foreign exchange and has been a prime mover in establishing mutually advantageous trading arrangements with many of the West European countries. Most assuredly the bulk of whatever oil the USSR is able to free for export to non-Communist countries will be directed to Western Europe, as it has in the past.[1] But the projected increments in sales are not likely to support any significant expansion in foreign exchange earnings and if the Soviet Union wishes to continue and expand its trade with the developed nations of the world, as it unquestionably does, then other, supplementary means of financing this trade must be found.

In attempting to satisfy its almost unquenchable thirst for Western technology, the Soviet Union can best offer raw materials which can be made available in competitive qualities and in amounts sufficient in size to generate significant, long-term earnings. One such raw material is natural gas. The role of natural gas in meeting Western European energy demands has been relatively modest, largely because local supplies have not been available, and not because of the absence of potential markets. This is to change somewhat in the coming years, as gas from the Groningen fields in the

Netherlands is developed for large-scale exports to France, Belgium, and West Germany.

The Soviet Union has recognized the potential for natural gas sales to Western Europe, it has the gas reserves to back up any reasonable sales contracts which might be arranged, and it can deliver the gas overland by pipeline, an advantage not available to others except to Groningen. The prospects for such sales are discussed below.

Before 1970

The possibility that the Soviet Union might become a major exporter of natural gas to Western Europe in the near future has aroused considerable speculation and consternation. It is known that negotiations in one form or another have been or are being conducted with Austria, France, Italy, Finland, Japan, West Germany, and Sweden.[2] Of these, agreement has been reached only with Austria, a tribute to the geographic proximity of the Austrian market to Soviet gas. Indeed, the availability of Soviet gas to Austria was accomplished upon construction of a gas pipeline only about 20 kilometers in length, simply by extending the Soviet-Czechoslovakian gas pipeline, known as Bratstvo (Brotherhood), built in 1967, into Austria to the town of Baumgarten.

In late January 1968 the Austrian state oil company, OemV (Oesterreichische Minearloelverwaltung A. G.), agreed in principle to import natural gas from the Soviet Union, marking the first sale of Soviet gas to a non-Communist country.[3] In a purportedly separate arrangement, Austria is to supply the USSR with 520,000 tons of steel pipe, of which 120,000 tons will be 40-inch and the remaining 400,000 tons will be 48-inch. This pipe will replace Austrian cash payments for the gas for a period until the price for the steel and pipe is paid up. Austria will not roll the 48-inch pipe but will provide West Germany with the necessary steel plate.

This purchase of pipe is the largest ever made by the Soviet Union and it is the first large-diameter pipe purchase from the West since the removal in November, 1966 of the North Atlantic Treaty Organization (NATO) embargo on sale

Figure 18

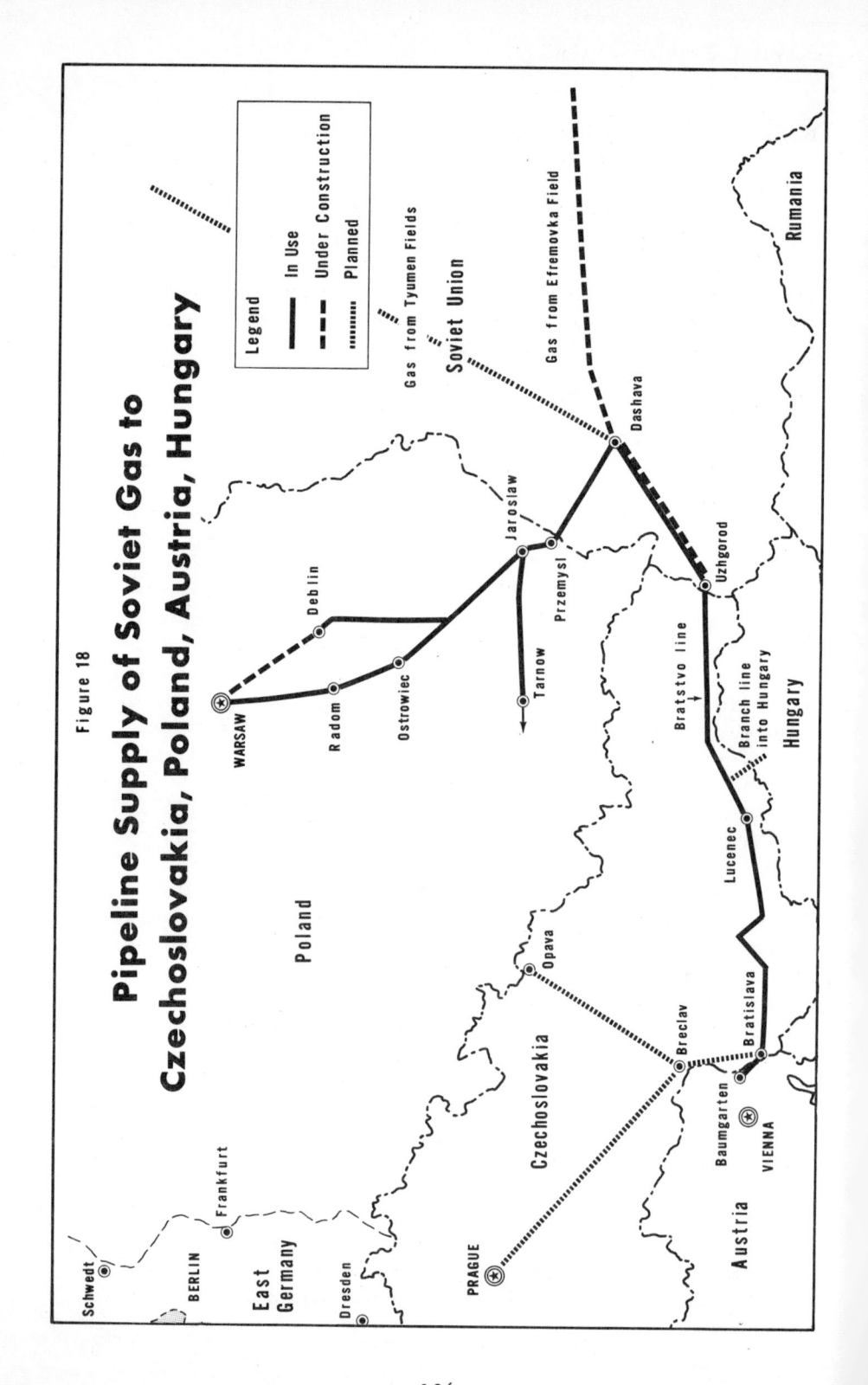

Pipeline Supply of Soviet Gas to Czechoslovakia, Poland, Austria, Hungary

of large-diameter pipe to the Soviet bloc.[4] It will considerably relieve the strain on domestic 48-inch pipe-making capacity, which admittedly falls short of requirements. Indeed, the 400,000 tons of 48-inch pipe from Austria would be sufficient to build three-quarters of the Ukhta-Torzhok sector of the Northern Lights gas pipeline, a distance of some 1,300 kilometers.

The first sales of pipeline natural gas by the Soviet Union to the West began in 1968, upon the completion of a pipeline spur into Austria from Czechoslovakia. Prior to that time exports had been limited to Poland through 1966 and were expanded to include Czechoslovakia, beginning in 1967 (see Table 48).

The first Soviet gas arrived in Austria in September, 1968, with delivery to total 200 million cubic meters during the year, increasing subsequently to a ceiling of 1.5 billion cubic meters a year by 1971. The contract covers twenty years in all and calls for a cumulative delivery of 30 billion cubic meters.

That no gas other than those quantities to be delivered to Austria probably will be sold to the West by 1970, even should agreement on price and interest rates be reached with, say, Italy or Japan, reflects the probability that a pipeline network between uncommited gas deposits in the USSR and importing countries in Western Europe could not be constructed in the time remaining. All other gas exports from the Soviet Union through 1970 will be directed to Poland and Czechoslovakia and possibly Hungary.

Under these limitations the total export of gas from the USSR in 1970 most likely will reach about 3.8 billion cubic meters (Table 49), of which roughly 30 per cent will be directed to the West (Austria).

After 1970

Forecasts of the probable levels of Soviet gas exports after 1970 are intrinsically tied to the prospects of consummating negotiations with Japan and finding some form of accommodation with sizable markets in Western Europe.

TABLE 48

Soviet Exports of Natural Gas,
Selected Years 1950-68[a]

Year	Million Cubic Meters[b]
1950	77.1
1955	138.7
1956	136.2
1957	170.0
1958	205.6
1959	222.0
1960	242.0
1961	271.7
1962	300.3
1963	300.9
1964	295.3
1965	391.5
1966	827.9
1967	1,290.0[c]
1968	1,780[d]

[a] Except for 1967 and 1968, all exports of gas during these years have been directed to Poland.

[b] Except for 1967 and 1968, data have been derived from Vneshnyaya torgovlya SSSR, statisticheskiy sbornik 1918-1966 (Moscow, 1967), pp. 81-83.

[c] Vneshnyaya torgovlya SSSR za 1967 god (Moscow, 1968). Distributed as follows (million cubic meters):

Poland	1,025
Czechoslovakia	265

[d] Estimate, broken down as follows (million cubic meters):

Austria	280 (Pravda, September 18, 1968, p. 1)
Czechoslovakia	500 (Tass, September 9, 1968)
Poland	1,000 (Estimate)

TABLE 49

Probable Exports of Natural Gas
by the Soviet Union, 1970

(in billion cubic meters)

Destination	Amount
Austria	1.0[a]
Czechoslovakia	1[b]
Hungary	0.3[c]
Poland	1.5[d]
Total	3.8

---

[a] Vyshka, June 19, 1968, p. 2.

[b] Radio Prague, January 12, 1968.

[c] Estimate.

[d] Polish Press Agency, December 19, 1967.

Italy as a Market

The Soviet effort to sell gas to Italy has attracted considerable publicity in the oil industry trade press, as could be expected when a commodity of Communist origin attempts to break into a new market, in volumes large enough to impact substantially upon future sales by established suppliers.

Italy, recognizing that the great Po Valley deposits, discovered shortly after World War II, will be exhausted some time in the 1970's, has expended considerable effort in developing new sources of supply, sufficient not only to make up for the eventual loss of Po Valley gas but also to meet anticipated future increments in demand. The demand for gas in Italy in 1969 was to be on the order of 10 billion cubic meters and requirements are expected to almost double in the coming decade.

When Italy set out in search of external supplies of gas, four countries--Algeria, Libya, the Netherlands, and the Soviet Union--could be considered likely sources. With the prospect of a number of suitors on hand, Italy has been content to play a waiting game, ready to enjoy the benefits of the competition among potential suppliers for a place in its gas market. At present, only one contract has been signed. Starting in 1969, Esso International, a wholly-owned affiliate of Standard Oil Company (New Jersey) is to begin delivery of liquefied natural gas from its fields in Libya in quantities of 3 billion cubic meter per year for a period of twenty years. Having this contract in hand and being assured of a flow of imported gas equal to 30 per cent of today's demand has relieved somewhat the immediate need to strike an agreement with other suppliers.

It would seem, based on preliminary information, that this waiting game has at least partially paid off. In September, 1968 (ENI) Ente Nationale Idrocarburi, the Italian state-owned oil company, announced that natural gas reserves, estimated at 60 billion cubic meters, had been discovered in the upper Adriatic Sea. Such reserves, according to the announcement, are sufficient to meet Italy's needs for six years at present rates of consumption. Most certainly these finds have considerably improved Italy's position in price bargaining with the Dutch and the Soviet Union. Yet to develop these finds to levels of substantial annual production is apt to be expensive and time-consuming, to the extent that having discovered gas in the Adriatic does not necessarily imply an end to negotiations with the Dutch and the Soviet Union.

More recently, the prospect of yet another source of supply of natural gas for Italy has appeared. The International Egyptian Oil Company (IEOC), a subsidiary of ENI, made a major gas find on the Nile Delta in 1967, and subsequent exploration indicates that the delta, both onshore and offshore, may develop into a major producer of gas.[5] Export of this gas would serve the interest both of Egypt, which would welcome the opportunity to earn foreign exchange, and Italy, if only to use the possibility of importing gas from Egypt to exert further downward pressure on the price for Dutch gas.

A number of factors would seem to mitigate against Italy importing natural gas from the Soviet Union. First, there is considerable--essentially correct--argument against the Italian economy's enlarging its dependence upon a source of energy of Communist origin which for political purposes might be denied at any time. Second, under any circumstances, to make the long-distance delivery of Soviet gas to Italy economically feasible, and drawing upon the economies of scale in pipeline construction, would require the building of a pipeline probably 40 inches in diameter. A pipeline of this diameter can move about 10 billion cubic meters of gas annually. But it is doubtful whether Italy would wish to import this much gas from a single supplier, for, in so doing, it would place the economy in a position of being imprudently dependent (perhaps as much as 50 per cent in 1975) upon an outside source. Indeed, it is unlikely that the Italian market in 1975 could accommodate that much Soviet gas. More likely, a quantity of between 4 and 6 billion cubic meters could be absorbed. The Soviet Union accordingly would then be forced to line up additional markets for its gas or to operate the pipeline at less than capacity.

And perhaps most important, with the finds of gas in the Adriatic and in Egypt, will there be room for the import of gas from the USSR?

With regard to the latter point, expert opinion is that while there may be no market for Soviet gas in Italy at present, by the early 1970's demand will have outstripped available increments to supply. Although warnings of the dangers inherent in too high a degree of dependence on Communist energy are not ignored, the Italian government is genuinely interested in diversifying its sources of gas supplies and certain officials regard the Soviet Union as just another supplier in this respect. Similarly, ENI, the prospective buyer of Soviet gas, has enjoyed a profitable trading relationship with the USSR. It is eager to expand this relationship and has been urging the Italian Government to come to terms with the USSR. Expenditures incurred in the liquefaction of natural gas and the relatively high costs of its transportation over comparatively short distances between North Africa and Italy favor the development of pipeline delivery of gas if the latter is at all feasible.

Finally, the USSR is not likely to become too concerned about having to find other markets to substantiate the construction of large-diameter pipeline. There are a number of attractive markets for Soviet gas in Western Europe, considering the route a pipeline to Italy might take, including Austria and West Germany[6] as well as all of the countries of East Europe, with the exception of Rumania which has domestic supplies of gas more than adequate to meet local requirements.

The possibility of deliveries of Soviet gas by pipeline to France has been mentioned, but French interest has declined noticeably following the conclusion of an agreement to import the equivalent of 4 billion cubic meters annually from Algeria in the form of liquefied natural gas (LNG).[7]

Assuming the construction of a 40-inch pipeline and assuming the allocation of half of its carrying capacity--or 5 billion cubic meters per year--to the Italian market would leave another 5 billion cubic meters annually to be placed elsewhere. This quantity could be absorbed in East Europe alone. Recent evidence indicates the intent of CEMA member nations to create a natural gas pipeline network to draw upon Soviet resources to meet East European energy requirements. The completion of the Bratstvo gas pipeline in 1967 bringing Soviet gas into Czechoslovakia may be regarded as the first major step in the initiation of such a network. Pipelines carrying gas from the western Ukraine into Poland already are in operation; a short branch pipeline off Bratstvo into Hungary was to be built in 1969; gas is to be delivered to East Germany beginning in 1972 and to Bulgaria in presumably the same year.[8] By 1975 Bulgaria is to be importing natural gas from the Soviet Union in the amount of 3 billion cubic meters annually. The scheme for delivery of this gas is depicted in Figure 19. The pipeline, which is to exit the USSR at Izmail, will cut across the narrow width of Rumania in an area whose terrain will not pose any particular construction problems. Although not so indicated, it is possible that Rumania might offtake gas from the pipeline, to serve Constanta. It is

Figure 19

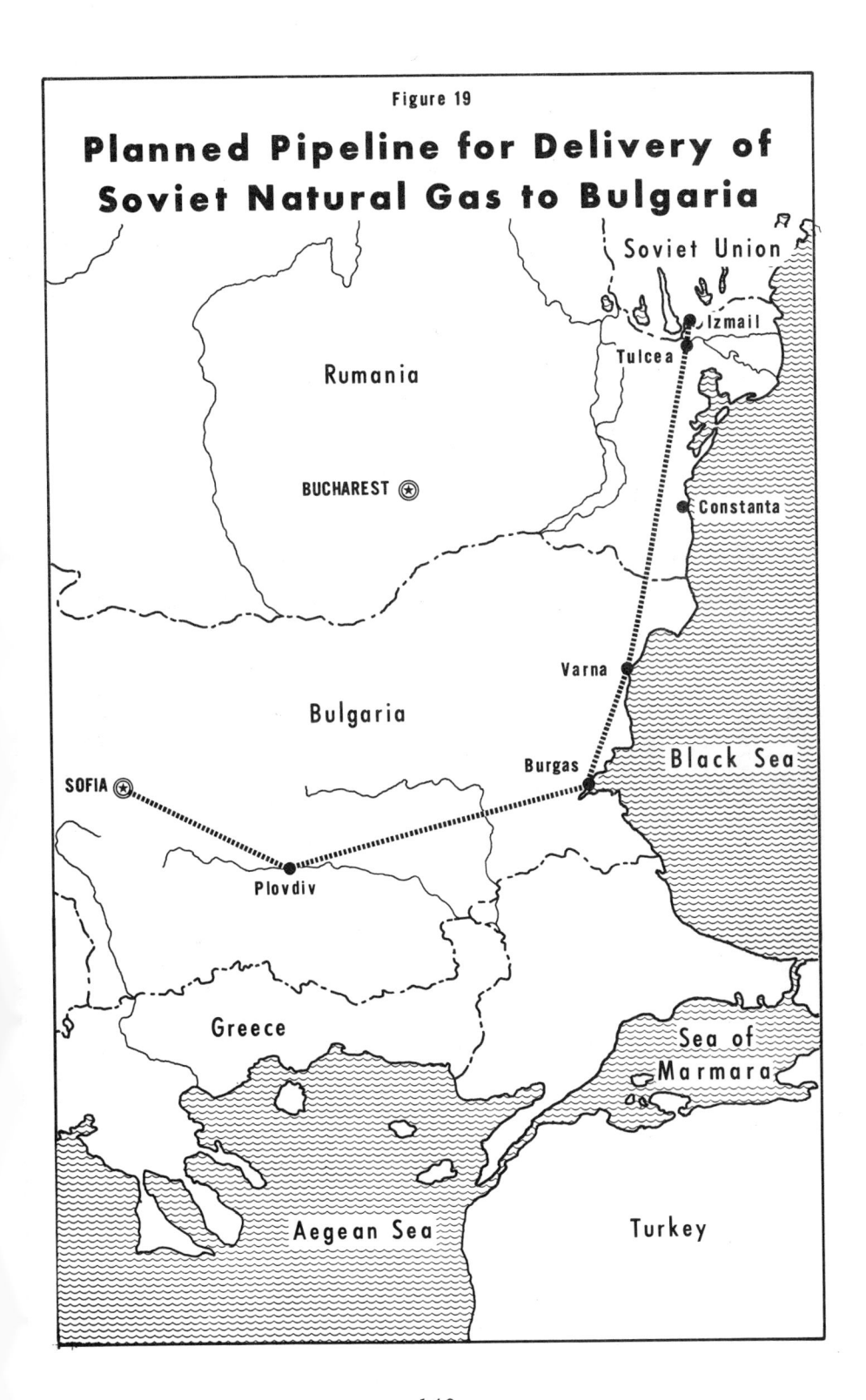

Planned Pipeline for Delivery of Soviet Natural Gas to Bulgaria

Figure 20

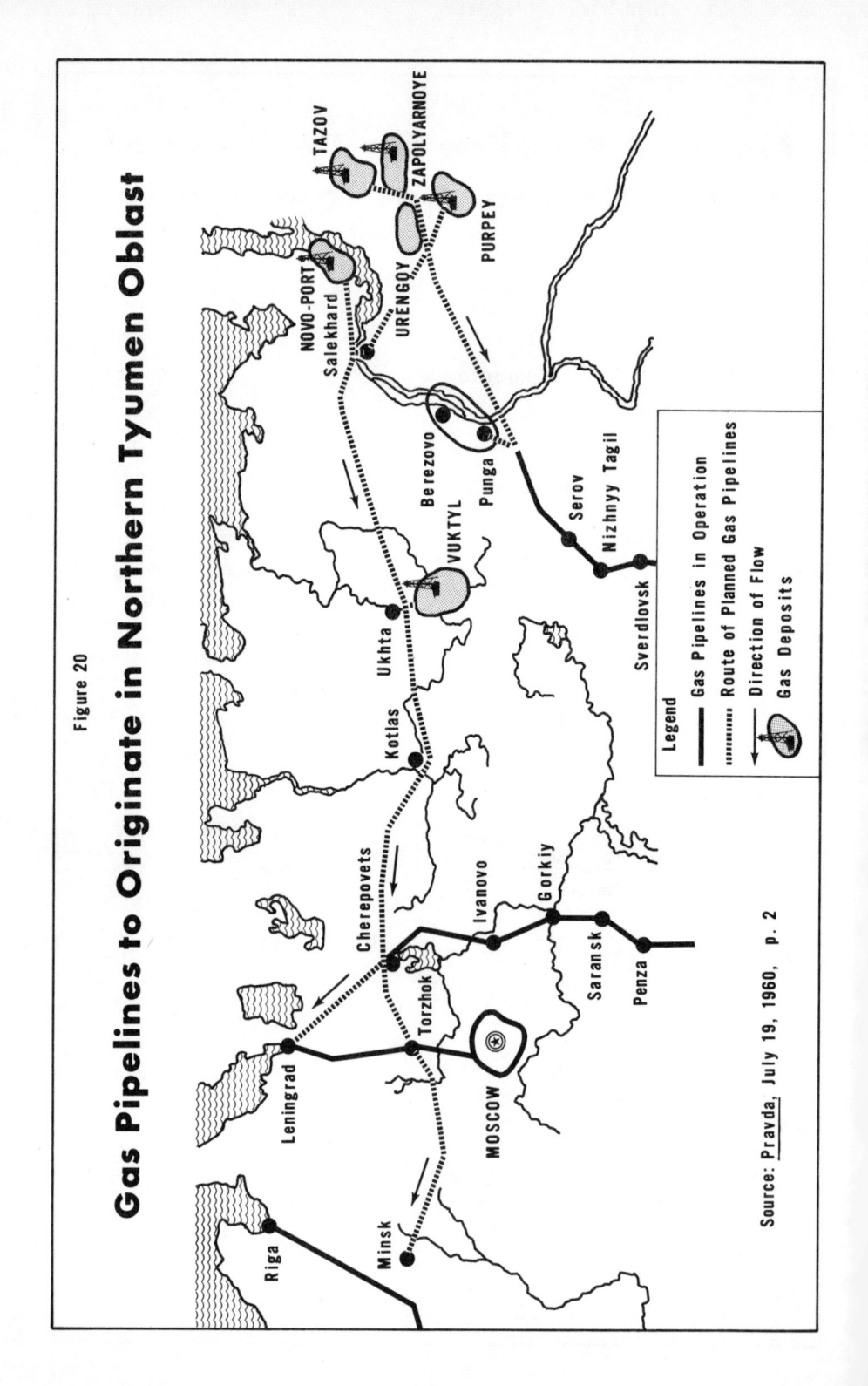

Gas Pipelines to Originate in Northern Tyumen Oblast

Source: Pravda, July 19, 1960, p. 2

Figure 21

The "Northern Lights" Natural Gas Pipeline System: Route of the Vuktyl-Ukhta-Torzhok Sector

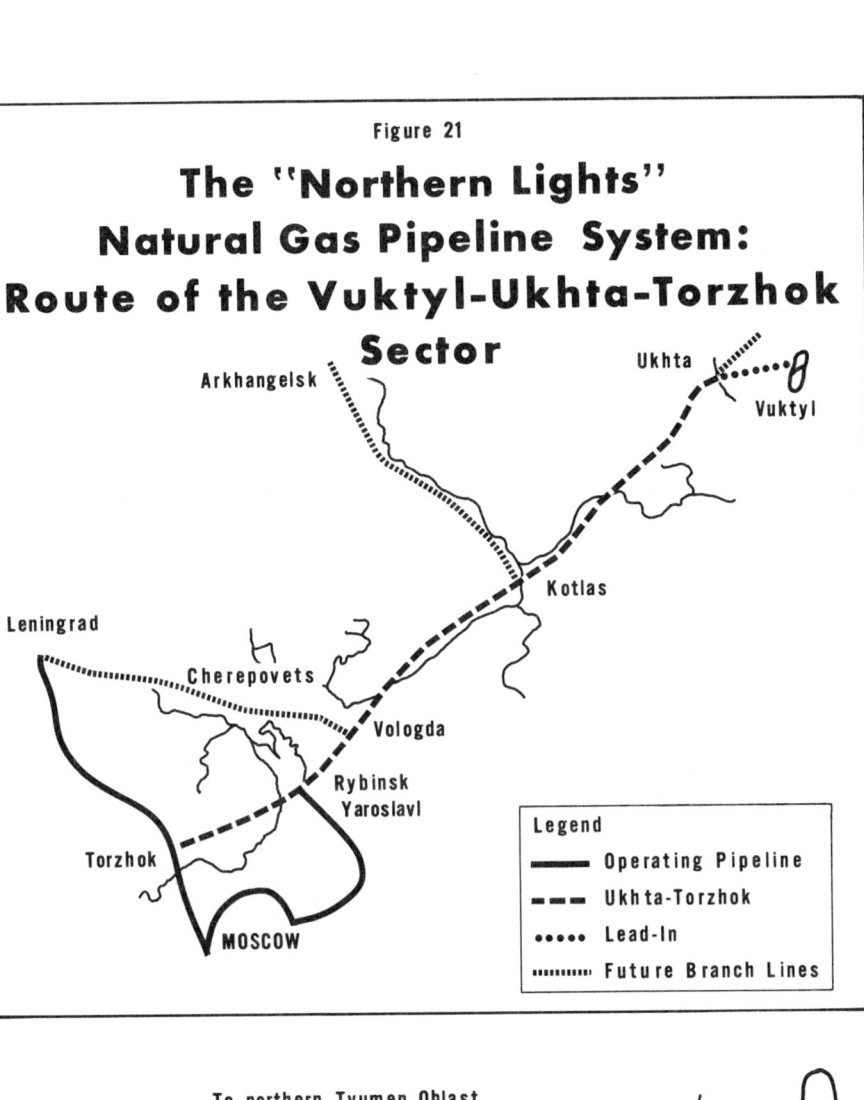

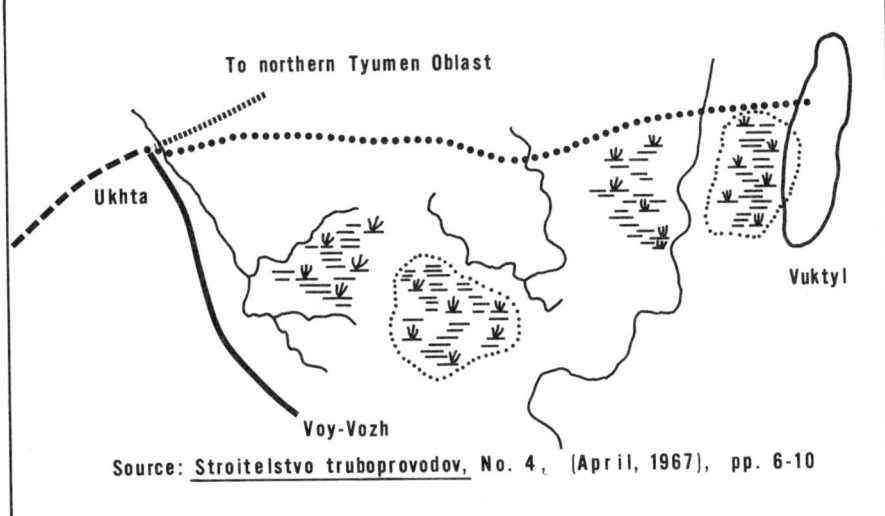

Source: Stroitelstvo truboprovodov, No. 4, (April, 1967), pp. 6-10

probable that the gas destined for Bulgaria will originate at the gas fields in the eastern Ukraine.

The Soviet Union at the present time has two major undeveloped sources of natural gas--in West Siberia and in Central Asia. Of these, long-term exports are most likely to come from the West Siberian fields but exploitation of these fields is dependent upon the construction of a pipeline system linking the fields with consumers in the central and western portions of the country. This system probably will not be completed in its entirety until the 1970's. As part of the much publicized Northern Lights gas pipeline system, a 48-inch line is now under construction between the Vuktyl gas fields, about 200 kilometers east of Pechora in the Soviet North, and the city of Torzhok, roughly 200 kilometers northwest of Moscow, and is the first step in bringing Siberian gas to the Moscow area. It is also something of a stopgap, in that it will tap gas resources located considerably closer to the consumer than the new fields in the northern part of Tyumen Oblast, found around the Ob River Bay. After completion of the Ukhta-Torzhok sector, construction will then be undertaken eastward from Ukhta toward the Ob River Bay and between Torzhok and Minsk.

Gas from the new Central Asia deposits apparently has been committed to the Moscow industrial area and to a lesser extent to the Urals, and is not likely, in itself, to be exported. The first of ultimately three large-diameter pipelines between Central Asia and Moscow was completed in late 1967. This pipeline, 40 inches in diameter and 2,750 kilometers in length, which was finished one year ahead of schedule, is now being paralleled by a 48-inch pipeline. Ultimately, a third, 56-inch pipeline, is to be laid along much of the same route.

The Northern Lights pipeline is to terminate at Minsk, in the republic of Belorussia. Minsk is but some 700 kilometers from the Soviet-Czechoslovakian border and extension of the pipeline beyond Minsk to the border for the purposes of exporting gas would be a relatively simple undertaking. Beyond that point, the respective countries through which the pipeline would pass en route to Italy would be responsible for construction of that portion of the pipeline within their own boundaries. It can be expected that the Soviet contribution to construction of the pipeline outside its own boundaries would be very small indeed. In fact, these other countries might be asked to participate in construction of that portion between Minsk and the Soviet border.

Figure 22

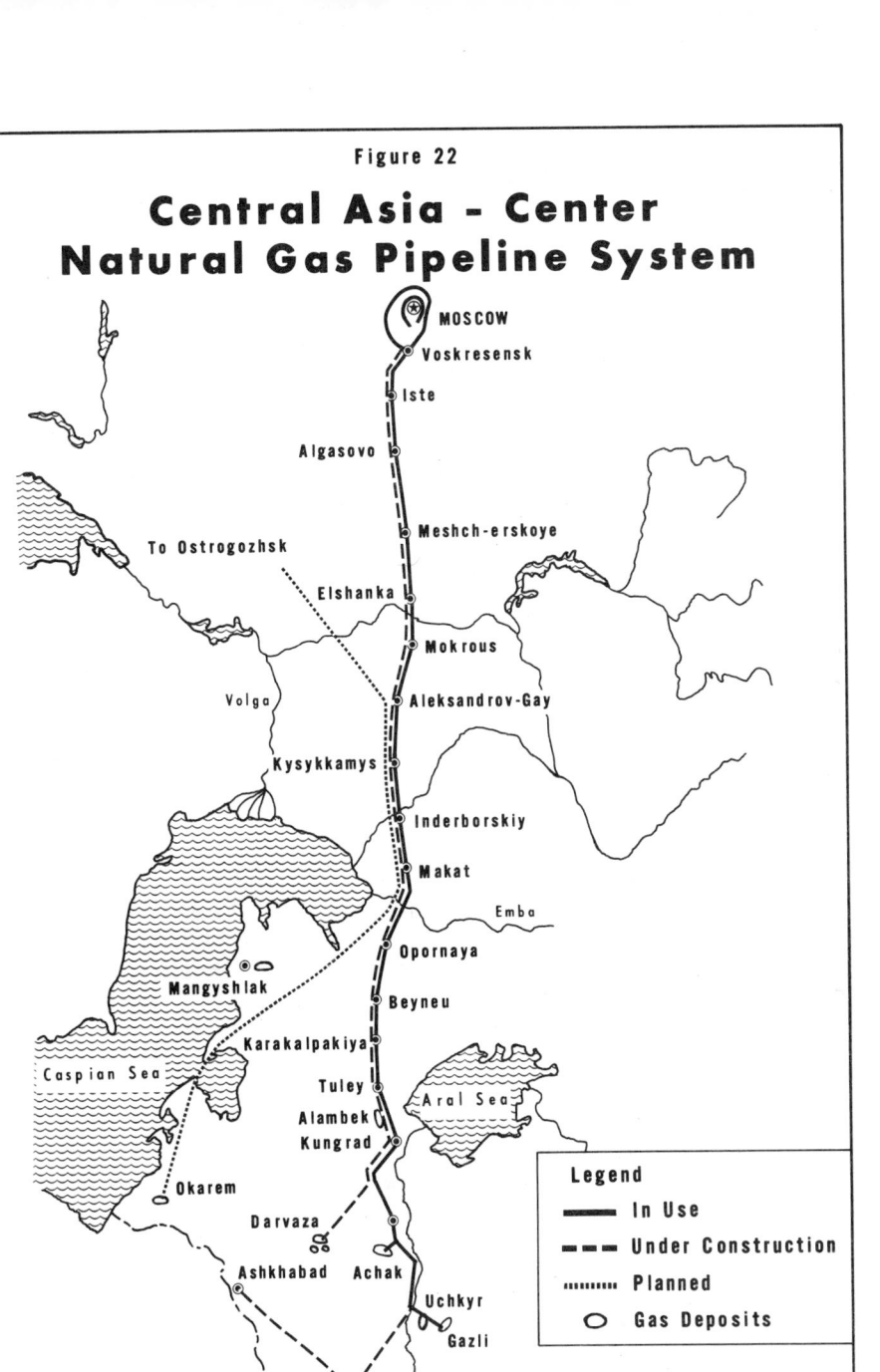

Central Asia – Center Natural Gas Pipeline System

The first line, 40 inches in diameter, was completed in 1967; the second line, of 48-inch pipe, is under construction. The third line will tap gas deposits on the Mangyshlak Peninsula.

Source: After A.K. Kortunov, Gazovaya promyshlennost SSSR, Moscow, 1967, p. 109

Such was the approach in building the Friendship crude oil pipeline; it worked satisfactorily then and probably would be applied again.

Care must be taken in an assessment of the reasoning behind the offer of Soviet gas to Italy. Should such exports materialize, it would appear that the USSR had taken a significant step forward in its search for means to increase foreign exchange earnings and that it had made a major addition to its barter basket of items attractive to European buyers.

The planned pipeline to Minsk is designed first of all to satisfy the Soviet consumer in the energy-deficient European part of the country. Between Minsk and Italy lies East Europe and these countries also are deficient in energy. Plans are being formulated, as noted earlier, for the USSR to supply East Europe with gas along much the same lines as Soviet crude oil is supplied to East Europe.[9] If a pipeline were to be built through East Europe en route to Italy, it is reasonable to expect that perhaps as much as 5 billion cubic meters of gas, equal to 50 per cent of the ultimate carrying capacity if 40-inch, eventually might be withdrawn in East Europe to supply local needs. Thus, the consumers in East Europe would benefit just as much, in absolute terms, as the customer in Europe and more important, the USSR would have found an easy way to meet its obligations to its socialist neighbors to the west.

This is not say that the Soviet Union attaches only minor importance to the sale of gas to Italy. To the contrary, the development of a new export commodity, especially one which can compete in quality with those of Western origin, would be extremely gratifying. In this particular instance, however, it should again be emphasized that the ultimate delivery of gas to Italy probably ranks third in relative importance to the other accomplishments which could be attributable to the construction of an export pipeline.

There are alternatives of course to the Northern Lights gas pipeline system as a source of export gas. The most reasonable alternative in terms of geographic proximity to the Italian market and to markets in East Europe as well are the gas fields in the eastern and western Ukraine. This alternative has become more promising upon the discovery of the Efremovka gas field in the eastern Ukraine, with reserves second in size in the republic to the giant Shebelinka field. Based on current pipeline construction plans, it is apparent that the USSR has recognized the feasibility of this alternative. Current plans call for the building of a 40-inch pipeline westward from Efremovka to Kiev and the concomitant construction of a separate, 40-inch pipeline beyond Kiev to the Soviet-Czechoslovakian border. Barring unforeseen construction difficulties, gas from Efremovka should be available at the border by the close of 1970, in quantities large enough for the USSR to follow through on any gas sales to the West which might be consummated in the next several years.[10]

Shortly thereafter, gas from West Siberia should also become available at the Soviet-Czechoslovak border, in apparently equal volumes, and the USSR will then have a gas export capability on its western border approaching 20 billion cubic meters for allocation between East Europe and markets in Western Europe.

Japan as a Market

Several offers to sell natural gas to Japan have been put forward by the USSR. The original Soviet proposal to Japan envisaged the construction of a large-diameter gas pipeline from the Okha gas field in the northern part of Sakhalin Island, southward to Nevelsk, a distance of some 850 kilometers. At Nevelsk the gas would be liquefied and carried to the Japanese port of Naoetsu by special tanker. Steel pipe, compressors, and other ancillary equipment for the pipe would be supplied by Japan. In addition, Japan would build the liquefaction plant and the special tanker. Quantities envisaged under this proposal were about 2 billion cubic meters annually.

In November, 1967 Teikoku Sekiyu, the potential Japanese customer, withdrew its offer to buy Sakhalin LNG, considering that the asking price of $12.90 per 1,000 cubic meters was too high.

The Japanese interest in buying gas from the USSR is very much analogous to that of Italy. Gas resources in the Niigata area of Japan are expiring and, lacking a domestic substitute, outside sources of supply must be located. In March, 1967 an agreement, covering fifteen years, was signed by Marathon Oil and Phillips Petroleum with the Tokyo Electric Power Company and the Tokyo Gas Company for the delivery of Alaskan LNG to Japan. Deliveries were to begin in July, 1969 and were to be made at an annual rate of about 1.42 billion cubic meters.[11]

Subsequently, in January, 1968, the Soviet Deputy Premier N. K. Baybakov requested continuing negotiations on the proposed LNG sales, to which Japan agreed. The Soviet Union has since revised and somewhat enlarged its proposal. One revision involves construction of a gas pipeline from Sakhalin Island to the Soviet mainland, and thence southward to Vladivostok. This would terminate the pipeline at a year-around ice-free port, more attractive than Nevelsk, which is not free for navigation throughout the year. Another proposal would involve Japanese participation in construction of a gas pipeline from the Yakutsk gas fields in eastern Siberia to the Pacific Ocean, a distance of 1,800 miles. Baybakov, in presenting this suggestion to the Japanese, implied that the fields were of sufficient potential to supply roughly 10 billion cubic meters of gas annually to Japan, but conversation centered around the export of 4 billion cubic meters per year.[12]

A plan of late 1968 called for the construction of a 1,600-kilometer gas pipeline, originating at Ikha in northern Sakhalin, running the length of Sakhalin Island, then extending out 50 kilometers across the La Perouse Straits, which separate Sakhalin and Hokkaido, and transversing Hokkaido to its southern regions to supply a proposed 2 billion cubic meters annually to a variety of consumers.[13]

This proposal was discussed in Tokyo in December, 1968 at a joint meeting of the Soviet-Japanese Economic Committee. Reportedly, this plan if promulgated would represent a considerably cheaper means of delivering Sakhalin gas to Japan than the long-standing proposal of natural gas liquefaction for shipment by tanker. Moreover, the question of selecting a suitable ice-free port on Sakhalin for tanker shipment of LNG would be avoided. Although no agreement was reached at the conference, areas of cooperation which could lead to some forms of joint venture in the future in the exploitation of the natural resources, including gas, of Siberia and the Soviet Far East were outlined.[14] While the acquisition of foreign exchange may be the prime mover behind the offer to supply natural gas to Japan, the Soviet Union would stand to benefit in a variety of ways should such a sale be consummated. The USSR has few markets in eastern Siberia and its easternmost regions capable of absorbing large quantities of natural gas; moreover, reserves are more than adequate in other, more attractive regions of the USSR. On this basis, exploitation of

Figure 23

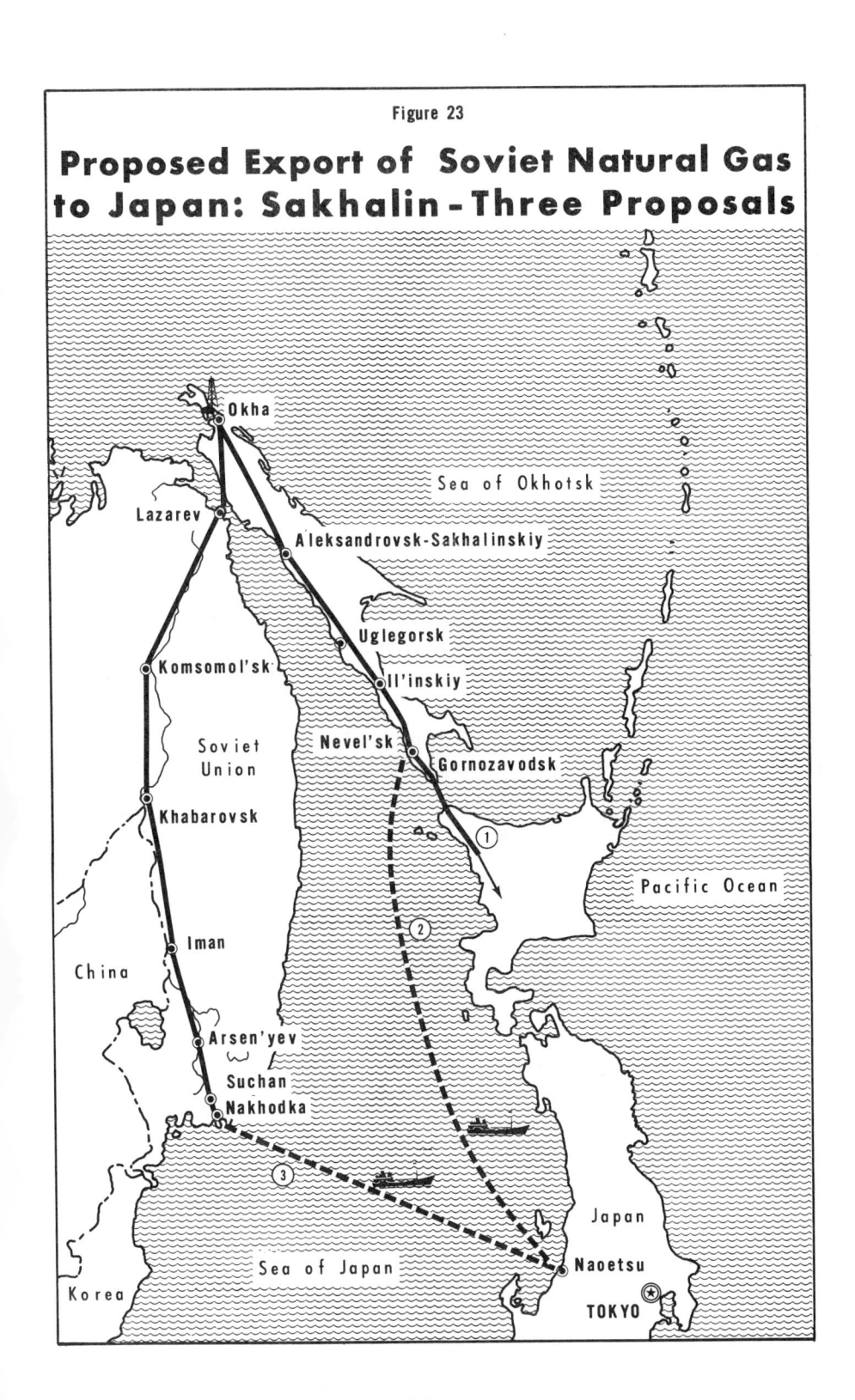

Proposed Export of Soviet Natural Gas to Japan: Sakhalin-Three Proposals

Figure 24

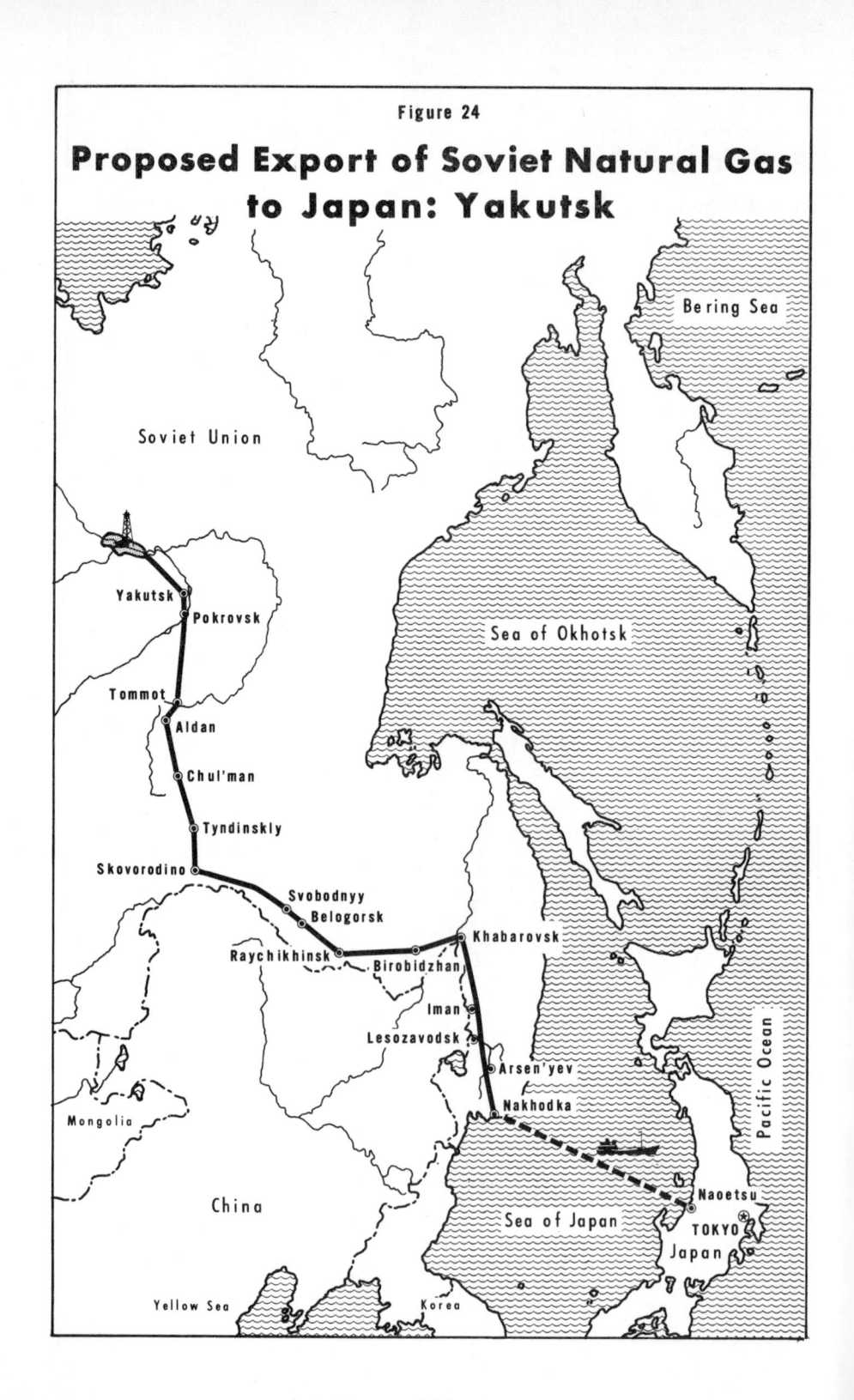

Proposed Export of Soviet Natural Gas to Japan: Yakutsk

the reserves currently available in those regions would be
postponed for a number of years to come. Yet it is recognized that the presence of a cheap source of energy (and raw
material) is a prerequisite to attract and support the variety
of diversified industries which Soviet planning officials would
like to see move into this relatively undeveloped area. So
far industry has been very reluctant to expand eastward beyond the Urals and southward into Central Asia, citing difficulties in obtaining supplies, the absence of transportation
and communication facilities, and the lack of an established
labor pool. Similarly, workers are just as reluctant to leave
behind those amenities of life available to them in established
commercial centers.

A natural gas pipeline constructed to fit the most recent
proposal to the Japanese, that of tapping the gas reserves in
the Yakutsk ASSR, would provide the rationale for producing
those gas reserves known to exist near the confluence of the
Vilyuy and Lena Rivers and would ensure adequate supplies
of fuel and raw material to sites heretofore dependent on
imports from the western regions of the country.

West Germany as a Market

The possibility of Soviet deliveries of natural gas to West
Germany was first broached in late April, 1969 by N. S.
Patolichev, Minister of Foreign Trade for the Soviet Union,
with Karl Schiller, Economics Minister for West Germany. [15]
Viewed against the background of the prolonged, and now dormant, negotiations between the USSR and Italy, which have
yet to produce any tangible results, West Germany now appears
as a most likely prospect for the export of Soviet natural gas
into Western Europe.

The demand for natural gas in West Germany is rather
sizable and growing, to an anticipated 27 billion cubic meters
by 1975. Prevailing estimates place one half of this demand
being satisfied from indigenous sources, with the Groningen
field in the Netherlands providing the balance. [16]

Most of the production of natural gas in West Germany
is concentrated in the northern portion of the country; moreover, Groningen natural gas is moved directly into the northern
area by pipeline. Thus, the immediate and future market for

Soviet natural gas would be limited largely to Bavaria. Production of natural gas has been advancing quite rapidly, increasing by about 55 per cent in 1968 to nearly 5.8 billion cubic meters. It is anticipated that by 1970 output will have risen to 9.8 billion cubic meters, but still far short of meeting requirements. Imports of gas from the Netherlands in 1968 were on the order of 2.7 billion cubic meters, roughly equivalent to one half of domestic production.

Because the West German economy will be heavily dependent upon imported gas, whatever the origin, during the coming years, the prospect of the Soviet Union as a source of supply has been welcomed in some circles, if only to the extent that the consumer perhaps may benefit from the presence of another seller in the marketplace. If Soviet gas were to be priced cheaper, as it normally would have to be to penetrate the market.

There is at present considerable political and economic pressure within West Germany to accept the Soviet proposal. This proposal foresees the delivery of between 3 to 5 billion cubic meters of gas annually, to be bartered for steel pipe, [17] most likely in the 48-to 56-inch diameter range. The political pressure derives from those who wish to capitalize on the improving relations between the Soviet Union and West Germany and those who seek low-cost energy necessary for regional industrial development. The economic pressure stems from the large diameter steel pipe manufacturers who are anxious to respond to the Soviet need.

In opposition are the domestic producers of natural gas who fear that large-scale imports of Soviet gas will impact negatively on their potential for future growth.[18] These producers are joined in their stand against the import of Soviet gas by operating oil companies that have a vested interest in the Groningen field. They judge that any reduction in the sales price of Groningen gas to West Germany, which the prospect of Soviet gas sales might entail, would have to be matched by comparable reduction in sales price quoted France and Belgium. Moreover, sharp increases in natural gas supplies are likely to bring about reductions in the demand for fuel oil.

Delivery of gas by the USSR to Bavaria could be accomplished by an extension of the pipeline now employed to bring Soviet gas into Austria, via Czechoslovakia. Alternatively, a pipeline could be built directly from Bratislava, Czechoslovakia, into southern Bavaria, crossing the frontier at a point, say, due east of Frankfurt.[19] Or gas could be provided to West Germany through extension of the pipeline to be built to East Germany in the early 1970's, if the political ramifications of such a route could be overlooked.

The Soviet Union undoubtedly views the prospect of selling gas to West Germany with considerable importance, first from the point of view of enlarging foreign currency earnings over the long run; second, the use of the gas to pay for steel pipe in the short run; and third, using the vantage of a foothold in the West German market to stimulate sales in other European countries. The next logical market for Soviet gas in Europe, once the West German market had been secured, would be France, where an opportunity for sales may become available in the 1970's.

Estimates of the probable export of natural gas by the Soviet Union in 1975 are assembled in Table 50. The total of 20.5 to 22.5 billion cubic meters presumes the successful conclusion of negotiations with Japan and Italy or West Germany. This total is allocated roughly equally between non-Communist buyers and other Communist countries and thus mirrors the relative allocation of Soviet exports of crude oil.

Should sales to Italy, West Germany, or Japan fail to materialize, the preponderant share of gas exports will be directed to the socialist countries of East Europe.

IMPORTS

From Iran

Since 1935 the Iranian government has attempted to arrange for the construction of an integrated iron and steel industry based on indigenous raw materials, first as a matter

TABLE 50

Probable Exports of Natural Gas
by the Soviet Union, 1975

(in billion cubic meters)

Destination	Amount
Austria	1.5[a]
Bulgaria	3[b]
Czechoslovakia	3[c]
East Germany	3[d]
Hungary	1[d]
Poland	2[d]
Subtotal	13.5
Italy/West Germany	5[e]
Japan	2-4[e]
Total	20.5 to 22.5

[a] *Vyshka*, June 19, 1968, p. 2.

[b] Bulgarian Telegraph Agency, October 24, 1968.

[c] *Izvestiya*, November 23, 1968, p. 4.

[d] Estimate.

[e] Estimates of exports of natural gas to Japan and to Italy presume the successful conclusion of negotiations with these countries.

of national prestige and second as a stimulus for further development of the economy. Various attempts to secure Western participation in such a venture proved unrewarding. The first Soviet overture to build a steel mill in Iran came in 1961, and the USSR offered to take payment through the sale of steel in Iran and in world markets. But because of close ties with the United States and the belief that a Western nation ultimately could be prevailed upon to build the mill, this offer was rejected.

In July, 1964 the USSR presented an informal offer to build a complete iron and steel plant for Iran as it had done for India. Iran, still hoping for Western assistance, was reluctant to accept the offer. However, because the availability of raw materials in Iran was not assured, private capital was willing to invest only in projects smaller than those envisaged by Iran. Therefore, if Iran insisted that a fully integrated mill was an absolute necessity for the prestige of the country, acceptance of the Soviet offer would be the only alternative.

No other offers were forthcoming and an agreement on the sale of Iranian gas to the USSR and the concomitant purchase of machinery and instalations by Iran was concluded in Moscow on January 13, 1966. The purchases by Iran included a heavy machinery plant and the much desired steel mill complex. The following January the Soviet Union extended additional credit to Iran to be used for the purchase of military equipment.

To repay the Soviet Union for this aid, Iran will export gas produced in association with crude oil from the Marun and Agha Jari fields in southern Iran. Because of the absence of local markets, this gas heretofore has been flared off, to the continuing annoyance of the Iranian government. Delivery of gas is to begin in 1970 at initial quantities of 6 billion cubic meters, gradually building to a maximum annual export of 10 billion cubic meters, to be reached in 1974. The agreement calls for a cumulative delivery of gas on the order of 140 billion cubic meters during the fifteen-year period 1970-84. The USSR will pay a price of $6.60 per thousand cubic meters for the gas, which will provide Iran with a total revenue of roughly $1 billion for a natural resource heretofore wasted, with no economic return to the country at all.

Construction of a pipeline to transport the gas from southern Iran to the Soviet border town of Astara, some 700 miles distant, presents a challenge to the builders. The terrain over which the pipeline is to be built and the extreme weather conditions which will be encountered en route will place a severe test on men and equipment. To overcome these obstacles the project, to be known as the Iranian Gas Trunkline (IGAT), has sought the latest techniques and know-how available, bringing together technicians, materials, and equipment not only from Iran and the USSR but from the United Kingdom, the Netherlands, France, Germany, Japan, and the United States as well.

The pipeline is to be built in two parts, a southern part roughly 400 miles in length and 42 inches in diameter and a northern part slightly in excess of 300 miles in length and 40 inches in diameter. Construction of the southern part, including the gathering system and spur lines, will be carried out by European contractors. The Soviet Union has assumed responsibility for building the northern part, between Qom and Astara, as well as a 75-mile, 30-inch branch pipeline to Tehran. Upon reaching full design capacity by the late 1970's, the pipeline will be capable of moving about 16 billion cubic meters per year, of which roughly 10 billion cubic meters annually will be allocated to Soviet markets.

The pipeline bringing Iranian gas into the Soviet Union will cross the Soviet-Iran border at Astara. From Astara a 220-kilometer pipeline will be laid to provide a link between the gas pipeline from Iran and the Transcaucasus gas pipeline Karadag-Akstafa-Tbilisi. Construction of this link was to begin in 1969. For the first 80 kilometers the system will consist of dual 40-inch pipelines; over the remainder a single 48-inch pipeline will be laid.[20] The line is to be completed the following year.

The major source of gas supply in the Transcaucasus (an area consisting of the republics of Azerbaydzhan, Georgia, and Moldavia) is found in Azerbaydzhan, as gas produced in association with crude oil. But production is on the decline and prospects for uncovering new deposits are not promising. Other gas is available, from the Stavropol and Krasnodar fields to the north--indeed, Stavropol gas has been fed to the Transcaucasus for some time--but these are old fields,

Figure 25

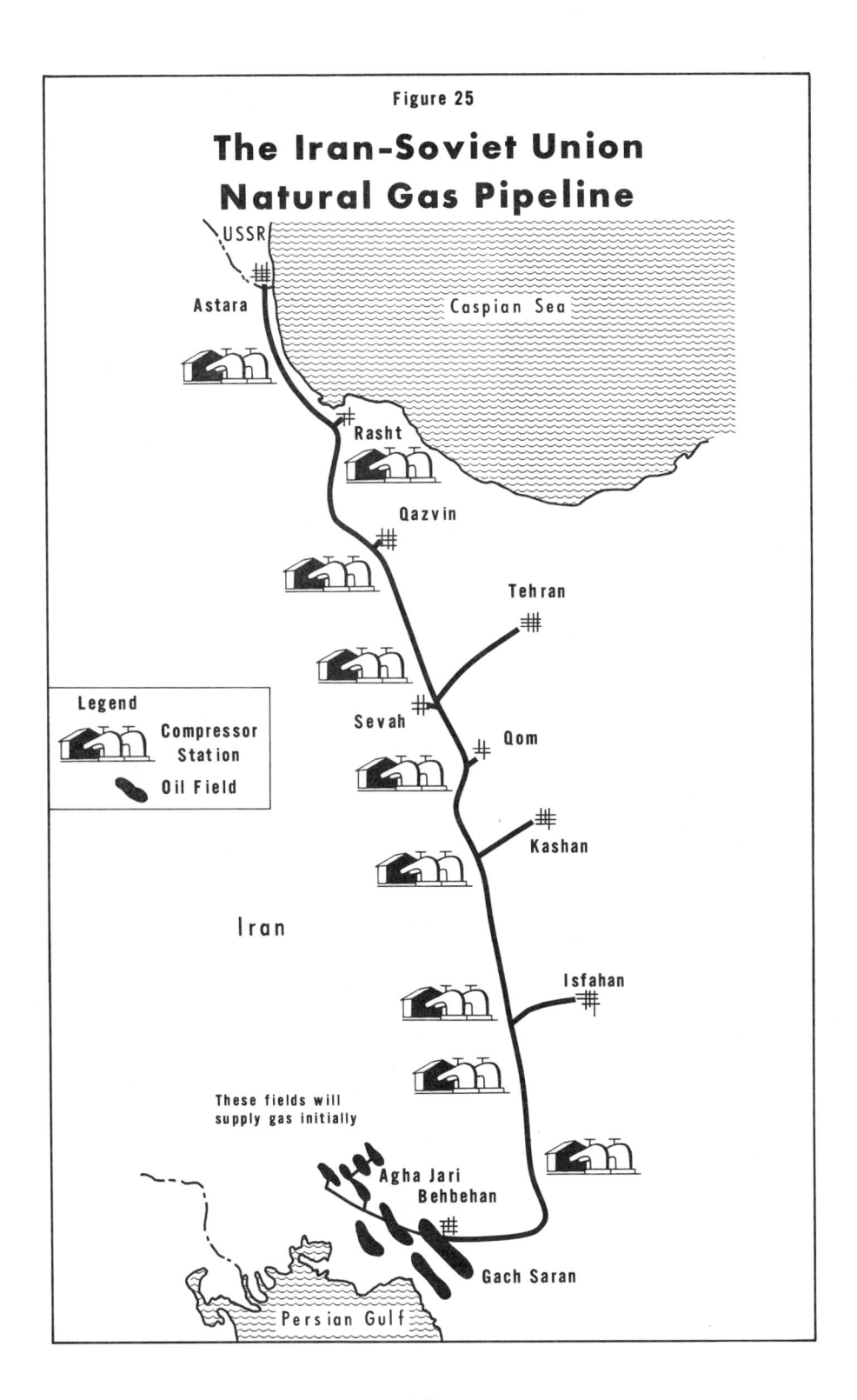

The Iran-Soviet Union Natural Gas Pipeline

USSR
Astara
Caspian Sea
Rasht
Qazvin
Tehran
Sevah
Qom
Kashan
Isfahan
Iran
These fields will supply gas initially
Agha Jari
Behbehan
Gach Saran
Persian Gulf

Legend
- Compressor Station
- Oil Field

largely committed elsewhere. Moreover, their output has been relatively static, and because the fields are faced with a production decline in the not too distant future, a substitute supply has to be developed.

It is probable that the quantities of gas imported from Iran will be in excess of local needs in the Transcaucasus, possibly by as much as 6 billion cubic meters per year, but this surplus gas, using the pipeline network now available, can be directed to a number of areas of the fuels-deficit regions in the European portion of the USSR.

The import of gas from Iran by the USSR can best be summarized as a marriage of convenience, with both partners to realize considerable economic benefits. For the USSR, it is perhaps more a marriage of geographic convenience, and represents probably the easiest solution to providing the Transcaucasus with supplies of gas more than adequate to make up the growing deficit in supply in that area.

While the agreement to import natural gas from neighboring Afghanistan is also a marriage of convenience, in this instance it appears that political gain on the part of the USSR, coupled with having presented Afghanistan with an opportunity to redress its negative trade balance with the USSR, stands out as the raison d'être.

From Afghanistan

In 1957 Afghanistan and the USSR signed a $15-million exploration assistance agreement. This exploration program realized little success until April, 1961 when natural gas was discovered at Hodja-Gugerdag, about 25 kilometers east of Shibarghan in northern Afghanistan. Further exploration in this area disclosed additional quantities of gas at Yatim Tag, about 16 kilometers east of Shibarghan.

The government of Afghanistan was particularly elated with the discovery of gas and was quick to propose the export of a portion of this gas to the USSR for the purpose of alleviating her unfavorable balance-of-payments position. An agreement reached in October, 1963 and ratified in June of the

Figure 26

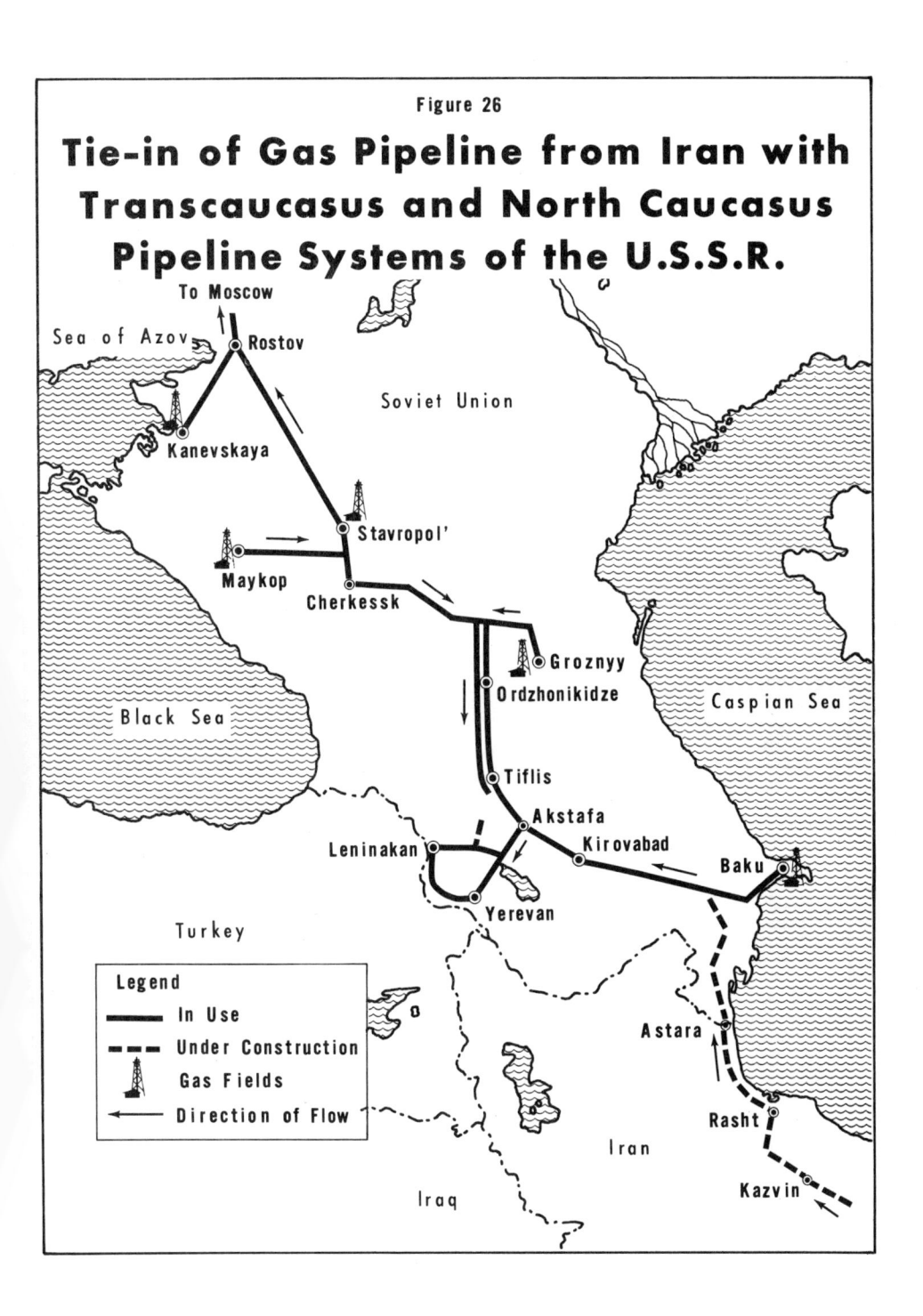

Tie-in of Gas Pipeline from Iran with Transcaucasus and North Caucasus Pipeline Systems of the U.S.S.R.

Figure 27

The Afghanistan-Soviet Union Natural Gas Pipeline

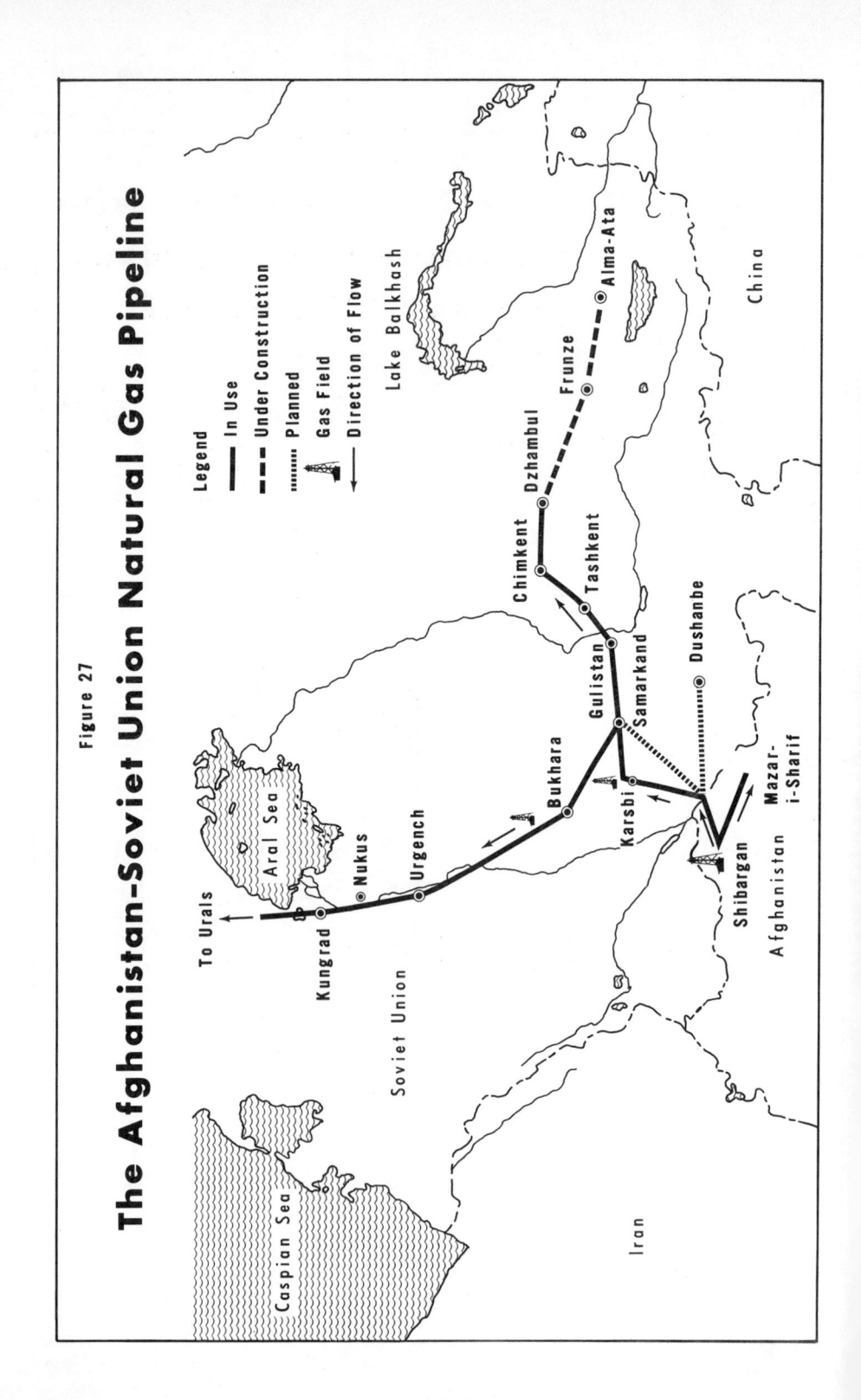

following year stipulated a Soviet loan of about $39-million, at 2.5 per cent interest, for the development of production at these gas fields and the construction of a pipeline to the Soviet-Afghanistan border. Beyond the border the pipeline would extend within Soviet territory to Mubarek in the republic of Uzbek. Exports of gas from Afghanistan would be at the rate of 1.5 billion cubic meters per year, increasing at some future date, to perhaps 3.5 to 4.0 billion cubic meters by 1975, at a delivered price of $5.60 per thousand cubic meters.

In addition to the pipeline to the USSR, plans called for building a gas pipeline to Mazar-i-Sharif to fuel an electric power station and to provide raw material for a fertilizer plant.

The Shibarghan-Kelif line is 98 km., 32 inches diameter, and has a capacity of 4 billion cubic meters; the Kelif-Mubarek (USSR) line is 270 km., 32 inches diameter, and has the same capacity; the Shibarghan-Mazar-i-Sharif line is 88 km., 16 to 12 inches diameter, and has a capacity of 1 billion cubic meters.

The Shibarghan-Kelif and Kelif-Mubarek pipelines were completed in 1967 and the export of gas to the USSR began that year. Cumulative delivery during the twenty-five-year agreement is to reach 58 billion cubic meters, all of which will be placed into the Bukhara-Urals and Central Asia gas pipeline systems. In contrast to gas imported from Iran, that from Afghanistan will become available in one of the centers of the Soviet gas producing industry and the annual imports will be almost lost among those quantities being produced locally.

Summary Imports

By 1975 both the gas pipeline from Afghanistan and the pipeline from Iran are to be operating at full capacity, meaning that total imports of gas by the Soviet Union from these two countries will fall within the range of 13.5 to 14 billion cubic meters. Only the import of gas from Iran will have any local significance, as noted, and total gas imports in 1975 will represent only some 4 per cent of projected indigenous output for that year.

NET TRADE IN GAS

What is considered to be a reasonable approximation of probable Soviet net trade in natural gas in 1970 and 1975 is presented in Table 52. Projections show a sizable increment in natural gas exports after 1970, largely reflecting presumed deliveries to Japan and Italy or West Germany. In terms of energy content, the possible export of 20.5 to 22.5 billion cubic meters of natural gas in 1975 would be equivalent to an export of roughly 17 to 18 million tons of crude oil.

But more important, as illustrated in Table 52, is the impact of imports of gas from Afghanistan and Iran upon the Soviet natural-gas trade balance. Indeed, the import of gas from Iran, to commence in 1970, and Afghanistan, which began in 1967, will in 1970 more than offset the volumes of gas exported. It is believed that in 1970 the Soviet Union will be a net importer of gas to the extent of 3.7 billion cubic meters, equivalent to the import of 3 million tons of crude oil. In the subsequent five-year period, the Soviet position in gas trade will improve, assuming sales to European markets and to Japan. Yet even should exports total 20.5 to 22.5 billion cubic meters in 1975, imports will sharply reduce this advantage to roughly 7 to 9 billion cubic meters. However, it should be kept in mind that exports to Italy and Japan can only be presumed for the purpose of constructing a natural gas trade balance. These are by no means "safe" estimates and the likelihood of gas sales to Italy or West Germany and Japan rests upon such a variety of political and economic factors that to attempt to develop trade estimates and to present them as an unyielding basis for making political and commercial judgments would be unwise. Thus, it is possible that the estimate of gas exports in 1975 is a major overstatement to the ultimate extent of the indicated sales to Italy and Japan.

Conversely, the estimates of the quantities of gas to be imported by the Soviet Union in 1975 have been clearly spelled out in trade agreements and are not subject to change, unless additional agreements are initiated.

TABLE 51

Imports of Natural Gas
by the Soviet Union, 1970 and 1975

(in billion cubic meters)

Origin	1970	1975
Afghanistan[a]	1.5	3.5 to 4
Iran[b]	6	10
Total	7.5	13.5 to 14

[a] Estimate.

[b] *Turkmenskaya iskra*, September 19, 1968, p. 4.

TABLE 52

Probable Soviet Net Trade
in Natural Gas, 1970 and 1975

(in billion cubic meters)

Indicator	1970	1975
Exports	3.8	20.5 to 22.5[a]
Imports	-7.5	-13.5 to -14
Net	-3.7[b]	7 to 9

[a] Assuming the successful conclusion of negotiations to sell gas to Italy and Japan (5 billion cubic meters projected for Italy and 4 billion cubic meters for Japan).

[b] Minus (-) sign denotes net imports.

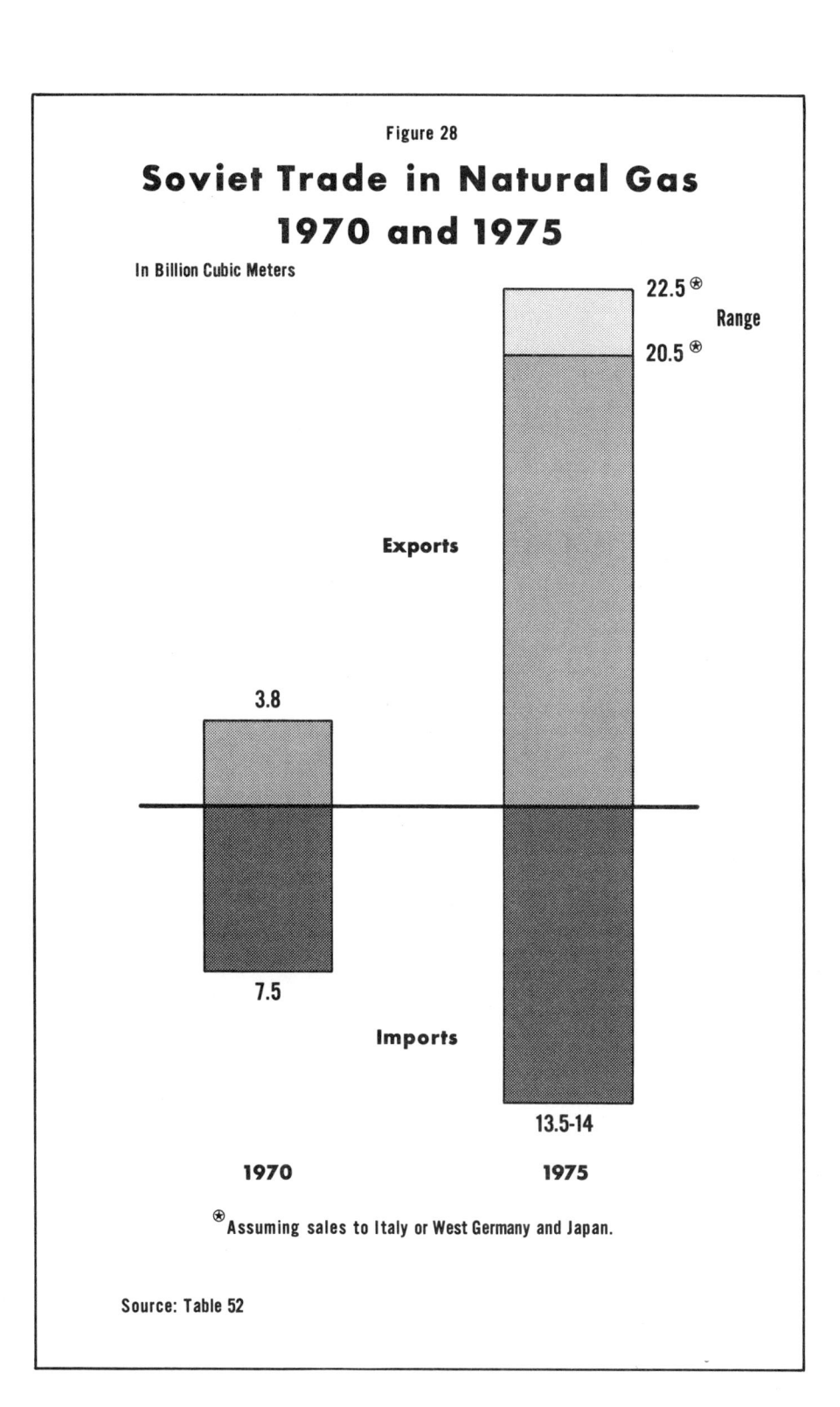

Figure 28
Soviet Trade in Natural Gas 1970 and 1975

In Billion Cubic Meters

Exports: 3.8 (1970); 20.5⊛ – 22.5⊛ Range (1975)

Imports: 7.5 (1970); 13.5–14 (1975)

⊛Assuming sales to Italy or West Germany and Japan.

Source: Table 52

Although it is believed that by 1975 the USSR may well be a net exporter of gas of only minor significance, if even that, the importance of the gross exports should not be overlooked. Imports of gas by a West European nation from the USSR (or by any country, from any source, for that matter) entail strategic and political vulnerabilities not present in oil imports, inasmuch as such energy supplies cannot be easily replaced nor quickly converted to other sources. Indeed, it is probable that a large imported supply of gas which was cut off for any reason could not be replaced in less than about two years. Thus the purchase of natural gas from the Soviet Union entails a much more sober responsibility than does the import of Communist oil, and prospective buyers should act accordingly.

NOTES

NOTES

NOTES TO CHAPTER 1

1. Christopher Tugendhat, <u>Oil: The Biggest Business</u> (London, 1968), p. 36. Standard Oil domination of the Far East was avoided through the initiative of Samuel in developing special tankers to haul kerosine, supplied by the Rothschilds, in bulk through the Suez Canal for delivery to markets in Japan, China, the East Indies, and India. The first such delivery was made in July, 1892 and permitted Samuel to offer kerosine at one-half the price of the Standard Oil product.

2. A. Beeby Thompson, in his <u>The Oil Fields of Russia</u> (London, 1908), p. 384, wrote that "no more marvelous example of the mineral wealth of Russia is needed than the Baku oil fields, where less than six square miles of barren desert yield nearly half the world's supply of petroleum; 2,000 wells grouped together within an area no larger than a London suburb yield as much petroleum as 20,000 wells spread over several thousand square miles of land in the United States."

3. Heinrich Hassman, <u>Oil in the Soviet Union</u> (Princeton, 1953), pp. 29-31.

4. The extent of the damage imposed upon the oil fields of the Apsheron Peninsula by these armed uprisings is vividly described in Luigi Villari's <u>Fire and Sword in the Caucasus</u> (London, 1906).

5. <u>Op. cit.</u>, p. 3.

6. "K 50-letiyu natsionalizatsiy neftyanoy promyshlennosti vsssr," <u>Neftyanoye khozyaystvo,</u> 9 (September, 1968), 1-7.

NOTES TO CHAPTER 2

1. Louis Fischer, <u>Oil Imperialism: The International Struggle for Petroleum</u> (London, 1926), p. 13. S. M. Lisichkin noted: "foreign monopolists did not wish to lose its rich sources of profits and such a strategic raw material to the Russian oil industry. The imperialists resolved to defend their interests

with open intervention. With the aid of internal counter-revolutionary forces they seized the rich Baku oil region in June 1918. Counter-revolutionary forces took possession of the Emba oil region. The Maykop oil fields, which became a battleground for the Red Army against the White Guards, was destroyed. In November 1918 agents of the English interventionists set fire to the Gronznyy oil fields. Production of crude oil in Central Asia ceased." ("K 50-letiyu natsionalizatsiy neftyanoy promyshlennosti vsssr," <u>Neftyanoye khozyaystvo,</u> 9 [September, 1968] 2.)

2. The subsequent claims for compensation have never been honored.

3. Interest in Soviet oil was not limited to international oil groups. Harry F. Sinclair, an independent of considerable financial and political wealth, made two unsuccessful attempts to gain oil concessions from the Soviet Union. The first, an oil concession covering northern Sakhalin Island, did not materialize because of failure to secure Japanese evacuation of the territory. In 1923 Sinclair and the Russians came to an agreement under which a joint company would be established to develop the Baku oil fields for a period of 49 years, with management and profits to be shared equally. Unfortunately, the involvement of Sinclair in the infamous Teapot Dome scandal the following year caused his contract with the Russians to be invalidated. One can only speculate as to what might have happened to the Soviet oil industry, and in particular its relations with the West, had Sinclair been able to proceed as originally envisaged.

4. This bid was based on the Standard Oil Company claim to the Baku oil fields, a claim obtained through an earlier purchase from the Nobel brothers. As had other international oil companies, Standard undoubtedly anticipated an early collapse of the Soviet regime, and thought this would be an easy means of acquisition of Russian oil properties. At that time, the Soviet government was interested in entertaining concession tenders, but not on those terms.

5. Leon M. Herman, "Export Ebb and Flow of Russian Petroleum," <u>Journal of International Economy,</u> April 7, 1945, p. 6.

NOTES 173

6. Amtorg Trading Corporation, Soviet Oil Industry (New York, 1927), p. 12.

7. S. M. Lisichkin, Ocherki razvitiya neftedobyvayuschey promyshlennosti (Moscow, 1958), p. 210.

8. In early 1927 a conference had been called between representatives of the Soviet Oil Syndicate and representatives of foreign oil interests. At the conference formulas were being worked out for the distribution of the rapidly growing oil exports from the Soviet Union. The conference broke up over the insistence of Deterding on a monopoly of Soviet oil exports and a limitation of Soviet exports of crude oil. See Soviet Union Information Bureau, The Soviet Union, Facts, Descriptions, Statistics (Washington D.C., 1929), p. 36. Deterding's opposition to Soviet oil unquestionably stemmed from the Oil Syndicate's having established a subsidiary -- Russian Oil Products, Ltd. (R.O.P.) -- in England. The temerity of the syndicate in challenging Deterding in his own market could not go unanswered. Deterding took his battle to the press and to official channels, and some charge, albeit without clear proof, that he was guilty of meddling in Soviet internal affairs.

9. Amtorg Trading Corporation, op. cit., p. 9.

10. In defense, the USSR later pointed out that the demoralization of the world's oil market could not be due to the part played by Soviet oil products which represented only 10 per cent of total world oil exports; see "The World Oil Conference," The Petroleum Times (May 28, 1932), p. 610.

11. The low prices for Soviet oil products were particularly inviting to the individual consumer, and for the American and British groups to be able to compete with these Soviet offerings in England meant involvement in a price war which neither relished. When it appeared that the USSR was in the market to stay, Deterding, out of economic necessity, was forced to grant to R.O.P. a large share of the British market. This apparent victory for Soviet oil, which in effect marked its emergence as a new factor in world oil trade, is discussed by Hans Heymann in "Oil in Soviet-Western Relations in the Interwar Years" in the American Slavic and East European Review, VII, 4 (December, 1948), 303-316.

12. Exports that year accounted for more than 28 per cent of crude oil production. As noted in Table 7, after 1926 the USSR exported at least 20 per cent of its output of crude oil to finance equipment imports.

13. The Petroleum Times, October 10, 1931, pp. 475-478.

14. Ibid., June 18, 1932, p. 700.

15. Ibid., October 24, 1931, p. 543.

16. Kessler had worked out a revised plan, following U.S. rejection of his first proposal, recognizing that U.S. laws did not admit of the application of the original plan. As revised, the plan did not differ greatly, except that production in the U.S. would not be regulated; only exports from the U.S. would be limited, and a number of other producing countries, including the Soviet Union, would be asked to subscribe.

17. The Petroleum Times, May 21, 1932, p. 580.

NOTES TO CHAPTER 3

1. The position of the Urals-Volga today as the major source of oil in the USSR is credited to Gubkin.

2. Vneshnyaya torgovlya SSSR, statisticheskiy sbornik, 1918-1966 (Moscow, 1967).

3. American Petroleum Institute, Petroleum Facts and Figures, 9th edition (New York, 1950), pp. 322-323.

4. John W. Frey and H. Chandler Ide, eds., A History of the Petroleum Administration for War, 1941-1945 (Washington, D.C., 1946), pp. 23-24.

NOTES TO CHAPTER 4

1. As later reported in Pravda (April 5, 1958, pp. 4-5) certain high government officials, most probably those having close ties with the coal industry, were exponents of the theory that explored reserves of oil and gas were quite small and that the structure of the primary energy balance, with its emphasis on coal, could be no different. The long-range plan had been to almost entirely eliminate petroleum fuels and by 1960 would have transferred all of the important branches of industry to the utilization of expensive solid fuels.

2. Probably the most complete study of the history of the Urals-Volga oil producing region is to be found in A. A. Trofimuk, Uralo-PoVolzhe--Novaya neftyanaya baza SSSR (Moscow, 1957).

3. A situation which could be traced back to the 1956-57 Suez crisis. Upon restoration of the Suez Canal to normal operations, domestic producers in the United States were faced with a major loss in export demand, to the extent of almost 1.2 million barrels a day, within a relatively short period of time. At the same time, the presence of a business recession depressed the growth in domestic demand. Because re-adjustments in supply were slow, an oil glut rapidly developed, followed by a general deterioration in the market. Some feel that the domestic oil industry has never really recovered from this setback.

NOTES TO CHAPTER 5

1. As of June 30, 1968 the world tanker fleet (10,000 dwt and over) was measured at 110,142,000 dwt, based on an inventory of 3,088 vessels. Of this tonnage, the tanker fleet of the USSR, 156 in number, represented only 3.3 per cent or slightly more than 3.6 million dwt. The surge in supertanker construction brought about by the closure of the Suez Canal in June, 1967 as an aftermath of the Arab-Israeli war was largely responsible for the 10.7 per cent increase in the world tanker fleet in the subsequent 12 months, but net new additions to the Soviet fleet were much less significant, to

the extent that the share of its tanker fleet in the world total was now less, if only marginally, than the 3.5 per cent share it held on June 30, 1967 (Petroleum Press Service [October, 1968], pp. 373-374).

However, it should be pointed out that the USSR flag tank ship fleet is about two-thirds comprised of vessels under 12,000 dwt. Nevertheless, measurement of the world tank ship fleet to include vessels of 2,000 dwt and over does not significantly improve the Soviet position. In such a measurement (Analysis of World Tank Ship Fleet, December 31, 1967, prepared by the Economics Department of the Sun Oil Company), the deadweight tonnage of the Soviet fleet is calculated at 4.04 million dwt, or 3.5 per cent of the world total. It is thought that the Soviet tanker fleet is sufficient in size to carry about 50 per cent of total oil exports to all areas, with the larger portion of the remaining 50 per cent handled by ships under Soviet control on a charter basis.

For a complete analysis of Soviet maritime policy, the reader is referred to The Soviet Merchant Marine, prepared by the Maritime Administration, U.S. Department of Commerce (Washington, D.C.: U.S. Government Printing Office, 1967). Several discussions of the development of the Soviet fleet which originally appeared in Russian-language sources are presented in Part II, "Soviet Trade in Oil."

2. International Affairs, 7, (July, 1960), 26.

3. November 13, 1967.

4. For the report of the first U.S. oil delegation to visit the Soviet Union, see Robert E. Ebel, The Petroleum Industry of the Soviet Union (Washington, D.C.: American Petroleum Institute, 1961).

5. The Oil and Gas Journal, LVIII, 47 (November 21, 1960), 154. The U.S. managed to drill more oil wells in a given year than total Soviet wells in operation at that time.

6. The New York Times, May 21, 1968, p. 14E.

7. Radio Moscow, November 22, 1968.

NOTES TO CHAPTER 6

1. Attempts to develop comparatave pricing in oil products trade between the USSR and East Europe and between the USSR and non-Communist countries are relatively meaningless. Trade in oil products can and does embrace a wide quality range within a single product category, for example gasolines of varying octane, and prices vary accordingly. Soviet statistics as a rule do not provide data on trade within a single product category, reporting only the total for, say, gasoline, kerosine, or diesel fuel, and the total value of all of the various grades of gasoline, kerosine or diesel fuel. Variances in the quality of crude oil can also justify a wide range of prices but it is believed that the quality of the crude oil delivered to Western Europe differs little from that delivered to Eastern Europe.

2. Members are the Soviet Union, Bulgaria, Czechoslovakia, East Germany, Hungary, Poland, Rumania, and Mongolia; Yugoslavia is an associate member.

3. The USSR has asked Czechoslovakia to invest in Tyumen (West Siberia) oil to guarantee increased deliveries after 1970.

4. Largely because of inferior quality.

5. Transmitted on March 4, 1968.

6. Radio Prague, May 16, 1968.

7. Ekonomicheskaya gazeta, 39 (September, 1968), 41.

8. Izvestiya, July 30, 1968, p. 2.

NOTES TO CHAPTER 7

1. U.S. Senate, Committee on the Judiciary, Eighty-Seventh Congress, First Session. Soviet Oil in the Cold War, (Washington, D.C.: 1961), p. 8; and Robert E. Ebel, The Petroleum Industry of the Soviet Union (Washington, D.C.: American Petroleum Institute, 1961), p. 166.

2. <u>Oil and Gas Journal,</u> November 28, 1960, p. 36.

3. Official Minutes, Third Arab Petroleum Congress, Alexandria, 1961.

4. <u>The New York Times,</u> December 9, 1967, p. 71; and <u>The Wall Street Journal,</u> June 19, 1967, p. 4.

5. <u>Platt's Oilgram,</u> July 6, 1967.

6. <u>Pravda vostoka,</u> June 21, 1967, p. 3. For a full exposition of the Soviet position regarding the relationship between the Western oil "monopolists" and Israel, see <u>Kommunist,</u> 12 (August, 1967), 109-117.

7. <u>Mideast Mirror,</u> September 14, 1968.

NOTES TO CHAPTER 8

1. <u>The New York Times,</u> September 13, 1955, p. 1.

2. Having complete and unquestioned control over the use of natural resources, including their application in foreign trade, the Soviet Union is in a position at any time to abrogate contracts or to interrupt supplies unilaterally and arbitrarily, should the political aspirations of the Government so dictate. Customers who find themselves dependent upon the Soviet bloc thus are vulnerable to such actions; although comparatively few in number, there are several outstanding examples where the USSR has used foreign trade as a means to secure purely political goals: (a) Largely as a result of an implied Soviet threat to curtail trade, the Conservatives in the Finnish cabinet, whose inclusion had been subject to criticism by the Soviet Union, withdrew from the cabinet and the cabinet was forced to resign; (b) A contract to deliver oil to Israel was summarily canceled following the outbreak of hostilities in the Middle East in 1956. Subsequent claims for damage were rejected by the Soviet side. Presumably such action was taken to reaffirm, for the benefit of the Arab nations, the support of the Soviet Union for their cause; (c) On February 5, 1969, Ghanaian police forcibly removed the crews of two Soviet fishing trawlers which had been seized some four months earlier, and placed

NOTES

them under arrest. Seizure had been on the presumption that the vessels might be engaged in arms-running. Shortly thereafter, the USSR warned Ghana they were unable to ship any crude oil in March (a contract, which began in September, 1968, called for the delivery of 700,000 tons of crude oil during the ensuing year). Ghana then was forced to appeal to Western suppliers to keep its refinery in operation. Although not so indicated, the cut-off of crude supplies, coming so soon after the trawler incident, strongly suggests Soviet retaliation.

3. *Pravda,* December 11, 1968, p. 2. Subsequently complicated by the severe winter of 1968-69, the effects of which may hold 1969 output to about 325 million tons. If so, the 1970 estimate may be an overstatement.

4. *Vyshka,* November 30, 1968, p. 2, in which Deputy Minister of the Oil Extracting Industry Kuvykin stated that output of oil would reach 345 million tons by 1970, but this total would exclude those amounts produced at fields under the responsibility of the Ministry of the Gas Industry.

5. In particular, see Lincoln Landis, "Soviet Interest in Middle East Oil," *The New Middle East,* and D. C. Watt, "Russians Need Middle East Oil," *The New Middle East,* 3 (December, 1968), 16-21 and 21-23, respectively. Landis does note that Russian interest in Middle East oil embraces both political and economic reasons and that the primary objective of Soviet strategy in the Middle East is "to be able to threaten the orderly flow of oil to Western Europe and the United States." Watt, on the other hand, is much more pragmatic about the Soviet _need_ for Middle East oil _per se._ He defines this need as encompassing domestic requirements, and export requirements as well, reasoning that by 1975 or so the Soviet Union could well become a net importer of oil, if it wished to maintain its position in oil trade.

Similarly, Edward Hughes reasons in "The Russians Drill Deep in the Middle East," *Fortune,* July, 1968, pp. 102-105, that while Russian interest in Middle East oil may have its political purpose, it is, nevertheless, also clearly founded in simple economics. Hughes observes that "Communist economists now estimate that the U.S.S.R. and the rest of the Eastern European nations could face a shortage of two million barrels a day by 1980." This statement, although not documented, can be attributed directly to the Albinowski article.

In its The Gulf: Implications of British Withdrawal "Special Report Series," No. 8 (Washington, D.C.: Georgetown University, February, 1969), p. 65, The Center for Strategic and International Studies apparently has utilized the same material when it writes that "Russian interest in the Gulf oil countries is increasing and it is probable that, with greatly increased industrial needs, the communist bloc will, in the near future, become a net importer of oil from beyond communist Eurasia and this may be presumed to result from the desire to control the oil economy of eastern Europe rather than to allow independent economic policies within the eastern European region of the Soviet perimeter." In other words the Soviet Union, faced with increasing domestic demand, the satisfaction of which will leave too little oil to meet growing East European requirements, very soon will be forced into importing large quantities of oil for redirection to East Europe in order to be able to retain control over the latter's economy. Again, the authors presumably have based their reasoning on Albinowski's findings which, as noted, overstate the situation.

6. Washington Evening Star, May 15, 1968, p. A-12.

7. Bakinskiy rabochiy, an Azerbaydzhan newspaper, referred in its September 8, 1968 issue (p. 3) to assertions in the Czechoslovakian press that as a result of restrictions in deliveries of iron ore, natural gas, and coking coals from the Soviet Union, metallurgical plants were not operating at full capacity. The newspaper then went on to point out that deliveries continued unabated from the Soviet side, but charged interference and sabotage on the part of certain Czechoslovakian groups and individuals of these deliveries in order to cause difficulties in the Czechoslovak economy which would then be blamed on the Soviet Union.

8. Reports of a Canadian team of oil men which visited the area in late 1967 attest to the validity of these claims, as discussed in Oilweek, February 12, 1968, pp. 20-22.

9. This long-range goal, and others like it, are meant to be only guides for planning officials, and revisions subsequent to the release of such goals should not be immediately taken to imply difficulties in the extraction sector. Long-range energy output goals are apt to be optimistic, a failing characteristic of most economies, and not just the Soviet.

NOTES 181

It can be expected that corrections will be instituted as requirements become more precise. Moreover, annual goals within a planning period such as a five year plan or a seven year plan are also subject to change; these changes, of course, are a direct reflection of recent successes or difficulties in energy production, distribution, and consumption. Table 33 points up the development of revised annual crude oil production plans as the Seven Year Plan (1959-65) unfolded, as well as reappraisals of longer-range goals brought about by more definitive knowledge of supply and demand requirements, as well as of the capability to produce.

Similar examples can be assembled for coal and for natural gas. Major downward revisions in plans for production of these fuels were made during the Seven Year Plan and in the subsequent Five Year Plan, as it became apparent that the original goals were far in excess of what could reasonably be expected. (See Table 34.)

10. Supporting in part a three-year contract to deliver 3.5 million tons of crude oil to Rumania, beginning in 1968.

11. <u>Oil Daily,</u> September 26, 1968. In practice, annual deliveries in the first several years under the contract will be less than this indicated average, and in the last years will be considerably more.

12. Radio Bucharest, June 6, 1968.

13. It is generally considered that the Soviet Union has been most anxious for the Suez Canal to reopen, for larger reasons. Its naval fleet in the Mediterranean, to gain access to the Persian Gulf, would have to sail around the Cape of Good Hope, and in the rapidity with which today's military operations must be carried out, this detour could have serious ramifications. Similarly, the closure of the canal has greatly complicated the delivery of Soviet military goods to North Korea and North Vietnam.

In 1966, the last full year of operation of the Suez Canal prior to its closure in June, 1967, a total of 2.9 million tons of crude oil and about 6.1 million tons of petroleum products were moved southward through the canal. Of this amount, 5.6 million tons, or more than 62 per cent, originated with the Soviet Union.

14. Shell subsequently negotiated a direct purchase of Soviet crude oil as reported in Petroleum Intelligence Weekly, October 17, 1968, p. 3. In an arrangement again made through Nafamondial, Shell will take delivery of about 250,000 tons of crude in Europe by the close of February 1969. Deliveries began in June 1968, at a c.i.f. price on the order of $13.50 per ton delivered to Italy.

15. Petroleum Intelligence Weekly, May 13, 1968, p. 1.

16. Izvestiya, April 5, 1968, p. 1.

17. Oil and Gas Journal, April 29, 1968, p. 48.

18. The Japanese effort to win a share in the development of Siberian natural resources has been at least persistent, if not successful. But what may be the first step toward what both sides hope to be a long and fruitful association was taken with the signing in Tokyo of a protocol confirming a timber development agreement, in which Japan will provide machinery and equipment to aid in the development of new timber resources in Siberia and will import an equivalent amount (in value) of timber in return (Economist, September 14, 1968, p. 94). This timber agreement, if it works, could clear the way for others.

19. Tass, August 12, 1968.

20. Radio Moscow, January 3, 1968.

21. The New York Times, January 11, 1969, p. C-39. Interestingly enough, neither the Soviet popular nor technical press reported in full on the remarks of Shashin. Internal publicity was limited to just a few generalizations, largely concerning the expansion in crude oil output through 1975 and the contributions expected from the new producing areas.

22. Robert W. Campbell, The Economics of Soviet Oil and Gas (Baltimore: Johns Hopkins Press, 1968), p. 231 ff.

23. Tass, December 27, 1968.

24. Stroitelnaya gazeta, January 12, 1969, p. 4.

NOTES   183

25. Kommunist, 2 (January, 1969), 84-87.

26. Ekonomicheskaya gazeta, 4 (January, 1969), 3-5.

27. What has been described as the possible prototype of arrangements under which East European countries may help the Soviet Union develop its natural resources proved most successful after completion of its first year. Under a ten-year treaty signed in 1967, Bulgaria was given permission to send construction workers and lumberjacks to the Soviet Union to cut timber for use by both countries. The treaty was an outgrowth of an inability on the part of the Soviet Union to meet Bulgarian requests to buy lumber, simply because there was no lumber to spare; local demand accounted for the entire supply. At the same time the USSR had no objection to Bulgaria cutting the timber itself. Under the treaty, four out of every seven carloads of timber cut would be handed over to the Soviet Union; the remaining three would be shipped to Bulgaria. (see The New York Times, February 9, 1969, p. 11).

28. As noted earlier, the USSR is to import 4 million tons of crude oil from Algeria during 1969-75 and is to take Iraqi crude oil in return for technical assistance rendered to their oil industry.

29. A rational basic to any argument for the expansion of East-West trade. For East-West trade to develop to meaningful proportions, the Communist countries must be able to generate hard currency. If payment for Western goods continues to be restricted to raw materials or selected items of equipment, significant East-West trade is not possible.

30. The view set forth by some that oil from the Alaskan North Slope in particular and the Arctic (both U.S. and Canadian) in general by 1980 may challenge the Middle East in terms of relative importance as a source of supply seems to be premature, based on what is presently known of the potential of this area and of the natural obstacles to moving this oil to market.

31. An exception has been made for lubricating oils.

32. Established in Antwerp in December, 1967 with a nominal capitalization of $60,000.

33. <u>London Daily Telegraph,</u> January 16, 1969.

NOTES TO CHAPTER 9

1.   The Minister of the Oil Extracting Industry V. Shashin has indicated that the export orientation of Soviet oil in the future would be largely westward (<u>The New York Times,</u> January 11, 1969, p. C-39).

2.   Markets for natural gas in Finland and Sweden are perhaps too small and too scattered to warrant serious consideration of imports from the Soviet Union at this time, although political considerations could set this logic aside. The USSR has been pressuring Finland to commit themselves regarding purchases of natural gas. Finland, to a degree, has acquiesced and expects to decide by 1973. If the decision is "yes," gas would start moving through the pipeline by 1977 at the earliest. Reportedly, imports would run about 2.7 billion cubic meters annually (See <u>Platt's Oilgram,</u> April 28, 1969, p. 1). On the other hand, France has the markets but to move Soviet gas to these markets would be a major undertaking and probably would have to be coordinated with the offtake of considerable quantities of gas along the way. Recently concluded negotiations between the USSR and France foresee a doubling of trade between these two countries by 1974 (see <u>The New York Times,</u> January 9, 1969, p. 6). Payment in part for plant and equipment to be delivered by France to the USSR through the sale of gas has been advanced by the Soviet side.

3.   A final version was signed in Vienna on June 1, 1968.

4.   Sweden, not a member of NATO, has continued to deliver 40,000 to 50,000 tons of 40-inch pipe to the USSR each year during and since the embargo. A discussion of this embargo, why it was instituted, and what it accomplished in a Soviet version, follows:

In November, 1962 NATO recommended that member countries not sell large-diameter (19 inches and larger) pipe to the Soviet bloc. However, those contracts in force at the time of imposition of the embargo were permitted to run their course. As a result, total deliveries of 40-inch pipe by the West (West Germany, Italy, and Sweden, plus very small amounts from Japan) to the USSR through 1965 reached about 1.1 million metric tons, sufficient to lay about 4,000 kilometers of pipeline. These purchases represented probably more than 40 per cent of total requirements for 40-inch pipe during the Seven Year Plan (1959-65). Indeed, all 40-inch pipe installed in the USSR through March of 1963 was imported.

The purpose of the NATO embargo was to delay construction of the Friendship crude oil pipeline being built to connect new refineries in East Europe with the major oil-producing areas of the Soviet Union and was undertaken "in the military interests of the alliance" (The New York Times, November 18, 1966, p. 13). It was believed that the Friendship pipeline was a vital link in the continuing supply of fuel to military forces in Eastern Europe. Any delay in the completion of this link, it was concluded, would seriously reduce the effectiveness of these forces.

Supply of 40-inch pipe in the USSR was sufficient to permit completion of this pipeline on schedule, that is, by the close of 1963. However, the USSR was in the midst of a rather ambitious program of development of natural gas production and transportation facilities, and the bulk of the available 40-inch pipe was directed to this end. As a result, the Friendship pipeline was not completed until late 1964, a delay of roughly one year. Moreover, it appeared that 32-inch pipe was installed along several hundred miles where original plans had called for 40-inch, although subsequent information tends to contradict this.

To offset those quantities of pipe denied by the embargo, a number of Soviet pipe mills were converted to the manufacture of 40-inch pipe. The production of considerable quantities of pipe in the range of 19-to-32-inch therefore was lost. Thus, perhaps the real impact of the NATO embargo was not on the supply of 40-inch pipe, whose availability very nearly matched requirements, but rather was in reduced supply of smaller diameter pipe. It is estimated that supply of pipe in the range of 19-to-32-inch probably was no more than 80 per cent of requirements.

Prior to the institution of the embargo, the Soviet Union appeared almost completely dependent upon imports for supplies of 40-inch pipe and apparently stood ready to conclude further contracts with the West. Indeed, at the time the embargo was imposed, an agreement had just been signed with a West German firm for the delivery of more than 160,000 tons of 40-inch pipe but the West German government, in a show of support of NATO aims and objectives, refused to approve the contract.

If the impact of the embargo upon Soviet pipeline construction plans can be measured by the bitter reaction of the Soviet government, then this impact was most serious. But Soviet pride demanded a show of strength on the part of the domestic pipe manufacturers and in March, 1963 Russian newspapers proudly displayed photographs of the first 40-inch pipe coming out of their mills. Although most of the early claims of self-sufficiency in large-diameter pipe following the crash program to develop indigenous capability were patently exaggerated, considerable progress was evident and the Soviet technical press was able to report that in 1965 domestic mills produced 900,000 tons of 40-inch pipe.

Following a review of the Soviet capability to manufacture large-diameter pipe and a comparison of this capability with forthcoming requirements, NATO lifted its embargo on November 10, 1966. The announcement of removal of the embargo noted (New York Times, November 12, 1966, p. 3) that the Friendship crude oil pipeline had almost been completed, after a delay of about the length expected in 1962, and added:

> As NATO's aim is not to place obstacles to East-West trade of goods which no longer have strategic significance, the council has decided that the current and prospective value of the embargo for the alliance no longer warrants its maintenance.

5. Oil and Gas Journal, October 7, 1968, p. 90.

6. There is speculation that the Soviet Union would be interested in extending the gas pipeline which presently terminates at Baumgarten in eastern Austria, westward through the whole of the country and into Bavaria.

NOTES 187

7. In truth France is already importing small volumes of liquefied petroleum gases from the USSR and these quantities are to be expanded to 100,000 tons yearly. These exports originate at a newly constructed facility in Riga on the Baltic Sea. Insofar as is known, the Soviet Union does not have the capability to produce, store, handle, and market liquefied natural gas in commercially significant quantities.

8. Izvestiya, December 12, 1968, p. 2.

9. The current intra-bloc exchange of natural gas involves relatively minor quantities. As noted, the Soviet Union has been selling small volumes to Poland for a number of years; exports to Czechoslovakia commenced in 1967 upon completion of the Bratstvo pipeline. Among the East European countries, Hungary has found it geographically convenient to import gas from Rumania. More recently, Hungary and Czechoslovakia have arranged to "swap" gas, an arrangement which reportedly will save the cost of instaling several hundred kilometers of pipeline. Czechoslovakia will deliver gas from its West Slovakian fields and from the Ukrainian gas terminal near Sala to the Hungarian industrial complex in the Gyoer area. In return, Hungarian gas from Haydubozlo will be fed into the gas distribution system being built in eastern Slovakia. This swap is to begin in 1970 (see Platt's Oilgram, May 21, 1968, p. 1).

10. A 40-inch gas pipeline by Soviet standards can move about 10 billion cubic meters annually. Reportedly all of the 10 billion cubic meters which the gas pipeline from Efremovka will be able to move up to the border will be available for export.

11. Petroleum Press Service, June, 1968, p. 222; Oil and Gas Journal, January 20, 1969, pp. 28-30.

12. Business Week, November 23, 1968, p. 60.

13. Petroleum Intelligence Weekly, December 9, 1968, p. 4.

14. Ekonomicheskaya gazeta, 52 (December, 1968), 42.

15. _The New York Times_, May 6, 1969, p. 2. Subsequent negotiations following this initial session between the Soviet Union and West Germany have centered around price and interest rates, where failure to reach agreement in the past has effectively precluded reaching any immediate agreement on gas exports by the USSR to Italy and Japan.

16. _The Oil and Gas Journal_, October 28, 1968, p. 58. It should be noted, however, that these estimates were developed prior to the Soviet offer. _Die Zeit_, July 11, 1969, p. 1 reported that contracts obligate West Germany to import up to 12 billion cubic meters per year of Dutch gas by 1975.

17. _Der Spiegel_, August 11, 1969, p. 27.

18. _Petroleum Intelligence Weekly_, September 8, 1969, p. 4. German producers argue that the extent of domestic reserves coupled with the projected imports of Dutch gas, make imports from the Soviet Union unnecessary.

19. Both of these possibilities imply that it will be gas from the eastern Ukraine that would be exported to West Germany.

20. _Vyshka_, August 8, 1969, p. 3.

PART II: SELECTED READINGS

INTRODUCTION

Part II presents a cross-section of pertinent material that has appeared in the Soviet and Eastern European press during the last several years. While by no means an exhaustive collection, it is representative of the coverage given the oil and gas industries and is considered adequate basis for independent analysis by other researchers. These published data are utilized by Soviet and Eastern European researchers themselves, and misleading information would be a disservice to the economy as a whole. Thus, the Western reader can safely accept what he finds in the open literature, provided of course that his analysis is made against a background of knowledge of the industry in general and especially provided that he has a clear understanding of the meaning of the various terms employed in the industry.

SELECTED READINGS 193

SOVIET TRADE IN OIL

"Iran and the CEMA-Member Countries"
Vyshka, January 12, 1969, p. 3

In recent years in Iran considerable attention has been given to the development of mutually-advantageous trade-economic and scientific-technical relations with the Soviet Union and with the socialist governments of Eastern Europe. This is an entirely new form of relations with other countries for Iran. It sharply differs from the relations with Western countries and assists Iran to stand up against the discriminatory trade and economic politics of the major capitalist governments. In the Iranian press quite a few examples have been carried to the effect that companies of the US, West Germany and other countries are not concerned about the national interests of Iran, compelling Iran to lower customs taxes on these goods imported, to lower purchase prices for Iranian products or have raised their prices for their own goods or have sold Iran knowingly food unfit to eat.

Beginning with 1962-63 the Iranian Government, with the national interests in mind, took a course of expansion of trade-economic and scientific-technical relations with the Soviet Union and with other governments of the socialist camp.

An important benchmark in the development of these relations has been the agreement to sell to the socialist governments, in addition to the traditional agricultural commodities, products of the Iranian oil industry. Thus, in 1965 the Iranian Government concluded an agreement according to which it will in the course of 10 years deliver to Rumania crude oil valued at 100 million dollars in exchange for Rumanian machinery, industrial enterprises and other goods.

The growth of trade between Iran and the socialist governments of Eastern Europe was an outgrowth of a trip in September 1966 by an Iranian economic delegation to Bulgaria, Poland, Czechoslovakia, Hungary and Rumania. As a result of this trip, in January 1967 a mutually advantageous agreement was signed with regard to the export of Iranian crude oil to Hungary,

Czechoslovakia, Poland and Bulgaria and the import by Iran from these countries of certain goods. Trade with the countries of Eastern Europe guarantees Iran a market for its goods. In this a trade balance is preserved, which saves foreign currency for Iran, which it needs. The Iranian press often has stated that the governments of Eastern Europe "carry on trade with Iran on an equal basis," while at the same time Western Europe and the US are not concerned about the national interests of the country.

Mutually advantageous financial relations exert much influence on the development of trade and technical cooperation of Iran with the socialist countries. In 1963-64 the USSR, Poland, Czechoslovakia, and Hungary extended long-term credits to Iran. These credits were given at a 2.5 percent annual rate, whereas at the same time the International Bank for Reconstruction and Development, in which the US plays the leading role, took 5.5 percent and then even 6.25 percent interest on its loans.

Using the 45 million dollars in credit given by three countries of Eastern Europe, Iran bought in Poland, Czechoslovakia and Hungary equipment for sugar, alcohol and other plants, an electric power station and also agricultural machinery.

Specialists from many socialist countries, together with local technical cadres, take an active participation in the industrialization of Iran. For example, in September 1967 a contract was concluded between Rumania and Iran on the construction in Tabriz of a tractor plant with an output of 5,000 tractors per year. Poland is delivering equipment for a number of sugar plants, several of which are already operating. Equipment for industrial enterprises in Iran are being sent from Czechoslovakia, Hungary and other socialist governments.

Economic relations between the USSR and Iran are expanding in particular. In conjunction with a number of Soviet-Iranian agreements the following important installations are being built in Iran with the assistance of the Soviet Union: a hydrotechnical complex on the Araks River; a metallurgical plant near Isfahan and a machine-building plant at Arak; the Trans-Iranian gas pipeline, from the southern oil fields of Iran up to the Soviet border; an electric power station, 33 elevators,

SELECTED READINGS 195

Other industrial construction is also under way. The share of the USSR in the total trade turnover of Iran is significantly increasing.

During the discussions of the leaders of the USSR and Iran which took place during the visit of Mohammed Reza Pekhlevi (September-October 1968) it was decided to develop further economic cooperation and to expand even more the mutual trade relations between the USSR and Iran based on an extended period of 12 to 15 years.

The development of trade and economic cooperation with the socialist countries should be examined from the point of view of that influence which will be exerted on the structure of Iran. Thus, heavy industry established with the aid of the countries of the socialist camp transforms the country into an economically independent government, supports a numerical growth in national technical cadres, and leads to an expansion in the system of higher educational institutions.

The development of trade and economic relations between socialist countries and Iran forces the imperialists to yield to Iran in a number of economic questions touching upon the national interests of the people. For example, the International Petroleum Consortium was forced to return to Iran 25 percent of (its) concessions and it agreed to hand over to the Iranian National Oil Company 20 million tons of crude oil for sale to the countries of the socialist camp. In the subsequent five years (1972-76) the Consortium will deliver to the Iranian National Oil Company 50 million tons of crude oil.

If not long ago foreign markets in many countries were closed to the products of the governmental Iranian National Oil Company because of obstacles placed by oil monopolies, then with the conclusion of oil agreements with the socialist countries Iran has entered the world market. The export of oil and gas offers the opportunity to the country to develop its oil and gas industry, its tanker fleet, which raises the whole economy of Iran to a new level.

In the countries of the socialist camp a vast market for the sale by Iran not only of oil and gas has been opened, but also increasingly for the products of mining and light industry and agriculture. Mutually advantageous trade - a powerful stimulus for Iranian progress on all sides.

The growth in economic cooperation with the countries of the socialist camp leads to a strengthening of the foreign relations of Iran with these countries, relations based on equality and non-interference in the affairs of one another. And this is highly valued by Iranian society. The imperialist countries, fearing a strengthening in relations between Iran and the socialist camp, attempt to drive a wedge into these relations. The journal "Khandanikha" wrote: "The unhappiness of the US relates to all spheres of relations between Iran and the socialist countries, particularly economic spheres. But here we prefer the East over the West." The relations of Iran with the socialist countries will be strengthened in the future, because this answers to the interests of the people of Iran and the socialist governments, aids in maintaining peace throughout the world and, in part, in such an unsettled region as the Near and Middle East.

"Billions for 'Black Gold'"
Sovetskaya Rossiya, October 10, 1968, p. 3

Information concerning the export of Soviet crude oil and petroleum products during the current year is still far from complete, so it would be better to discuss data pertaining to 1967.

In 1967 the oil industry of the USSR made available for export almost 80 million tons of crude and products which, it suffices to say, was about equal to the level of crude oil extraction in our country in 1956. All of this oil was sold. The return to Soyuznefteeksport, the All-Union Oil Export Organization, from the sale of this oil in foreign markets exceeded 1 billion rubles for the first time, reaching almost 1.1 billion rubles.

Soviet geologists, oil workers, railroad workers and sailors with full reason may consider that they played a direct role in the emergence of the "oil billion" on the world market.

SELECTED READINGS

Crude oil and petroleum products always have played an important role in the development of foreign trade relations of the USSR. Prior to the war, for example, this group of commodities accounted for 13 percent of the value of all Soviet exports, holding in certain years first place, second and third, interchanging with wood products and furs. From that time, however, the vast economic shift in our country placed machinery and equipment in first place among Soviet exports, which prior to the war we were able to market only in insignificant amounts.

Vast natural resources permit the Soviet Union a wide variety of exports. Along with industrial items we can sell certain quantities of raw materials.

Last year we sold 35 million tons of crude oil and petroleum products to socialist countries. Guided by the principles of proletarian internationalism, the Soviet Union cooperates with the brotherly socialist governments in establishing a series of new branches of industry. In part, the USSR cooperates in the setting up in CEMA member countries oil refining and petrochemical industries, industries without which contemporary economic and scientific-technical progress is unthinkable. Inasmuch as the majority of governments of socialism possess only the smallest of oil resources, the export of crude oil from the USSR to these countries is one of powerful and real levers for their progressive, general economic development.

Almost 45 million tons of crude and products were sold to capitalist countries, chiefly to West European countries, providing about 10 percent of the oil requirements of these importers.

As is known, the closure of the Suez Canal as a result of the Israeli aggression has led to a certain complication in the world oil market, under the influence of which prices for crude and products in West European and certain other capitalist countries have risen. According to estimates of Western experts, the capitalist countries of Europe already have paid about 1.5 billion dollars more for oil (than they would have under normal conditions).

The countries of socialism, purchasing crude and products from the USSR, need not overpay for these commodities, inasmuch as their prices are fixed. This is one of the advantages of the world socialist market. Here, mutually advantageous prices, agreed-upon for a number of years, are in effect and do not change under the influence of frequent commercial fluctuations in the world markets. Such stable prices prevent the overflow of means from one country to another under the influences of various disturbances, accompanying capitalism.

In the current year Soviet oil workers will provide the country with more than 300 million tons of "black gold". Successes of the domestic oil industry guarantee that the buyers of Soviet crude and products may depend upon timely delivery, in the necessary quantities.

"Oil: Profits and Losses"
<u>Izvestiya</u>, July 30, 1968, p.2

Last year's aggression by Israel brought damage not only to the Arab countries. Its aftermath even more so is being felt by those who tolerated the aggressor.

In the first weeks after the closure of the Suez Canal to shipping the Western press quite reluctantly spoke about the possibility of any damage to the West, not wishing to interfere with Israel reaping the fruits of aggression. But when the English pound sterling became feverish and was devalued, and the prices of petroleum products in West Europe countries threatened to rise, the West spoke of the closure of the Canal as a real catastrophe.

Still more: Now it had become necessary to move thousands of ships around Africa, flying the flag of more than half a hundred governments. Almost 250 million tons of freight, handled by these ships in a year, move now from ports of origin to ports of destination along a route which is 8 to 15 thousand kilometers longer than before. This quantity of freight includes more than 150 million tons of oil - the remainder is rubber, ores, raw materials and other goods upon which depends the operation of hundreds of enterprises in the various branches of industry. At present all of these goods have become much more expensive in the West European market.

SELECTED READINGS

The most critical situation is that concerning oil. Prior to the beginning of the Israeli aggression England imported a ton of Kuwait crude oil at an average price of about 15 dollars. After the route from Kuwait to England has been lengthened by 10 thousand kilometers, the price for the crude was raised to 18 to 20 dollars. Prices for West Germany, France, the Scandinavian countries and all other countries of Western Europe were raised by about the same degree. As a result the total losses for Western Europe during the year only in the import of oil is estimated currently at about $1.2 billion.

Despite this, the Israeli military leaders impede the removal of sunken vessels in the Canal, thus rendering a great favor for the American oil monopolists. The oil industry press does not hide the fact that the inactivity of the Canal is a powerful stimulant for a growth in the profits of the oil industry of the US.

The growth in profits of the American oil monopolies in 1967 was without precedent in the entire history of the oil industry. Such growth is continuing into this year. According to information by the "Petroleum Intelligence Weekly" the profits of the five largest oil companies of the US rose during January-March 1968 by 12 percent, whereas during the corresponding period of 1967 these profits increased by only 6 percent, although such a growth had been considered significant. A "majority of the firms" - the weekly hurries to gladden its readers- "anticipate this tendency to prevail for the remainder of the year."

The losses now being suffered in the capitalist oil market, and ultimately by millions of consumers, consist in themselves just another type of tax which the capitalist countries pay to the major imperialist monopolies.

In sharp contrast to the pricing nightmare that shakes the capitalist oil market and enriches a handful of magnates, there is price stability and soundness in the oil market of the socialist countries. No type of price increase is threatened this year, when such CEMA member countries as Bulgaria, Hungary, East Germany, Poland and Czechoslovakia will be importing from the USSR about 30 million tons of crude oil and petroleum products. If the socialist camp did not have its own stable oil market and if prices here would increase by the same degree, as they did for England, that is on an average of 4 rubles per ton, then these countries would have to pay an additional minimum of 120 million rubles a year.

In order to secure such an additional sum, these countries would have to additionally produce and export, for example, more than 6 thousand diesel locomotives, manufactured by Czechoslovakia and East Germany, or more than 2 thousand passenger (railroad) cars, which are exported by Poland, East Germany and Hungary, or more than 700 thousand tons of fruits and vegetables, grown in Bulgaria and Hungary, or 10 thousand Hungarian motor buses, or about 9 million tons of Polish hard coal.

As an unforeseen extra payment for oil, this vast sum would disappear as non-productive expenditure or, in other words, would be simply thrown to the winds, whereas with this sum one would be able to buy more than 50 thousand trucks and automobiles, or more than 10 thousand metal-working stands, or 1.2 million television sets, or more than 7 million cubic meters of good-quality construction lumber.

Stable prices, based on average world prices, and free from speculative increases and declines, are one of the advantages of the world socialist market. Such prices prevent the overflow of means from one country to another under the influence of economic and political disturbances, continuously accompanying capitalism. The establishment for an extended period and changed only upon agreement of both sides in consideration of mutual interests, these prices aid in the plan development of the economy, the equal movement (expansion) of socialist countries along the route of economic progress.

In part, Czechoslovakia, as an example, having planned to import during the years 1966-70 a total of 39 million tons of crude oil from the Soviet Union, set aside the required means for payment of this oil. These means may be overdrawn only if Czechoslovakia requires more oil. As concerns such a business factor as a possible increase in prices, then its influence has been removed earlier. And this not only makes the organizational side of long-range planning easier, but it guarantees the economy that it will be free from unforeseen business expenditures.

At the same time Soviet deliveries of crude oil through the Friendship pipeline, the carrying capacity of which is presently being expanded, provides Bulgaria, Hungary, East Germany, Poland and Czechoslovakia with the opportunity to build a fuels-energy balance on a progressive basis, to modernize and intensify the basic branches of production, in part the chemical industry and metallurgy. All this combines to raise the effectiveness of the national economy.

"Soviet Oil on the World Market"
<u>Ekonomicheskaya gazeta</u>, 39, September 1967, p. 44

The Soviet Union holds second place in the world in terms of the volume of oil output and first place in terms of the annual rate of increase in output. During the anniversary year, the nation's oil workers have accepted the obligation of producing 286 million tons of "black gold".

Approximately one-quarter of all the oil produced in the nation is exported. In terms of the total exports of the USSR, crude and oil products hold second place after machinery and equipment. In 1966, the earnings from the sales of petroleum and oil products for exports were 958 million rubles.

Soviet liquid fuel is presently delivered to 50 nations in the world. The USSR offers for export a broad range of crude and oil products. The high-octane automotive gasolines, the "water-white" kerosene, the low-sulfur diesel fuels with a low freezing temperature, various grades of mazut, base lubricants and other oil products are well-known abroad.

Our nation is a traditional supplier of oil on the world market. In 1921, when for the first time after the revolution the Black Sea ports were opened, the first batches of Soviet oil products were shipped off. Even in 1934, the exports of Soviet liquid fuel exceeded six million tons, which at that time was around 10% of all the oil products consumed in the world (minus the US).

Over the period between the First and Second World Wars, the USSR exported 50 million tons of crude and oil products. Of this amount, 78% went to the nations of Western Europe,

15% to the Asian nations, 6% to Africa, and 0.8% to America, while an insignificant amount of oil commodities was delivered to Australia and New Zealand. From the data given one can see that both in the past and at present the Western European nations have been the basic market for Soviet oil.

The war and the postwar reconstruction period led to a hiatus in the exports of Soviet crude and oil products. Since 1955, the exports of Soviet oil products began to rise and by 1966 had increased by more than 10 times.

A significant portion of the Soviet oil exports goes to the socialist nations. Four European nations -- Poland, East Germany, Czechoslovakia and Hungary -- receive Soviet oil which is pumped over the "Friendship" Oil Pipeline. It is delivered by tankers and by rail to the other socialist nations.

The deliveries of Soviet oil have played a major role in the independence struggle of the heroic Cuban people. After the property of the Anglo-American companies was nationalized in Cuba in 1960, the imperialists decided to blockade the young republic and refused to deliver oil to it. Since then each year the Soviet ports have shipped more than four million tons of liquid fuel for Cuba.

At the present stage such capitalist nations as Italy, France, Finland and Sweden are major purchasers of Soviet crude and oil products.

The commercial operations involved in the sales of crude and oil products are carried out by the "Soyuznefteksport" Association.

In considering the planned nature of our national economy, "Soyuznefteksport" since the beginning of the 1960s has changed over to the extensive practice of concluding long-term contracts which are a good basis for mutually advantageous trade. The contracts for the delivery of Soviet crude and oil products are concluded for a term of from one to six years. On the basis of such long-term agreements, the sales of Soviet oil commodities are being made to "ENI" (Italian), "Neste" (Finnish), "Union Generale des Petroles" (French), and "Indian Oil" (Indian) firms.

On the market of the capitalist nations, "Soyuznefteksport" has encountered strong resistance from the major Anglo-American companies, association and cartels. The oil cartel controls the output, transporting and distribution of petroleum in the developing capitalist nations as well as the developing lands.

In supplying a major strategic commodity, oil, the oil companies have established close ties with the governments of many capitalist nations. They have made use of these ties in their fight against Soviet oil exports, and as well they have resorted to putting pressure on the governments of the nations purchasing Soviet oil by using the American State Department as well as through NATO. In particular, such pressure has been put on Italy which annually purchases 6 to 7 million tons of Soviet oil. Without limiting themselves to this, the oil companies have repeatedly resorted to malicious slanderous campaigns in the reactionary press which has created the myth of the "flooding of the markets with Red oil" or the "dumping of Soviet oil."

The events which are unfolding now in the Near East are very closely linked with the activity of the powerful international oil companies. It is not accidental that this region of the world has repeatedly been an arena of armed clashes.

In order to mask their invidious role in the events in the Near East, the Western imperialist circles are again attempting, as they say, to shift the blame and are artificially stirring up a furor that the Soviet Union is trying to increase the sales of oil and oil products and to thereby capture markets which previously belonged to Arab oil.

The exports of Soviet oil are rising not at the expense of other nations, but rather due to the annual increase in the consumption of oil throughout the world, and the share of the USSR in international oil trade over the last several years has remained constantly at 4 to 5%. Prior to the war, over the period from 1925 to 1935 the Western European nations imported around 210 million tons of oil and oil products, and of this amount 30 million tons or 14.3% came from the USSR. In the postwar period, from 1955 to 1965, the nations of Western Europe imported 2 billion 228 million tons of oil and oil products, of which 143 million tons or 6.4% came from the Soviet Union.

Over the last several months there have been no fundamental changes in Soviet exports. In 1963, the USSR sold 73 million tons of oil and oil products, of which 41 million tons were delivered to the capitalist nations. This means that each week these nations received 800,000 tons of oil.

Thus, whom is the American weekly Petroleum Intelligence Weekly trying to frighten by announcing that during the period of the Near Eastern crisis in several weeks the Soviet Union sold around a million tons of oil and oil products? This attempt to accuse the Soviet Union of using the presently existing situation on the world oil market for its own purposes is clearly clumsy.

Equally awkward is the statement of several Western newspapers that in selling its oil and oil products, the USSR is presently trying to conclude long-term contracts. Trade on a long-term basis in fact is characteristic for Soviet foreign trade, since it corresponds to the planned system of our national economy. At the same time, all international trade in oil and oil products is carried out basically on a basis of long-term contracts due to the specific features of such a commodity as the liquid fuel is. But, over the period since the beginning of the military actions in the Near East up to the present moment not a single significant long-term contract has been concluded for the delivery of crude and oil products in the USSR.

One could not think of anything more unlikely than the statement of Western propaganda that Soviet oil has been delivered to Israel. The American publication Platt's Oilgram has published the completely unsubstantiated statement that the Israeli tanker "Haifa" has carried oil from the USSR to Israel.

Soviet oil and oil products have not been delivered to Israel since 1956 when the Soviet government condemned the aggressive actions of Israel and since then nullified all licenses for the delivery of oil to that nation. Since then not a single contract has been concluded for supplying Soviet oil to Israel. Moreover, in all of the Soviet contracts there is a clause providing that "the purchase does not have the right to transship or in any way dispose of the purchased commodities to any nation without the preliminary written agreement for this sale." This condition of the Soviet contracts is the guarantee that the Soviet commodities in no manner can get to Israel.

All statements of this nature are a fiction designed to drive a wedge between the Arab and Soviet peoples. This fact was pointed out by the USSR Minister of Foreign Trade, N.S. Patolichev, who stated that such false statements are directed at "sowing doubt among the Arab peoples vis-a-vis the position of their sincere friend, the Soviet Union."

Without being satisfied by slanderous attacks aginst the USSR in the pages of the Western press, the imperialists are trying to use the press of individual Near Eastern and African nations for spreading various sorts of slanderous notions on Soviet oil exports. However, the materials published in certain newspapers of these nations and directed against the USSR do not reflect the opinions of the broad popular masses in the Arab states which are fully aware of the noble role of the Soviet Union in establishing peace in the Near East.

The propaganda pen-pushers should be aware that the USSR does not need a great increase in oil exports or certainly oil dumping at the expense of others. The exports of Soviet oil will expand without this, and will remain in full keeping with the national economic plans of our nation.

"The Export of Crude Oil and Petroleum Products"
Vneshnyaya torgovlya, 8, August 1967, p.10-14; By
E. Gurov, head of Soyuznefteeksport

The commercial production of crude oil in Russia has been carried out for more than a century. In 1910 Russia occupied first place in the world in terms of crude oil output. However, in the subsequent years, the extraction of crude oil was reduced. Foreign capital, to which belonged one-half of the oil industry of Czarist Russia, was not interested in its development. The nationalization of the oil industry after the Great October socialist revolution delivered to the service of socialist construction this important branch of the national economy. However, foreign capitalists for a long time held to the thought that their influence in Russia would return. It is characteristic that the large American firm "Standard Oil Company of New Jersey" after the act of nationalization bought out the Nobel group, the largest of the foreign oil monopolies in Russia.

The desire of the capitalists to recover the oil concessions in Russia was one of the major reasons of the Anglo-French military intervention in the south of Soviet Russia, after the failure of which they inflicted on the oil extracting industry of our country serious damage. The American magazine "Business Week" on 7 May 1960 wrote the following: "When 40 years ago the red military forces expelled from Baku the last vestiges of the English and Turkish forces, the rich oil deposits of the Caucasus were in a state of chaos. Drilling had stopped, the fields were abandoned, and output did not exceed one thousand tons per day. After nationalization the young government of the Soviets had before them the task to establish the fields anew." Already by 1921, when after an interruption, provoked by military circumstances, the Black Sea ports were opened, the first exports of oil products from Batumi were made. In 1924/25 plan year, the export of Soviet oil products exceeded the level of the prewar years, and reached 1,372 thousand tons. At this time up to 10 percent of all of the oil products entering the world market originated with the Soviet Union. The major consumers of our products were England, Germany, France, Italy, and Turkey. Imports from the Soviet Union by these countries accounted for 11 to 24 percent of total oil imports.

The successes of the Soviet government in the establishment and further development of the oil industry permitted them already by the middle 1920's to secure a leading role among the major exporters of liquid fuel. In the period between the world wars Soviet oil and oil products were purchased by more than 40 countries. To Western Europe was delivered 78 percent of total pre-war oil exports of the USSR, to Asia - 14 percent, to Africa - 6 percent, to America - 0.8 percent, to Australia and New Zealand - 0.2 percent. Practically all countries in Western Europe bought Soviet oil products. Asian importers of Soviet oil were Japan, India, Turkey, Iran, China Afghanistan, Yemen, and a number of others. In Africa - Egypt, Algeria, Morocco, Tunisia, Congo, South African Union In America - Uruguay, the U.S., Argentina, Canada, Brazil.

SELECTED READINGS

The share of Soviet deliveries in the oil imports of a number of these countries was quite significant.

From 1926 through 1935 Western Europe imported a total of about 220 million tons of crude oil and products (excluding re-exports), including 30 million tons, or 14 percent, from the USSR. During 1929-33 the share of Soviet deliveries in the European oil imports reached 17 percent.

In the period of establishment of the Soviet regime and up to the Second World War crude oil and products were a major item of Soviet export. In the total foreign trade of the USSR from 1918 through 1940 the products of the oil industry accounted for 13 percent.

From the middle of the 1950's, when the Soviet people had recovered from the wounds inflicted on the country by the war, the export of crude oil and products began to be actively developed. It was necessary to establish communications with buyers anew. By the end of the 1950's the USSR again was a major exporter of oil.

The basic economic reasoning for this was the rapid development of the oil industry of the USSR in fulfilling the Five Year Plan goals. During these years (1951-55) major centers of crude oil extraction were established in the Urals-Volga ("Second Baku") from which emanated a rapid expansion in the oil extracting industry.

Imports of Crude Oil and Petroleum Products
by European Countries during 1926-35

	Million Tons Total	From USSR	USSR Share of Total (Percent)
Western Europe, total	220	3.0	14
Italy	15.1	7.2	48
Spain	6.4	2.5	39
Belgium	6.3	1.7	27
Germany	25.1	4.2	17
Denmark	4.8	0.8	17
Sweden	6.7	1.0	15
France	41.9	5.3	18
England	87.5	5.8	7

During this same period in a number of capitalist countries there took place a strengthening of the position of national governmental oil companies with the purpose of protecting the domestic oil market from the preponderance of foreign oil monopolies. Raising the role of governmental enterprises in such countries as Italy ("ENI"), France ("CFP"), the UAR ("DPO"), Argentina ("YPF"), Uruguay ("Ankap"), Brazil ("Petrobras"). They accelerated their activity and began to decisively emerge in the market private independent national firms.

Having encountered difficult circumstances in the mid-1950's in the capitalist market, and also a high degree of competition in the world oil market, the Soviet Union forced their entry into the oil product market.

In the 1950's as in earlier periods, one of the major importers of Soviet crude oil and products among the capitalist countries was Italy. From 1957 through 1962 Italian firms increased their purchases of Soviet oil from 0.5 million tons to 7.1 million tons. The share of the USSR in the total Italian import of oil and oil products increased during these years from 3 percent to 18 percent. The import of Soviet oil by Italy was connected with the realization of substantial economic advantages. American and English oil monopolies direct a significant part of their profits from the sale of crude oil and products to banks in the US and Great Britain, which has a significant negative influence on the balance of payments of the purchasing country. Soviet foreign trade organizations use the profit from the sale of crude oil and products to buy from Italy rigs, equipment, steel pipe, petrochemicals and other goods produced by Italian industry.

Considering the profitability of trade with the Soviet Union, the Italian concern "ENI" completed a contract on the purchase of Soviet oil over an extended period of time. In October, 1960 a major contract was completed involving 12 million tons of crude oil over a period of 4 years. Later, in 1963, an agreement on the delivery of 25 million tons of oil during 1964-70 was reached.

Business contacts with France, Finland, Sweden, are increasing. In the early 1960's commercial relations with a number of developing countries was established. Contracts were signed relating to the delivery of crude oil and products to young government corporations of India and Ceylon, and also with government companies of Brazil and Argentina.

From year to year the circle of buyers increases and the popularity of Soviet oil products grows in the world oil market. In terms of quality, Soviet crude oil and products does not lag behind the best of the world and our buyers value this.

At the end of the 1950's the export of Soviet crude oil and products sharply increased to other Socialist countries, chiefly CEMA members. From 1955 through 1960 the delivery of Soviet oil and products to Socialist countries increased from 4.2 million tons to 15.2 million tons. In 1964 the crude oil pipeline "Druzhba" was completed, through which the delivery of crude oil to CEMA countries sharply increased and reached 18.3 million tons in 1965.

The development of the economics of the Socialist countries generated in these countries a rapid growth in the consumption of oil, largely covered by imports from the Soviet Union.

Export of Soviet Crude Oil to CEMA Countries
(Thousand tons)

Country	1963	1964	1965
Bulgaria	464	1,799	2,146
Hungary	1,497	1,758	2,046
GDR	3,060	3,936	4,923
Poland	1,416	1,703	3,213
Czechoslovakia	4,222	4,760	5,964

Among the socialist countries one of the major buyers of Soviet oil and products has been Cuba. In the summer of 1960 revolutionary Cuba nationalized the oil refineries of the American and English monopolies. In answer to this step the latter attempted to subject the heroic island to an "oil hunger", leading to a boycott on deliveries of oil to Cuba. The Soviet Union, true to the principles of proletarian internationalism, responded to the request of Cuba to supply it with crude oil and products under mutually advantageous conditions. In subsequent years Soviet oil deliveries almost completely satisfied the demands of the country for liquid fuels. In 1965 Cuba imported from the USSR more than 4.7 million tons of crude and products.

For the postwar structure of Soviet oil exports the sharp growth in the share of crude oil and the reduction in the share of products is characteristic. This change reflects the major structural shifts in the whole international trade in crude and products.

Up to the middle of the 1930's for international oil trade the overwhelming share of oil products was characteristic. The majority of the countries exporting liquid fuel then preferred to export already prepared oil products and did not consider it necessary to encourage the construction of oil refineries in importing territories. In 1937, for example, the oil import of Western Europe was made up of 70 percent oil products and 30 percent crude oil. A different situation is observed in the postwar years, especially in the last ten years. The consumption of petroleum fuels grew to such levels, that it became more expedient to build oil refineries in the major regions of consumption and to import crude oil for these refineries. This was dictated by the necessity to save foreign exchange, the ease of transport of comparatively single-type crude oil compared with various types of petroleum products, the possibility to more fully utilize the products of oil refining in the major regions of consumption which, as a rule, are highly industrialized regions. Large Western European countries have been transformed into exporters of oil products. As a result European imports of oil are now made up largely of crude oil and only one-fourth is in the form of oil products. In recent years in the oil imports of all capitalist countries of the world the share of crude oil approaches more than 70 percent and about 30 percent for products. Thus, structural changes in Soviet oil exports in general correspond to these changes, which have taken place in the past 10 years in the whole of the international trade in crude oil and products.

Throughout the entire history of development of Soviet oil exports the trading organizations of the Soviet Union (initially "Neftesindikat", then "Soyuzneft", and beginning in 1931 "Soyuznefteeksport") have come up against the competition from the major oil monopolies of the world. Prior to 1956 more than 90 percent of the world oil market belonged to 8 monopolies, the largest of which was the American "Standard

Oil Company of New Jersey", and the Anglo-Dutch group "Royal Dutch-Shell". Subsequent events indicated the danger, associated with the dependence of the capitalist countries in the question of delivery of oil from the so-called international monopolies. In connection with this, at the end of the 1950's there occurred a sharp acceleration in the activities of the national firms and simultaneously the process of establishment of state oil companies. In this period many countries of Asia, Africa and Latin America entered into a struggle for their economic independence, without which complete political independence and liquidation of the remains of colonialism are not possible.

No less important events took place in the developed capitalistic countries. The major consuming countries, such as the European countries and Japan, not having their own oil and having to import this oil from other countries, purchased oil from the basic sources of crude oil and products, the American and English monopolies. With the growth in consumption of oil in these countries, their dependence on the foreign monopolies turned into a different time. Nevertheless, the national capital of these countries was such that they felt themselves capable of entering into a struggle with such powerful American and English oil trusts as "Esso", "Shell", "Mobil", "Gulf", and others.

In connection with the appearance of new firms in the market place which, differing from the monopolies and their affiliates, were given the title of "independent" firms, the competitive battle in the world oil market was sharpened. Conditions in the market were particularly complicated in the early 1960's, when the western monopolies, and especially the American, turned from the slander of the purposes and aims of Soviet oil exports to the organization of a struggle against the export of crude oil and products from the USSR. In this struggle the most widely varying means were used: from the commercial "price war" to political pressure on buyers of Soviet oil through the "Common Market", NATO, and other imperialistic groupings. In November, 1962 the U.S. Government proposed to the NATO countries a limitation on the import of Soviet oil, and also on the export of equipment from the western countries to the USSR. After such a "proposal" the NATO Council recommended to the countries of the "Common Market" a limitation on the allowable volume of imports of Soviet crude oil and products of 10 percent. Another discriminatory measure against Soviet oil exports was the laying down of a so-called black list of ships, carrying Soviet oil goods. Ships, listed hereby, were deprived of the possibility of carrying freight from the U.S. or to the U.S. In the same period the NATO Council recommended an embargo on the sale of large diameter pipe to the Soviet Union.

In 1963, when the government of Ceylon decided to reduce the share of Anglo-American companies in the market and to increase the import of oil products from the USSR, the U.S. State Department directed a note of protest to the government of Ceylon and sharply cut back the scale of economic "aid".

Not being confined to economic and political pressures on the countries buying Soviet oil, the monopolies organized an ideological aggression against oil exports from the USSR. On the pages of the burgeoise press there appeared a multitude of articles on the political motives of the Soviet export of oil. The American oil companies prepared a two-volume research effort on the theme: "Impact of Oil Exports From the Soviet Bloc". The American press did not hide the fact that these two volumes were designed for the gratification of the monopolies "conditioning" higher instances and for the governments of a number of capitalistic countries.

The aggressively constructed circles in the world oil market several times attempted to clash the interests of the Middle East and other oil producing countries with the interests of the Soviet Union. In part, the reactionary propaganda often claims that Soviet oil is displacing Arabian, Iranian, and Venezuelan oil in the international market.

The Soviet oil industry is developing at a rapid tempo. From 1955 through 1965 the extraction of oil in the USSR increased from 71 to 243 million tons. However, the greater portion of Soviet oil is consumed in the USSR and in other Socialist countries. A certain share is directed to capitalist countries, where the total consumption of oil products during the past ten years doubled and in 1965 reached about 1,300 million tons (including bunker fuel, asphalt, and other products of refining). With such a growth in consumption of oil in the capitalist countries, demand may be covered by imports from all sources.

SELECTED READINGS

Facts lead to the conclusion that the increase in the export of Soviet oil took place not on the basis of displacement of other suppliers in the market, but as a result of the rapid growth in the consumption of oil throughout the whole world. Thus, during the period 1955 through 1965, the export of crude oil and products from the Soviet Union to capitalist countries increased by 31 million tons. During this same period the growth in export of crude oil and products from the Arabian oil producing countries reached 236 million tons, including 59 million tons from Libya, 53 million tons from Kuwait, 52 million tons from Saudi Arabia, 32 million tons from Algeria, and 31 million tons from Iraq.

Utilizing recent events in the Near East, the western press has attempted to accuse the Soviet Union of endeavoring to take the place of the Arab countries in the world oil market. But, as explained the USSR Minister of Foreign Trade, Patolichev, "These reports are blind invention, the purpose of which, as with the many other insinuations in this same plan, is to create doubt in the Arab peoples relative to the position of their true friend - the Soviet Union."

The Soviet Union always supports the developing countries in their just struggle against any attempts to disturb their freedom and independence.

Despite the counteraction of the monopolies, "Soyuznefteeksport" has been able to assume one of the leading places among the major oil companies of the world. Soviet crude oil and products are sold at present to more than 50 countries. The largest buyers of our products are the Socialist countries and developed capitalist countries such as Italy, France, West Germany, Finland, Japan, and others. The Soviet Union offers crude oil and products on a mutually advantageous basis to any country, independent of social-economic make-up. The Soviet Union, in the interest of development of planned economy, is interested in trade on the basis of long-term agreements and contracts. At the present time significantly more than one-half of all Soviet deliveries of crude and products are carried out under terms of long-term contracts and agreements. This answers not only to the interests of the Soviet economy, but is also completely advantageous to the buyer.

Long-term contracts under agreed aims and conditions of delivery are an excellent guarantee of necessary and stable supply of oil over a long period of time. Today, great tasks stand before the Soviet export of oil concerning the increasing of its effectiveness.

Oil export in the future will play an important role in the development of foreign trade of the USSR. This is guaranteed by the rapid growth in the domestic oil industry and by the interest in Soviet oil in the world market place.

"Oil Monopolies--Inspirers of Aggression"
Ekonomicheskaya gazeta, 27, July 1967, p.43

The blood of Arab soldiers, fellahin, women, old people, and children was shed on the Sinai desert, on Syrian land, and at the walls of ancient Jerusalem. The Israeli aggressors succeeded in reaching the Suez Canal and created a threat to the vitally important center of the UAR. The territory they seized is more than three times as much as the area of Israel itself.

The bloody hand of the aggressor has been stopped by efforts of peaceloving peoples, primarily the entire Soviet Union and other socialist countries and states who had freed themselves from colonial rule. A sharp political struggle developed at the emergency session of the UN General Assembly between the forces of progress headed by the Soviet Union and other socialist countries and the forces of imperialist reaction headed by the United States. Progressive world society demands the immediate and unconditional withdrawal of Israeli troops from the territories they seized, political regulation of the Middle East crisis, and compensation for damage inflicted on the Arab states by military aggression.

The decree of the June 1967 Central Committee Plenum states: "Israeli aggression is the result of a conspiracy of the most reactionary forces of international imperialism, primarily the United States, directed against one of the detachments of the national liberation movement and against progressive Arab states on the path of progressive social-economic reorganization in the workers' interests who are carrying out anti-imperialist policies."

SELECTED READINGS 215

Analysis of the events shows that the aggressors operated according to carefully worked out and prepared plans of attack on the Arab states which provided for destroying their airfields and basic areas of troop and military equipment concentration.

The political goals of the aggression are clear: overthrow or at least weaken the progressive regimes established in Arab countries which, having achieved political independence and using it, were taking active steps to be free from economic dominance by imperialist countries and their monopolies and which were carrying on a struggle to create their own national independent economies.

The present period of the national liberation movement is characterized by an active struggle for economic independence and for achieving equal participation of the developing countries in international economic ties. The area of the Arab East is the center of the most sharp economic contradictions between the largest imperialist countries and countries which have freed themselves from colonial rule.

The plan of military attack on the Arab countries which was implemented by the Israeli aggressors had as its task setting straight the shaky position of the huge Western monopolies in the economy of these countries by using military pressure.

The economy of Arab countries at present depends and is based upon the oil-extracting and oil-refining industry. Receipts from oil exports make up from 20 to 90 percent of the income of Arab countries.

Known reserves of this important strategic raw material in Arab countries of the Middle East and North Africa are valued at 30 billion tons -- more than 60 percent of the entire known reserves of oil in the capitalist world. Nearly 470 million tons were extracted here in 1966, or 35 percent of the entire output of oil in capitalist countries. The largest oil-extracting countries are Saudi Arabia (117 million tons), Kuwait (114), Libya (72), Iraq (67) and Algeria (32).

However, the oil wealth of these countries still goes into the hands of the Western oil monopolies who obtain billions of dollars of profit. Eighty percent of oil output in these countries

was shared by 8 foreign companies in 1966. Among them are
five US companies: Standard Oil (New Jersey), Texaco, Gulf
Oil, Standard Oil of California, and Socony Mobil; British
Petroleum, a British company; Royal Dutch/Shell, a British-
Dutch company; and Companie Francaise de Petrol. The share
of these companies is correspondingly different in different
countries. Thus, 95 percent output is concentrated in US oil
companies in Libya, 80 percent -- in the hands of French companies in Algeria, 47 percent is controlled by the British in
Iraq, 29 by the French, and 24 by US oil monopolies.

The world crude and products market in recent years has
been characterized by a chronic overproduction and by supply
excess to demand. Imperialist countries are using this situation as a means of economic blackmail of the oil-extracting
countries, of exerting pressure on them to lower prices, and
of (achieving) a corresponding reduction in government receipts.

In connection with the continuing worsening of the economic
and financial conditions in the developing countries and the constantly growing sum of profits being pumped out of them, the
countries of the Arab East in recent years adopted a number
of serious measures directed toward establishing state control
on the activity of the oil monopolies and on reducing the share
of profits going to the oil magnates and increasing their own
profits from oil exports. The first steps were taken to create
a state sector in the oil-extracting industry.

Thus, oil regions in the "third world" became the arena
for a bitter struggle between imperialist countries to maintain
their supremacy in the economy of the Arab East and the Arab
countries for their natural rights as sovereign states in the
area of extracting, refining, and exporting oil.

The Arab states, relying on the support of socialist countries and on a united front of all countries who had freed themselves from colonial rule, achieved major victories in this
struggle. The chief victories were the creation of OPEC
(Organization of Petroleum Exporting Countries) and the increase in the share of profits received by oil-extracting countries from the oil monopolies. Five Arab countries joined
OPEC, as did Iran, Indonesia, and Venezuela. This organization has as its task the establishment of control on the level

SELECTED READINGS 217

of output and export of oil by introducing a quota for oil output for each country binding for Western oil companies.

Until 1950, Arab countries, making concession to the oil monopolies and according to existing agreements, secured only 20 percent of the profit from oil export. From 1950 on, as a result of the bitter struggle which took on different forms in different countries, the share of profit obtained by oil-extracting countries increased to 50 percent. From 1956 on, the large international oil companies had to allot up to 58 percent pure profit to the oil-extracting countries.

Expanded profit sharing and oil exports led to a significant growth in the profits of the oil-extracting countries at the expense of a corresponding reduction in the oil monopolies' profits. Thus, profits for Arab countries from oil exports nearly doubled in the last five years: from $1.16 billion in 1960 to $2.252 billion in 1965.

Events following Israeli aggression showed how vulnerable the interests of the oil monopolies in this region are. Arab countries, still very much dependent on the capitalist market, demonstrated their ability for new large-scale action in the struggle against the dictates of the oil monopolies in the name of Arab unity and punishment to the aggressor and his accomplices.

The 5 June conference in Baghdad of 10 Arab states (Saudi Arabia, Iraq, Kuwait, Lebanon, Algeria, Syria, Libya, Abu Dhabi, Qatar, and Bahrain) adopted and declared a decision that, in connection with Israeli aggression, Arab states would cut off shipments of oil to any country taking part in this aggression or rendering aid to the aggressor. Conference participants warned the Western oil monopolies that they would declare a collective boycott on any of them hindering the implementation of this decision.

What are the consequences of the Israeli aggression for the imperialist countries? Let us turn to the figures on the geographic distribution of oil exports from countries of the Arab East (in millions of tons):

	1965	1966
Britain	44	49
United States	15	13
Italy	57	58
Japan	48	54
France	48	52
West Germany	44	50
Netherlands	19	25
Total	369	418

At present the boycott on shipping oil applies to Great Britain and the United States. The figures cited show that Great Britain and the United States imported 49 and 13 million tons of oil, respectively, from these countries in 1966. Oil imports from countries of the Arab East comprise nearly 20 percent of the total volume of imports for the United States. Great Britain covered 70 percent of its import needs for crude oil annually by importing it from Arab countries. Western Europe daily received nearly 1 million tons of oil from countries of the Arab East which covered 65 percent of its daily consumption.

According to reports of the official Western press, commercial and strategic oil supplies in capitalist countries are sufficient to cover requirements for 2 to 4 months.

In connection with Israeli aggression and its consequences, the extraction and export of oil from Arab countries has been reduced, and delivery of oil to Western Europe from this area is less than 30 percent of the normal level.

If one takes into account that, in connection with the transition from coal to oil, in the last five years the need for oil has tripled and oil now covers 50 percent of Western Europe's energy needs, then the problem of guaranteeing oil for this portion of the world is extremely critical. The New York Times on 29 June wrote that Western Europe is threatened by an "immediate oil deficit of critical proportions."

A rise in prices for crude oil and oil products has been noted on the world market as a result of Israeli aggression, chartering of oil tankers has become more active, and freight charges are rising.

British oil companies have already raised the price of gasoline. The British government has denied reports that rationing gasoline will be introduced in the country in the near future; however, according to reports available, coupons for gasoline have already been printed in England in case its sale is controlled. It is supposed that distribution of gasoline to owners of private cars will be limited to 200 miles of travel by car per month.

The situation of supplying oil to Western Europe has been aggravated by closing of the Suez Canal, an act evoked by Israeli aggression; the Canal formerly was the main oil artery. Of the total ships passing through the Canal, three-quarters were oil tankers carrying oil to Europe and the Western hemisphere. According to The New York Times, an additional 1,440 tankers would be needed to transport the needed quantity of oil around Africa and the Cape of Good Hope. This means that the entire tanker fleet of the capitalist world must be increased by 22 percent. But there are no possibilities for this.

The Israeli aggressors and their accomplices placed the world face to face with aggravation of international tension and with economic problems difficult to solve. It is not a coincidence that political intellectuals even in capitalist countries sensibly connect the problem of armed conflict in the Arab East with the general problem of the struggle against the policy of aggression and economic pressure. President De Gaulle and his minister of foreign affairs stated that the Middle East conflict cannot be considered apart from the aggressive war in Vietnam.

Regulating the armed conflict in the Middle East demands the adoption of important political decisions in order not to let the presumptions aggressors ruin people's peaceful lives even further. The Arab people's struggle for strengthening their political and economic independence has reached a new stage. Outcome of this struggle will depend on many factors. One thing is clear: the imperialists will not succeed in turning back the wheels of history. The aroused Arab East, having thrown off the yoke of political domination, never will submit to military threats or economic pressure. Forces fighting for peace and against colonialism and neocolonialism are on the Arabs' side. All progressive mankind is on their side.

The message of leaders of the party and government of the Soviet Union to the emergency conference of the Afro-Asian People's Solidarity Organization states: "An important condition for the success of this struggle is the closest unity of the Arab peoples with socialist countries and with all social and political forces fighting against imperialist aggression."

"The Transport Fleet of the USSR Is Growing Rapidly"
Bakinskiy rabochiy, January 13, 1967, p. 1

We talked about the new year and about those who were at sea at night, when it was born. The world looked at us from a wall of multi-colored shreds of continents and blue seas and and oceans along which "went" small toy ships. And the fact that they did not fall, but simply splashed toward the blue squares of the network of meridians and parallels, had no relation to the laws of the earth's gravity. It was replaced by microscopic magnets at the bottom. These toy ships will now be rearranged every 10 days on a new map in the office of Viktor Georgiyevich Bakayev, Minister of the Sea Fleet. He familiarized me with the position of the Soviet ships on 1 January 1967.

"On this map you see only those ships which on New Year's night were either enroute or in ports of the world being loaded or unloaded. Thus, we had 137 ships in the Pacific Ocean, 58 in the Indian Ocean, 282 in the North Norwegian, Baltic, and Barents seas, 254 in the Black, Mediterranean, and Red seas, and 102 in the Atlantic Ocean, including some on the shores of the American continent. Look at the specific peculiarity of the New Year horizon. The Soviet ports were crammed with ships -- the crews have hurried home to greet the holiday. But here are the totals. On New Year's night, 833 ships are in the oceans and ports of the world."

"This figure indeed agitated UPI correspondent Robert Bakhorn, who wrote an article in the first days of the New Year on the Soviet commercial fleet. He begins by saying that 'the Soviet Union plans a revolution on the sea. If things continue as the Kremlin hopes, in the 1970s the Red Flag could fly over the largest commercial fleet in the world, putting the American commercial fleet in second place."

SELECTED READINGS

Bakayev smiled:

"This American journalist is simply rendering his due to the fashionable idea in the West of representing the growth of the Soviet commercial fleet as an aspiration on the part of the USSR for domination of the seas and expansion. The emergence of our new ships on the waterways, undoubtedly, is changing the correlation of forces in world navigation. However, the growth of the Soviet fleet is explained not by a desire to 'outgallop' anybody but by urgent requirements for developing the national economy of our country on the whole, as well as by its growing international ties. We do not want to depend on anybody. The companies of the largest capitalist states, which are accustomed to dictate their conditions of sea shipments, are now compelled to take into account the Soviet commercial fleet. They, understandably, are being subjected to despondency by the successes of the USSR in developing navigation. They are afraid of losing immense profits from freight shipments and of losing opportunities for influencing freight rates. The mood of the companies is sometimes conveyed to the journalists."

"Our fleet is indeed growing. Do you want a visual comparison?"

Bakayev went to the cabinet and got a photograph glued together from two pieces. It was not a professional picture taken for "history." The minister acknowledged in an embarrassed manner that he had taken the photograph and that it showed the position of the ships on 1 January 1957. Then the country was also preparing to mark an anniversary, the 40th of Soviet power.

And this was 10 years ago. The map is somewhat smaller and there are not ships with magnetic bottoms. In their place are tags with the names of ships. There were 76 of them. Yes, only 76, and their names easily fit on the map. In the Pacific Ocean on New Year's night of 1957 there were 8 ships, there were 10 ships in the Indian Ocean, 12 off West Africa, 34 in the Mediterranean and Black seas, and 12 in the North Atlantic. And this was all.

The comparison is truly impressive: 76 at that time, and 833 just 10 years later.

"Otherwise we could not have developed our commercial relations with foreign countries so impressively," said the minister.

"During the Five Year Plan, the tonnage of Soviet ships must increase by 150 percent. Now, as is known, the Soviet commercial fleet is in sixth place in the world. What place will it hold after 1970?"

"I will not venture a guess. I have already said that we are not competing for a place but are proceeding from our urgent needs. Moreover, the allocation of these 'places' will depend also on who will further develop their fleet and how they will do it, without counting, of course, Liberia. How paradoxical that this country now carries the title of 'master of the seas'. It owns more than 1,400 ships with total tonnage of more than 20 million. This is almost twice the tonnage of the US fleet."

"How did this happen?"

"Very simple. In the language of the merchants, it is called 'hiding under a convenient flag.' The favorable tax policy and national legislation of a number of countries (including Liberia) permit foreign ship owners, under a convenient cover, to derive additional profits by leasing the flag of a foreign country."

"But let us return to our fleet. Each year it will increase by about one million tons. Until the country is completely provided with its need for sea transport."

I called the minister's attention to another expression of the American journalist: "Along with a campaign to ship its freight on its own ships," wrote Bakhorn, "the Kremlin plans also to ship the freight of other countries. It is not impossible that in the not too distant future, the US perhaps will have to count on the Soviet Union to transport their raw materials."

"What do you say about this?"

"Such thoughts have not crossed my mind. Evidently the American journalist has consulted with knowledgeable people

SELECTED READINGS

on these questions. They are more prominent. And I do not intend to dispute their opinion."

"The year 1967, an anniversary year for the Soviet Union, has begun. What are the plans for sea transport for this significant year?"

"More than 140 million tons of freight will be shipped. The fleet will receive more than 100 new ships of various designation, new kilometers of piers will appear in the ports of the country, and thanks to mechanization, there will be a considerable acceleration in the time for loading and unloading the ships."

"As for participation of the Soviet commercial fleet in international sea shipments," said Bakayev in conclusion, "the USSR has favored and continues to favor cooperation with foreign organizations on the basis of equal rights and mutual advantage. By the way, the many years' experience of business relations with foreign shipping companies has earned the Soviet commercial fleet the reputation of a reliable partner who consistently supports the spirit and letter of the conditions and agreements."

"Petroleum: The Policy of Social Progress, and the Policy of Neocolonialism"
Neftyanik, 1, January 1967, p. 10-11

With the birth of the first Soviet socialist state in the world, there were also born new principles of international relations, one of which is the establishment of relations of equal rights and fraternity with the peoples oppressed by imperialism. Back in 1916 Lenin stated that the future socialist Russia would make every effort "to draw close to and amalgamate with the Mongols, Persians, Indians and Egyptians" and "Help them in the transition to the use of machines, to lightening of labor, to democracy and socialism" (Collected Works, Vol. 23, p. 55).

The treaties with Mongolia, Iran, Turkey and Afghanistan were an example of the friendly, equal-rights relations between the country of the Soviets and the formerly oppressed peoples. Now, following the Lenin policy, the Soviet Union has established such relations with most of the young states of Asia, Africa, and Latin America. This became possible thanks to the fact that in the postwar period there took place the collapse of the colonial system of imperialism, out of the ruins of which have already emerged about 70 independent states.

A characteristic feature of the modern national-liberation movement is, on the one hand, the striving of the young states to create their own independent national economies, and on the other, that of the imperialist countries, by neocolonialist methods, to prevent them from doing this, and to keep them in capitalist dependence.

It should be noted that the economies of the former colonial countries which have gained their political independence are, as a rule, one-crop economies, which unquestionably deprives them of the possibility of immediately rejecting economic relations, and especially foreign trade, with the capitalist powers.

As a result of this, in most of the countries which have won their political independence, the capitalist monopolies, just as formerly, exercise control over the economy and the national resources, and mercilessly rob the young states.

In the countries of the Near and Middle East, for example, there are vast proven reserves of petroleum, but the owners of these riches are not the peoples of these countries, but foreign capitalist monopolies. In 1965 alone American companies took out of these countries 210 million tons of the black gold; British companies, 130 million; French and others, 50 million. Having invested less than 800 million dollars, in just the last three years these monopolies have received 6 billion dollars of net profit. Is this not robbery of the national wealth of these nations? This is neocolonialism in action.

US imperialists are especially active in the policy of neocolonialism. In carrying it out, they depend on their experience, built up over decades. As we know, the countries of Latin America freed themselves from the Spanish yoke back in the 19th century, and most of them became independent. But this independence, as a rule, continued to be just a matter of form. The countries of the continent did not create their own national economies, and were converted into appendages to the economy of the United States, the monopolies of which exploited them without restraint. According to data of General Clay's commission, created by the President in 1962 to study problems of "aid," in the course of 17 years the US granted to Venezuela, for example, "aid" in the amount of $273.8 million. At the same time the annual profits of the US oil monopolies alone, running things in this country, exceeded $450 million.

The US imperialists are carrying out a similar policy with respect to all the other countries of Latin America. Only socialist Cuba has broken the chains of this bondage.

The US is trying to extend the experience of colonialism in Latin America to other continents, and especially to Asia and Africa.

Robbery by the imperialists of the national natural wealth of the young states deprives the latter of the possibility of developing their own national economies.

However, it must be noted that, especially in recent years, new and important processes are taking place in the national liberation movement. In a number of young countries -- the UAR, Burma, Guinea, Mali, Algeria, Congo (Brazzaville), Syria, and others -- important social reforms have been accomplished on paths of non-capitalist development.

To be sure, the depth of these processes in each country is not the same, and they are manifested in different forms.

The young developing countries have established the closest economic relations with the Soviet Union, which are being carried out on the basis of complete equality and mutual advantage. Soviet credits to these countries, as distinguished from those from capitalist countries, are granted without any political conditions whatever, and with favorable economic conditions -- for ten or fifteen years, with interest at from 2.5 to 3 percent a year, while the imperialist countries take up to six percent and grant the credits for shorter periods, and, as a rule, from private lenders.

Credits from the Soviet Union today already exceed 4 billion rubles, by means of which there have been built or are being built more than 600 industrial, agricultural or other projects and over 100 educational, medical, and scientific institutions, in 30 countries of Asia and Africa.

Striving to maintain their control over the extraction, refining, and supply of oil in the young countries, the capitalist monopolies have sabotaged oil prospecting in these countries in every way. The Soviet Union, true to its international duty, has come to the support of the young states against the insolent dictates of the oil monopolies. Economic cooperation of our country with the young states has developed along many lines.

The Soviet Union, expanding and strengthening economic cooperation with the young states, is directing its efforts to rendering aid and cooperation to the peoples of these states in the solution of problems of over-all national importance, and especially in satisfying the growing demands of the developing countries for ferrous metals, machine building, electric power, and the training of personnel.

The Soviet Union is giving cooperation to these countries in the solution of the important petroleum problem. With the help of the USSR, 20 refineries and chemical plants have been built, or are being built, in the young states.

In seven countries, with Soviet cooperation, large modern refineries have been, or are being, built, which will increase the production of petroleum products by 60 percent.

Of 40 different industrial enterprises being built in India with the help of the USSR, and called upon to strengthen substantially the Indian national economy, two petroleum refineries have already gone into operation. In 1966 the petroleum refining capacity of India's state industry exceeded 7 million tons. And today India is able to rebuff the insolent claims of the western oil monopolies. Speaking at the ceremony opening the second unit of the refinery at Koyala (state of Gujarat) in October 1966, President Radhakrishnan of the Republic of India expressed gratitude to the Soviet Union for cooperation in the building of this plant, and said, "The plant is one more example of the varied assistance which the Soviet Union is rendering the industrial development of India."

In the spring of 1965 the construction of a petroleum refinery in Ethiopia began. Emphasizing the great importance of Soviet help in its construction, Emperor Haile Selassie, at the ceremony of laying the cornerstone, stated that it could be truly called the first heavy industry plant, and its construction was striking evidence of the coming of a new era in the history of the economic development of Ethiopia.

The imperialists, in order to keep the young states in the status of raw material appendages, have in every way blocked prospecting of the mineral resources in their territory, and have used their own deliveries of petroleum products to these countries for economic and political blackmail. Soviet petroleum experts here, too, came to the aid of the young developing countries. They helped to discover ten oil deposits of commercial importance in India, on the basis of which state refineries are being built. Search for oil begun several years ago in the UAR and Syria by Soviet specialists has resulted in the discovery of a number of deposits, the development of which will not only permit these countries to free themselves from having to import liquid fuel (but) the UAR in the near future should become a major exporter of petroleum, while Syria, where four large deposits were discovered with the help of Soviet specialists, likewise will be able to export part of the oil extracted. Thus has been dispelled the myth of western experts as to the poverty of the petroleum resources of India, the UAR, Syria and other countries. As the Indian newspaper, Patriot, writes: "Having discovered rich oil deposits in Kambey, Ankleshvar, Kalola, and other regions of the country, Soviet petroleum specialists have not only dispelled the myth, persistently propagated by western 'experts', about the lack of petroleum in India, but have also helped the Indians to lay the foundation of a national petroleum industry."

Soviet petroleum experts have helped discover promising structures for oil and gas in Mali, and are conducting exploration in Pakistan, Congo (Brazzaville), and other countries. The petroleum people of Mexico, for example, invited our specialists to come there, and the latter conducted slant drilling of a well with a Soviet turbodrill. Thus have been established practical professional contacts between Soviet petroleum people and those of other countries. Our experts have helped the Afghan people discover important reserves of natural gas.

The Soviet Union has rendered great assistance to the young countries in the training of their own petroleum personnel, both in Soviet educational institutions and by the construction of training centers in these developing countries. In the 1964-65 school year, in the Baku Institute of Petroleum and Gas alone about 500 students from 29 countries of the world were being trained. With our help a petroleum institute and a petroleum technicum in Algeria have begun to train personnel for their own country and many other countries of Africa. Algeria, as is known, has very rich reserves of oil and gas, but so far they are being developed by foreign, mainly French, monopolies. The leaders of the country are putting forth as one of their principal demands that their energy resources serve the cause of industrialization of the country, particularly the establishment of a chemical industry on the basis of Sahara petroleum. Of 24 million tons of petroleum extracted in 1964, only 1.5 million were refined in Algeria itself, while the rest was exported in crude form to Europe. In March 1966 there was a ceremonious opening of the first state oil pipeline, with a length of 800 kilometers and a throughput capacity of 10 million tons a year. Step by step the Algerian government is pushing the foreign monopolies out of the economy of the country.

The western press has been writing a great deal in recent years about the "Russian oil offensive" in the young countries. But it deliberately passes over in silence the fact that this "offensive" is being conducted in alliance with these countries against the imperialist petroleum monopolies. While the colonial powers are trying to keep the countries of Asia, Africa and Latin America in the status of raw material appendages, the Soviet Union, building economic relations with them on a basis of equal rights and mutual advantage, is helping them to develop diversified national economies.

Cooperation on the basis of equal rights is helping the rise of the national economies of the young states, promoting their industrialization, and strengthening their political and economic independence. The peoples of these countries more and more understand that the way of real economic and cultural progress lies not through the "bounty" of imperialism, but through their own labor, resting on mutually advantageous cooperation with the countries of the socialist world. This is delivering the peoples from the imperialist path of neocolonialism and opening up for them the non-capitalist path of development.

This path has become possible in the era since our October revolution. The existence of the Soviet Union and a mighty world system of socialism, and the great successes of the peoples of the socialist countries in the building of socialism and communism, have created favorable international political, economic and ideological conditions not only for the downfall of the colonial system, but also for the movement of the liberated peoples along the path of social progress.

The postwar years, and especially the last decade, show convincingly that the policies of the Soviet Union and the other countries of socialism with regard to the young states are diametrically opposed to the policies of the imperialist states. The imperialists are doing everything to fasten on to the young liberated countries the fetters of neocolonialism. But the wheel of history cannot be turned back. The Soviet Union by its example and its Leninist policy is promoting the social progress of the peoples of these countries.

Following the wise counsels of Lenin, the Communist Party and the Soviet state have always supported and will always support peoples struggling against colonial oppression and neocolonialism; they will develop comprehensive cooperation with the countries which have achieved national independence; they will by every means promote the strengthening of the anti-imperialist front of the peoples of all continents; and they will expand relations with the communist and revolutionary-democratic parties of the young national states.

"Lessons of an Ill-fated Embargo"
Vneshnyaya torgovlya, 12, December 1966, p. 50-51

The Permanent Council of NATO was compelled to change its ban introduced in 1962 on the export to socialist countries of large-diameter pipe. Thus another instructive lesson was added to those which history regularly offers to the inspirers of the "cold war". The inevitability of such a lesson was foreseen by many 4 years ago, when the Washington politicians and the NATO generals began to be led by the oil monopolies.

Introduction of the embargo was one of those far from original measures with the help of which the most reactionary imperialist circles strive to hinder the development of foreign trade ties of the USSR with the capitalist countries. In particular, after beginning in the middle 1950s to purchase Soviet oil and oil products in an ever greater amount, the West European and other countries became convinced that they could pay for this vitally important commodity through the export of traditional products and they were not at all obliged to pay gold or sacrifice their political sovereignty as was actually demanded by the monopolies of the US and England. The mutually advantageous conditions in the sale of Soviet and Rumanian liquid fuels debunked the myth of unselfishness of the large Western oil companies, united in a so-called international oil cartel.

The system created by the cartel for supplying the capitalist countries with "black gold" turned the world oil market into a bundle of very complex contradictions. A sober approach toward these contradictions could have engendered many acceptable ways for their resolution. However, the Western and particularly the American oil companies took the slippery paths of adventurism. One after another they throw out appeals to save the Western world from the "Red oil threat," dictated not only by greediness but also by the desire of the cartel to enlist for protection of its dominating position in the capitalist economy everybody who instantly responds to any variety of the "Red danger."

In late 1960, the presidents of the Rockefeller Standard Oil monopoly and Mellon Gulf Oil openly demanded that the American government use military discipline within the framework of NATO to organize a "united repulse of the Red oil threat." Diplomatic protests and economic sanctions were placed against West European countries, Japan, and other importers of Soviet oil. The NATO generals did not wait. Declaring that the "Friendship" pipeline can "increase the mobility of Soviet divisions in East Europe," they demanded on 21 October a ban on the export of large-diameter pipe to socialist countries.

Still fresh in their minds is the storm of indignation aroused in this connection not only among the broad public and the trade unions, but also among business and official circles of England, Italy, Japan, and many other countries. Only official Bonn, ignoring the protests of the large West German metallurgical concerns and the strikes at the plants, demonstratively supported the NATO embargo at that time.

"In this case," said the statement of the Ministry of Internal Affairs, USSR of 22 March 1963, "the interests of the large oil monopolies were obviously served in the US policy and, under its influence, NATO policy. However, even if the US succeeded in limiting somehow the trade of other countries of the West with the Soviet Union, the ones to suffer here would not be the Soviet Union, which with its powerful, thoroughly developed economy can manage with its own resources, but above all those countries and their industrial companies which would be deprived of beneficial orders and markets and would find it necessary to reduce their production accordingly and to dismiss a portion of their workers."

The sober evaluation of the NATO demarche made from the Soviet side has completely been confirmed. For the second year, the "Friendship", the largest pipeline in the world, is operating along its entire great length, easily supplying oil to East Germany, Poland, Czechoslovakia, Hungary, and Bulgaria. "According to the assertion of economists in Washington," the New York Times acknowledges, "the Soviet Union has increased its own production of 40-inch pipe from a small amount in 1961 to 600,000 tons in 1965; now the Russians do not depend to a large extent on deliveries of pipe from abroad." On the other hand, the business press of West Europe and Japan reports on increasing difficulties in the sale of pipe and foresees a gloomy necessity of resorting to the closing of new shops and to the dismissal of thousands of more workers of the metallurgical industry.

In October 1966, it became known that Bonn, which had zealously supported the embargo on the export of pipe to countries of socialism, requested that Washington revoke this discriminatory measure. The enlightenment of Bonn engendered in business circles of capitalist countries a genuine sensation. Commentators counted several reasons which could have induced Bonn politicians to ask for mercy. Besides the rapid development of the Soviet pipe-rolling industry, it was also noted here that competitors of West German firms are not taking into account the NATO embargo and are prepared to increase the sale of pipe to the socialist countries, that the

Erhard government, with the help of this gesture, would like to placate the trade unions, etc.

However, the main reason for the retreat which has resounded in NATO should be noted. It is undoubtedly the fact that 4 years ago certain Western politicians evidently were overzealous in yielding to a false alarm of the American oil business. Would it not have been advisable for them then to read those lines of the Soviet statement which stated that "it is not the strategy or safety of the West European or other countries, but the 'strategy' and 'safety' of the high profits of American oil companies which determine trade discrimination and limitations in this policy and which the US tries to improve on other countries, increasing thereby economic instability in the West and undermining faith in relations among states."

The US oil monopolies can still continue in some places their economic blackmail of countries exploited by them. However, they are already powerless to stop the development of mutually advantageous economic ties between the two world systems. It is characteristic, for example, that the business press of the West, anticipating new orders, is more and more often writing in calm tones about plans for the construction of new large gas pipelines in the USSR. As for the fantasies about the influence of the new construction on the "mobility of Soviet divisions," they now engender only a skeptical smile. The peaceful aspirations of the Soviet Union are being doubted less and less by the business circles of West European and other capitalist countries.

One of the main tasks of foreign economic ties of the USSR is to develop further foreign trade with the industrially developed capitalist countries which display a readiness to develop trade with the Soviet Union. The mutual benefit of such trade is a good antidote for discriminatory policy in international economic relations.

"Oil and the President"
Bakinskiy rabochiy, November 13, 1966, p. 3

The readers of Bakinskiy Rabochiy are familiar with the article by V. Berezhkov from the journal Novoye Vremya

SELECTED READINGS 233

entitled "The Unrevealed Secret of the Kennedy Murder." The author delves in detail into the recently published book by the US publicist Justin on the Dallas tragedy of 22 November 1963. Justin, just as many other publicists, argues that the "crime of the century" was the result of a secret plot which was carefully prepared by the extreme right-wing circles in Dallas.

Dallas is the oil capital of Texas, which extracts about half of the US "black gold." It is the same Texas where the multi-millionaires from the oil business adjust everything. Considering that relations between the clan of Texas oil industrialists and the Kennedy house were about the same as those between the Montecci and Capoletti families, the conclusion unwittingly arises that the oil magnates of the "Lone Star State" could easily have been the organizers of the plot which took the life of the 35th president of the US.

27.5 Percent, Which Makes Multi-Millionaires

US tax legislation has become a gold vein from which the oil barons extract annually at least one billion dollars besides their regular tax profits. On the whole not in one branch of big business is such a large fortune scraped together so easily and quickly as in the oil extraction industry.

This began in 1926, when a group of oil businessmen, by bribing some congressmen and intimidating others with the "great risk of an oil suit," achieved an unprecedented tax indulgence, a so-called deduction for the depletion of resources. According to the accurate expression of US observers, the Texas oil industrialists pumped from the resources deductions for depletion which constituted as much money as did the barrels of oil from the resources.

What is the meaning of this "great charity" which loaded with money many oil businessmen at the expense of the rank-and-file taxpayers? The businessmen who "made America happy" because they engaged in such an "improper" business as the extraction of liquid fuel have the right to pocket 27.5 cents of every dollar of gross income without taxation. He pays taxes to the state only from the remaining amount, and then after deducting expenses for test wells which produced no oil.

Harvey O'Connor, a progressive American publicist, writes in his book "Oil Empire," which discloses the mechanism of enriching the oil pirates, that "when 27.5 percent is deducted from the gross income received from operation of existing wells, and then all the losses are deducted from the dry wells, Uncle Sam finds that in some instances he is in no condition to take even one penny from the hands of these newest nouveau riche."

Senator Douglas, a democrat from Illinois, showed that in 1954 one oil firm with a net profit of 4 million dollars paid a tax of 404 dollars, which is less than the amount taken from the lowest paid worker. Another corporation with an income of 5 million dollars did not pay one cent, and a third with a gross income of 12 million received a tax advantage of 500,000 dollars. Over 5 years (1953-57), 12 oil industrialists-millionaires paid an income tax of 22.5 percent of the gross profits, while in any other branch they paid at least 52 percent. According to calculations of the National Association of Oil Dealers, the US government in 1953 could have had tax receipts of one billion dollars more if, instead of the 27.5 percent deduction, it had used a new, more appropriate and true depletion of resources. In 1950, the Budget Commission of the House of Representatives tried to present a plan for reducing the magic deduction to 15 percent. However, Sam Rayburn was concerned that the child would be killed long before it saw daylight. And here is what Chairman Dauton of the commission said: "The Speaker of the House of Representatives objected sharply to any changes in the conditions regarding deductions for depletion of resources in the oil industry. I know that he openly said this to other members of the commission."

Considering that Sam Rayburn is a Texan, everything falls in its place. He simply carried out the wish of Hunt, Cullen, Murchison, and other Texan multi-millionaires and millionaires from the 27.5 percent deduction.

John Kennedy Against Harold Hunt

The injustice of the tax deduction set forth for the conquistadors allegedly in the interests of the well-being of the US oil industry is quite evident. President Roosevelt tried to deprive the Texas magnates of the unfounded advantages. However, the oil barons, who were helped by World War II, were able to discard the plans of the President. Roosevelt only managed to make an inveterate enemy of his very close assistant, Vice President John Garner, a Texan and henchman of the oil circles of Dallas. Dallas noted Roosevelt's death with cocktails.

Harry Truman acknowledged that the "perspective program for developing natural resources does not require that we give each year hundreds of millions of dollars in the form of freedom from taxation of the privileged few at the expense of the "majority." But Truman stopped with this protest.

Dwight Eisenhower also did not think of encroaching on the tax advantages of the oil industrialists. On the contrary, no US government was under such influence by the oil kings as was the cabinet of Eisenhower.

Only John F. Kennedy, who considered the 27.5 percent deduction a most "unnecessary and unjust advantage," firmly decided to change the tax legislation. The president's attention was attracted to the fact that the domestic US prices for crude oil considerably exceeded the world prices. As a result, the consumers of oil products each year overpay by about 3 to 4 billion dollars, and the competition capability of US goods of the refining industry where oil is used as raw material is decreasing.

Even before he took office, Kennedy asserted that he would review the entire tax system, and if he "observed any discrepancies in oil or any other goods I will advocate closing this hole."

After becoming President, Kennedy charged a special committee with developing measures to eliminate the extreme advantages for oil industrialists. In early 1963 he introduced for examination by Congress a draft law on increasing taxes in the oil industry, which would reduce the deductions for depletion of resources by about one-third. Acting further in the interests of the businessmen in the refining industry, Kennedy published the draft of a future US oil policy which specified the gradual deduction of domestic prices for oil to the level of world prices.

Of course, the Texas oil businessmen were hostile to Kennedy's plans. They counterattacked in an anti-communist spirit. Agents of Dallas oil magnate and billionaire Harold L. Hunt accused the President of under-estimating the threat engendered by the "export of oil from the Soviet bloc." They asserted that Kennedy's policy was hindering US companies from "competing with the Soviet oil industry not for life but to the death." The newspapers and radios financed by Hunt yelled powerfully that the President's measures in limiting profits of the oil industrialists constitute a "blow on Texas," "a rush a mile below, toward socialism," "a copy of the Soviet Five Year Plan," etc.

Thus a genuine war flared up between the Texas billionaires headed by Hunt and John Kennedy. This has reflected the contradictions between the "old dollars" (the Kennedy family is one of the richest such examples in the US) and the "young dollars," the "upstarts," who became rich during the past few score years (including the Texas oil businessmen and the California war industrialists). In October-November 1963 it reached its culmination.

Five weeks before the Dallas shooting a very close friend of Hunt, T. Dailey, who owned the newspaper Dallas Morning News, had breakfast at the White House, where he acted impudently. He sharply condemned Kennedy's policy and slanderously stated: "You, Mr. President, instead of being a leader on a horse, are dragging along in a dirty cart."

On 7 November 1963, the heads of three large oil companies visited the White House. This was the first and last meeting of President Kennedy with the great oil businessmen. The talks lasted only about 25 minutes. The journal Oil and Gas, in commenting on this meeting, wrote: "The representatives of oil circles who visited the White House left there without any assertions on the part of the President that henceforth relations between the government and industry would be strewn with roses." Four days later in Chicago the Congress of the American Petroleum Institute met. It was attended by 6,000 industrialists and bankers. At this "council of oil gods," violent accusations and unconcealed threats were made against Kennedy.

On 22 November, the day of the President's murder, Dailey's newspaper encircled with a bold mourning border the following statement: "Welcome, Mr. Kennedy, to Dallas." Several hours later John Kennedy was deprived of the possibility of continuing the war against the Dallas oil billionaires. He was killed.

Two details were later revealed. Dailey, after signing to the press the issue with the provocative statement, remained in his office, where he met with Jack Ruby. The statement had been generously paid for by three Texas businessmen, including Nelson Hunt, son of Harold L. Hunt, the very first of the Texas billionaires from the oil business.

Another fact was disclosed. On 14 November, when Ruby's "Carousel" was in full swing, a secret meeting was held in one of its rooms, which was attended by Ruby, a certain Weisman, the policeman Tippit (he was killed soon after Kennedy's assassination), and one other incognito "wealthy oil industrialist." This, according to reports of the US press, is what the latter was called by Earl Warren, chairman of the US Supreme Court, who headed the commission which investigated Kennedy's death.

The report of the Warren Commission calls all these facts "chance events." But are there not too many "chance events" in this "crime of the century" which directly indicate that the oil magnates of Dallas could have been the inspirers and organizers of a secret plot against President Kennedy, who was hindering them?

"Prices in the World Socialist Market"
Vneshnyaya torgovlya, 11, November 1966, p. 19-22

The problem of the scientific development of a system of price formation is one of the most complex in the political economy and economic practice of socialism. Among the economists of socialist countries there are various points of view on such questions of principle as the methodology of calculating in the price the net income, cost of fixed assets used, and the consumer cost of goods. In developing prices in trade among the socialist countries, it is necessary to establish a concept of cost and socially necessary expenditures for the production of goods, show the price-formation factors in conditions of the world socialist system, resolve the question of the role which prices must fulfill in mutual relations among countries, etc.

An important landmark in development of the system of price formation in the world socialist market was the 9th Session of CEMA, which was held in July 1958. Generalizing current practice, the session resolved that in trade among member countries of CEMA the prices must be established on the basis of prices of the major world trade markets, they must be stable for a long period and common for all countries, and they must consider the quality of goods and the expenditures for their delivery. Such a system of price formation with individual variations in certain cases has also been adopted by the socialist countries of Asia in their trade with member countries of CEMA. At the 9th Session it was agreed to specify the "major commodity market" as the market which is leading in the international trade of a given commodity and which is geographically attracted toward the appropriate socialist countries.

The decisions of the session pointed to the necessity of further improving the system of prices in trade in the world socialist market. It was also resolved to study and develop possible ways of transferring to their own base the prices in trade among socialist countries. It was considered that to transfer prices to their own base, specific economic and organizational premises are needed: the development of specialization of production among countries, regulation of their domestic prices, drawing together of the methods of determining costs of production and price formation used in each country, etc.

Prices virtually did not change during 1958-64 in trade among all these countries. At a meeting of the CEMA Executive Committee in December 1962, it was decided to make corrections in those prices for which there were considerable deviations from the world prices, as well as to change accordingly transport tariffs on the basis of the principles of development of the new contractual prices.

SELECTED READINGS

The use of world prices as a basis for developing a system of price formation in trade among socialist countries arose historically, on the strength of objectively existing conditions. It is based on the fact that the two world markets are linked with each other and that world prices in the end are formed under the influence of the law of cost in its international application. The foreign trade of socialist countries is beginning to influence notably the formation of world prices for many commodities. The share of this trade in the entire turnover of international trade is still not great. However, for certain commodities, such as timber, oil, cast iron, grain, and certain types of machinery and equipment, this share is considerable.

On the whole much work is now being done in member countries of CEMA both to improve the existing system of prices in their mutual trade and to study and clarify the opportunities for using their own price base.

However, certain economists propose a simplified resolution of these complex problems which constitutes breaking away from accumulated experience of price formation in the world socialist market and contradicting the economic principles. Because, in their opinion, their socialist cost has formed on the world market detached from worldwide conditions of production, it has now become possible to develop their own base for prices. It is necessary, they suggest, only to select the appropriate indicator which characterizes the national cost of a commodity and to resolve the "technical" question on the method of averaging such indicators. They propose to take the internal wholesale prices of each country as a basis. However, some feel it necessary to reform wholesale prices for this in order to draw them closer to the national socially necessary expenditures. Others propose to clear wholesale prices of the turnover tax, as well as of certain national peculiarities of price formation. There is also a third point of view in accordance with which, in developing their own base for prices, they should proceed from the national cost of commodities, cleared of net income at all stages of production. To form prices in the world socialist market, it is proposed to add again some share of net income to the cleared cost of production of each commodity.

Some propose that we must adopt the regional cost of production of a commodity in the world socialist system, cleared of elements of net income, as the initial base for prices in trade among socialist countries.

They feel that price proportions, and not their level, are important in the world socialist market. Therefore, the inclusion of net income in the price, in their opinion, is not expedient.

However, all these proposals are made without a sufficient analysis of the foreign economic relations in the world socialist market. In essence, they ignore the question of how and to what degree the implementation of a particular proposal will affect the genuine economic interests of each state and how it will be reflected in the course of world economic competition with the capitalistic countries. They lose sight of the fact that selection of a system of prices in trade among socialist countries is far from a technical question. It is a great economic problem connected with the sharp national interests of each of the countries. Without considering and ensuring such interests, the plans for developing their own base of prices are impracticable.

In all probability they should start not with the selection of cost indicators and methods of averaging them, but with a study of the initial conditions for developing their own base for prices and with the development of objective criteria to ensure the correctness of these prices.

It is evident that in contemporary conditions in the economic relations among socialist countries, economic accountability, the basic principle of socialist economic management, must be observed. And the prices of the world socialist market should be developed on the basis of this principle. The method of economic accountability does not exclude mutual assistance among the countries or even their gratuitous assistance to one another.

It is also evident that this method on the scales of the world socialist system must differ in its use within each country. In our opinion, economic accountability must signify here a quantitative common determination in cost indicators of the material values and services transferred by the countries to one another, in which such indicators must be determined on the basis of operation of the law of cost in its international application with calculation of certain other factors. It follows from this that calculation of socially necessary expenditures of labor is the main function of price in the world socialist market. This price also has a stimulating function. However, it must fulfill the stimulating function in such a way as to be advantageous to all countries without harm to any of them.

For the common determination of the costs of goods delivered to one another, the countries must have a system of indexes the criteria of which are determined on the basis of international cost in one or another form. Each country can compare the cost of commodities shipped and that received in return also in its internal prices. Proceeding from the results, it can change the structure and direction of its trade.

As was indicated in the "Main Principles of the International Socialist Division of Labor", approved by the June 1962 conference of representatives of communist and workers parties of participating countries of CEMA, the calculations of the effectiveness of foreign trade are an important, but not the single, criterion for the basis of rational ways of deepening the international socialist division of labor. Therefore, final decisions on its development can be made only on the basis of the complex calculation of the economic interests of individual countries and the socialist system on the whole, as well as political factors. It would be incorrect in determining the export-import prices to proceed only from their own expenditures incurred in the production of particular types of products. In establishing such proportions it is necessary to consider in what degree in a given country opportunities have been used for decreasing the cost of production of given commodities, and also to pay attention to the conditions of their production in socialist and capitalist countries, the interest of all countries in importing such commodities, the possibilities of deliveries from non-socialist countries, etc.

From the point of view of inter-state relations, the question of prices in trade among socialist countries is a question of increasing or decreasing the national income created in each of them. It is important that the prices not engender a redistribution of this income among countries to the detriment of others. However, the advantages of foreign trade can be different for each country depending not only on the level of export and import prices but also on the correlation of the amounts of national and international cost of exported and imported commodities. In this question the problem arises of the equivalence of the exchange among countries.

Equivalent exchange on the world market infers in principle an exchange which is made in accordance with prices based on the international cost of a commodity with calculation of its capital intensity, i.e., calculated expenditures (the theoretical basis for calculating such expenditures was given in the book by V.P. Dyachenko entitled "Theoretical Principles and the Method of Calculation of Price-Forming Factors in Planning Prices," Moscow, 1964, p. 19-20). In many cases, however, especially under conditions of the world socialist system, such a determination is insufficient. The structure of commodity turnover among socialist countries is determined to a large extent by means of the law of systematic and proportional development which is in effect throughout the entire system. Proceeding from the principle of mutual assistance, the socialist countries were forced in a number of cases to deliver to one another commodities, the conditions of production and transport of which were less favorable than the average world conditions. Therefore, the exchange of these commodities can occur in accordance with prices determined on the basis of international cost, but with the insertion of corrections in them. Price in such cases can, at the consent of the interested parties, deviate from the international cost with calculation of the specific nature of the production conditions of the socialist countries.

In connection with the fact that the national conditions of production and cost of commodities differ from their worldwide, international conditions of production and cost, each of the countries in equivalent exchange can receive through foreign trade an amount which consists of a greater measure of national cost than that which it gives to another country in exchange. As a result of such an exchange, the national income of importer-countries will increase in quantity and value. However, this does not harm the exporter and it does not decrease its national income.

Thus, price in mutual trade of socialist countries will begin to carry out the functions of stimulating the rational division of labor among them without leading to an overflow of national income from one to another. Therefore, the assertion of certain economists that price in foreign trade must fulfill without fail the redistribution function is incorrect. In accordance with the decision of the 9th Session of CEMA on stimulating prices, this should be resorted to only in unusual cases, at the mutual agreement of the interested parties.

In examining the cost of a commodity as a basis for its price, K. Marx differentiated between national and international cost. However, because the socialist countries are not separated from the entire world economy, in our opinion we cannot speak of a developed separation of international cost in the world socialist system.

The process of forming a regional cost for socialist countries for various commodities is to be found at various stages of development. It depends on the share of the socialist countries in world production and trade, on the development of economic ties among countries of the socialist and capitalist systems, and a number of other factors. Therefore, at the present time, in developing a system of prices in the world socialist market, it would be incorrect to give up the calculation of international cost on the level of the entire world economy.

Developing in accordance with their specific laws, the world socialist market at the same time is an integral part of the worldwide market. Proceeding from an abstract concept of the universal development of the cost of a commodity, one can also speak in an abstract manner about cost in the world socialist market as regional, and not international, cost. In reality, this regional area is not separated economically from the entire world economy, in which the level of socially necessary expenditures and cost changes in connection with the changes in the processes of the scientific-technical and social-economic development of mankind.

The existing cost indicators of the socialist countries are unsuitable for use as a basis of prices among them also because, as a result of the methods of developing them, they do not correspond to the socially necessary expenditures for the production of commodities. The domestic prices in each socialist country are given the functions of redistributing the national income among the spheres of the national economy, as a result of which they are knowingly diverted from their cost basis. Therefore, the weighted average from the domestic wholesale prices of each of the socialist countries for a particular commodity will not correspond with the average level of socially necessary expenditures of the world socialist system.

The situation changes little if, instead of wholesale prices, there is an averaging of the cleared cost of production of the commodity which reflects the complete expenditures of wages for its production. The fact is that wages, just as prices for commodities, include the redistribution of elements which are different in various countries. The levels of wages in various branches of production of each of the countries do not correspond with the amount of socially necessary expenditures for output of these branches. Much work must also be done to improve prices and regulate wages in each of the countries before their cost indicators begin to correspond with the levels of socially necessary expenditures for the production of commodities.

The development of their own base of prices by averaging the domestic wholesale prices would result in unsuitable conditions for countries which export a large portion of commodities which require little processing. In Soviet exports to CEMA countries, such goods constitute about 60 percent, and in imports from these countries they constitute about one-third. Such a structure of commodity turnover developed because a shortage of raw materials and fuel is felt in the European socialist countries. Under the existing systems of development of wholesale prices in socialist countries, the level of prices for this group of commodities is considerably lower than for commodities requiring a high degree of processing (due to inclusion in the price for each degree of production of net income as a specific percent of the cost of production). Thus, the use of the above method of developing their own base of prices would lead to the situation whereby about one-half of the USSR exports to other socialist countries would be, without any reason, considerably undervalued in comparison with the prices of our import.

The level of labor productivity in a number of branches of the economy of socialist countries is still lower than in the industrially developed capitalist countries. In these conditions, the world prices constitute a unique guideline in the process of eliminating such lags. The development of prices in mutual trade of the socialist countries by averaging the wholesale prices or cleared cost of production could slow down the decline in the cost of production of products and the increase in labor productivity in these countries.

The following conclusions also speak against the use of cleared cost of production prices. This indicator is not used directly for price formation on the national levels. It reflects neither the consumer cost of a commodity which must be considered in the price nor the influence of supply and demand. Its use for developing their own base of prices would result in a separation of this base not only from world prices but also from commodity-monetary relations which actually exist within each country. As a result, all objective criteria and real indicators of the correctness of prices of the world socialist market would be lost. Moreover, the calculation of cleared cost of production in a broad circle of goods with the necessary degree of accuracy virtually involve great difficulties.

Consequently, objective conditions have not yet ripened for developing prices of the world socialist market by averaging the cost indicators of each of the socialist countries. We need a large degree of leveling of economic levels, a further drawing together of them, and regulation of the methods of price formation and the systems of wages in each of these countries. We also need the development of suitable factors for the mutual compilation of cost indicators, a comparison of the levels of labor productivity and wholesale prices of socialist countries with world prices, a study of the practice of developing wholesale prices for basic commodities in each country, etc.

In opposing the use of world prices in developing the system of price formation in trade among socialist countries, certain authors allude to the fact that world prices are basically determined by the conditions of production and exchange in capitalist countries, where production relations exist which are alien to us. Such an argument is not convincing. From an analysis made by K. Marx on the circulation of commodities "produced with various means of social production," one can conclude that the nature of the social-economic structure in which a commodity has been produced cannot have decisive significance for formation of its international cost (K. Marx, Capital, Volume II, page 107).

In determining the prices in the world socialist market, the concept of international cost cannot be limited by the framework of this market or of several countries. In practice, this would have a negative effect on the development of the economy of socialist states and their mutual relations. In our opinion, the international cost of commodities circulated in the world socialist market must in principle be determined on the basis of the level of labor productivity achieved within the framework of the entire world economy. Here, as has been said above, it is possible to make corrections to this level at the consent of the interested parties.

An improvement in the existing method, which is based on the use of world prices which reflect international cost as their basis, should now be considered the chief method of improving price formation in the world socialist market. This method, despite the shortcomings inherent in it, reflects to a greater degree the true conditions of the present stage of development of the socialist countries. With its help, we can stimulate the rational distribution of labor, an increase in its productivity and a reduction in the cost of production of commodities; i.e., we can ensure fulfillment of the tasks which stand before the system of socialism under conditions of peaceful economic competition with the capitalist system.

One of the chief methods of improving the existing system of prices of the world socialist market which are based on world prices is a further calculation of the specific nature of the social and natural conditions of production.

This is why, in using world prices for developing a system of price formation in the world socialist market, these prices should be corrected not only by a "clearing" of speculative factors. In those cases where there are economic grounds for correcting world prices, it should be done also with calculation of the difference in production conditions for each commodity in the countries of the socialist and capitalist systems. Evidently it becomes necessary with such correction to resolve the question of expediency of increasing the purchase of particular commodities on the capitalist market.

We should particularly delve into methods of stimulating inter-state specialization of production within the framework of CEMA with the help of the system of prices for specialized products in the processing industry. The creation in any of the socialist countries of large-scale production calculated to meet the demands of other countries for specific commodities makes it possible for it to decrease sharply the cost of production of these commodities. Such a decrease in cost of production is often possible only if other countries reduce their own production of specialized products and import them from the country in question. Thus, international specialization of production is actually a collective measure of several countries. Therefore, the price for the output of such specialized enterprises must acquire the character of an internal price of the world socialist system as the "single cooperative" of the peoples. This price in the future will deviate more and more from the world prices which should be viewed as its upper limit.

To create material interest in specialization among all countries, they must jointly enjoy its advantages. To do this in certain cases it would be sufficient to reduce the export prices for specialized products in comparison with the level established on the basis of "cleared" and corrected world prices.

At the present time, it would be incorrect to establish prices in the world socialist market for specialized products of each country by directly recalculating its cost of production. A reduction in prices for such products for mutual deliveries could be established by proceeding from the specific portion of the actual decrease in expenditures for the manufacture of the products received as the result of specialization. Concrete proposals on this question must be drawn up with calculation of the conditions of specialization of production of each commodity. However, a common approach toward determining such a saving must consist of calculation of the cost of production and the level of capital investment in the products.

It should be noted that inter-state specialization, in strengthening the tendency to decrease prices in the world socialist market in comparison with the prices in the world-wide market, cannot in all cases ensure such a decrease in the near future. There may be cases whereby expenditures cited for the manufacture of specialized products will be higher than the world price on the basis of which the export price is established. In these cases, increased exports will inflict losses on the exporter. As a means of eliminating such losses, there should be an increase in the technical equipping and an improvement in the organization of labor at specialized enterprises.

Certain economists of socialist countries feel that with inter-state specialization of production there should be no withdrawal from the existing system of establishing export prices in order not to complicate this system. In their opinion, the country which does the specialization should enjoy the advantages from specialization of production of a given product and from its export. Other countries, in turn, will enjoy the advantages from specialization in the production of other commodities. Such an opinion deserves serious attention. It is not impossible that in practice, depending on actual conditions, both variations of resolution of the question can be used.

Thus, the theoretical aspect of the question and the common practical interests of the socialist countries indicates that at the present stage of development of the economy of socialist countries, it is necessary to consider in trade among them the international cost and world prices. The world price must be considered at least inasmuch as it is determined by the level of labor productivity, based on the use of modern equipment and organization of production.

The fundamental base for prices of the world socialist market must express the socially necessary expenditures for the production of commodities, and not the average amount from the available cost indicators of each socialist country. Such a base of prices can be built by improving the system of prices on the basis of the world prices. The latter should be corrected in certain cases with the consent of the interested countries not only on the amount which is determined by the attitudes of exploitation in the capitalist world, monopolization of the market, market and cyclical fluctuations, etc., but also on the basis of the differences in national and other conditions of production and transport in socialist and capitalist countries. Such differences must be considered first for those commodities which the socialist countries cannot now purchase in the capitalist market (due to the policy of embargo, the lack of currency, etc.), as well as for those commodities the production of which is maintained at a specific level in the world socialist system due to reasons of economic independence from the capitalist market. Moreover, the prices of the world socialist market must be deviated from the world prices by stimulating the specialization of production, its rational distribution, and the development for each of the countries of an export and import structure which would ensure maximum effectiveness of its foreign trade for the entire national economy.

"Where Is the Oil to Come From?"
Polityka, September 24, 1966, p.
by Stanislaw Albinowski

The problem of energy supply for a modern economy is one of the most difficult and complicated economic problems. However, the great amount of research work already done makes it possible to formulate one of the two following conclusions without being afraid of erring; -- the demand for fuel and power will grow in all CEMA (Council for Economic Mutual Assistance) countries, but the geographic distribution of the raw material resources is extremely uneven; and -- on a large scale, crude oil is currently the most economical power fuel, but it is precisely what Poland is short of, whereas we have immense resources of hard coal.

With such a situation, an indirect way out is the only practical solution to the energy problem. We cannot give up coal, because this would entail losses amounting to billions of Zloty resulting from the incomplete utilization of existing mines; whereas, on the other hand, we should not, I believe, give up the use of oil in long-range planning, for this would amount to condemning ourselves to economic backwardness. I must say I was greatly disturbed to learn from Professor Secomski's recent book entitled, Podstawy Planowania Perspektywicznego (The Foundations of Long-Range Planning) that in 1980 crude oil is to account for a bare 3.2 percent (0.3 percent in 1960) of Poland's consumption of basic fuels. If we add to this the 4.1 percent share of natural gas and methane, the index will be something like 7 percent, whereas in the USSR and in the European Common Market it will reach a level of 60 percent by 1975!

Someone may feel inclined to say: alright, but how are we to increase the share of liquid fuels without our own oil, while supplies from the Soviet Union cannot be unlimited? Well, there is only one answer: we have to consider the possibility of importing crude oil from certain countries of the so-called Third World.

I would like to consider this problem by including all of the CEMA countries and against the background of the trends prevailing in the most industrialized capitalist country -- the United States. Now, in the US per capita consumption of crude oil was about 2.3 tons back in 1963, whereas in 1980 it is supposed to rise to 2.9 tons. In the USSR it increased during 1960-65 from 600 kilograms to 870 kilograms and according to the most reliable estimates it will rise to the level of 2 tons by 1980. This index varies among the other CEMA countries, but the minimum we should average in 1980 is 1,500 kilograms. This is essential if we consider the requirements of technical progress, the savings possible by the more extensive use of crude oil (in place of coal) as a source of power, and our plans in the fields of motorization and diesel traction in railway transport.

Here are just two examples. All CEMA countries, and especially the Soviet Union, Czechoslovakia, and East Germany, are expanding their automotive industry. Only five years from now, the production of automobiles will reach millions of units. In the course of the subsequent decade this tendency will grow, and very likely the remaining CEMA countries will also have large automobile industries. Hence, the consumption of fuels for such purposes will increase many times in the period under consideration. And now for another example. All CEMA countries are planning to change over to electric and diesel power in railway transport. For instance, it follows from Professor Secomski's book that in Poland the share of diesel fuel is to rise from the 1960 level of 0.1 percent to 50 percent in 1980. Considering that railway freight is to double at least (to about 130,000 million ton-kilometers, not counting passenger traffic), it is not difficult to imagine (experts could easily calculate it) what tremendous amounts of fuel will be consumed by the railways alone. Let us add that in the two decades from 1960-80, our merchant marine is to expand to almost six times its present size. And nobody builds steamers any more.

The same reasoning can be applied to the CEMA countries. Consequently, the 1980 per capita crude oil consumption of 2 tons in the USSR and 1.5 tons in the other CEMA countries, which was reached a while ago, is by no means exaggerated.[1] And if so, it is easy to calculate our total demand for crude oil in 1980. Assuming that at that time the population of the USSR will be 280 million, and that of the other CEMA countries, 115 million, we arrive at the following conclusion: in 1980, the total demand for crude oil will be 560 million tons in the USSR and 170 million tons in the other CEMA countries for a total of 730 million tons.

And how can this demand be met? Soviet crude oil output will probably be about 630 million tons, and will thus exceed domestic consumption by 70 million tons. The total output of the other CEMA countries can amount to -- according to current estimates -- from 23 to 33 million tons. Let us be optimistic and take the higher figure. Furthermore, let us assume that of its 70 million ton surplus, the USSR will earmark only 20 million tons for export to capitalist countries, leaving 50 million tons for the other CEMA countries (since the ratio is now 50:50, this assumption is again optimistic from our point of view). With such a situation, the other CEMA countries will show a 1980 crude oil deficit of about 90 million tons. This roughly corresponds to the entire 1980 deficit in the fuel and power balance of CEMA.

Theoretically, this deficit may be covered by coal, since the USSR and Polish reserves are adequate. But this would make a farce of a rational economy. Hence, the only alternative left is to import, including from a non-socialist area. I do realize that the prospects of importing 90 million tons of crude oil, which, at only $15 per ton, presumes an annual expense in foreign currency of $1.4 billion, may arouse opposition. I also realize that in practice this concept will represent an immensely difficult task. But, I believe, we have to face the facts. I wish to present a line of thought which -- when considered briefly -- will prove two theses. First, that all of this is practicable, and second, that it is profitable.

First, let us consider the practical aspect. Let us dot the "i" right from the beginning. Oil is a strategic material not only militarily but also economically. It determines the tempos of a nation's economic life. Its sudden disappearance

1. According to the existing estimates on development, Poland's 1980 consumption of crude oil is to be 20 million tons, which is not quite 600 kilograms per capita.

could result in tremendous difficulties. Therefore, the first question is: can the socialist countries make themselves largely dependent on oil imports from an area outside our socialist bloc? And here is a tentative answer: first, one might point out that so far our dependence (here I mean Poland) on imports from the developed capitalist countries of such important commodities as rolled products (steel sheets, regular and high-quality steel bars, seamless and welded pipes, etc.,) is very great. Capitalist countries cover from 40 to 90 percent of our demand for imports of these commodities and of a number of others, such as copper, copper wire, mercury and marine diesel engines. Second, let us recall for the sake of comparison that already in 1964 European Common Market countries imported almost 200 million tons of oil from abroad -- chiefly from the Near and Middle East and North Africa. The British economy relies completely on imported oil, and this from areas of their former colonies, i.e., from countries having none too friendly political feelings for their former rulers. The fact that this oil is extracted and supplied to the British market by monopolies in which British capital has either a controlling or a significant interest does not really mean anything. For there is the possibility of nationalization of the property of such consortiums in former colonial countries, and well-known instances in the past have shown that this is what happens when the situation demands it. One of these is the problem of a market. If Iraq or Iran nationalized the foreign oil companies right now they would "choke" on their oil. A boycott on the part of the monopolized markets in the western countries would be more than likely.

At this time I pass on to the third argument. It could be an important thing for the socialist countries to become importers of large quantities of crude oil. There are a number of countries where oil deposits are already nationalized, or controlled by the state (Algeria, Egypt). Furthermore, we may suppose that in many poorly developed countries which are only now beginning to explore and develop their crude oil resources, the active presence of socialist countries as large-scale importers and partners conducting a policy of mutual benefit might guide social developments in these countries along a path concordant with the interests of anti-colonial forces.

Therefore, the problem which has just been posed should not, I believe, be looked upon as one-sided dependence. It is true to say, Great Britain is dependent on the supplies of oil from the Middle East, but it is also true that such great purchases entail a reverse dependence. By analogy, this reasoning could be applied -- mutatis mutandis -- to our future relations with individual countries of the Third World.

Up to now we have been discussing the political problem. Is it practical to import 90 million tons of oil annually into CEMA countries? The expert then immediately visualizes a whole fleet of tankers, a complex of port facilities (which would have to be built from the start), thousands of kilometers of pipelines, and a whole system of storage facilities. This is a tremendous problem, but, if we have been able to build by joint effort the "Friendship" pipeline in 3 years, whose northern "thread" alone pumps 14 million tons of crude oil annually, if Yugoslavia is currently assembling a pipeline with a capacity of 30 million tons annually, I do not doubt that we would be able to prepare within the next 15 years for the import of 90 million tons of crude oil.

Now to consider the financial and commercial aspect of this problem: what are we going to pay for this import? Oil is a foreign exchange commodity, and an annual expenditure of 1.4 billion dollars is not a trifle even for 5 or 6 CEMA countries together. But in reply to this question one may say that even now plans are being formulated for a new type of trade between the socialist countries and the Third World. They involve long-term agreements whereby the socialist countries supply economically undeveloped countries with capital goods in return for crude oil. Such an agreement, covering a period of ten years, was signed between Rumania and Iran late in 1965. In January 1966 Czechoslovakia also signed an agreement with Iran whereby she receives crude oil and farm products in return for a complete foundry and a machine tool factory. The agreement is based on a 10-year credit. Similarly, Hungarian officials are contemplating major purchases of crude oil from Algeria and the National Oil Company. Obviously these are only the first steps, but they show that such an approach is feasible.

And now to look at the other side of the coin. Is it worthwhile? It has to be, because without adequate quantities of crude oil the proper economic development of our countries would be impossible. The alternative solution, i.e., increased oil prospecting in the CEMA area, even if geologically a success, would not, I believe, be an economic success. This follows from a simple calculation: every million zlotys invested in the fuel industry has so far yielded for Poland an added output of 146,000 zlotys, as opposed to 663,000 zlotys in the ferrous metallurgy industry, and 2,805,000 zlotys in the machine-building industry. Obviously, the ratios may be different in the other CEMA countries, but it remains a fact that under our geological conditions, investments in the extraction of energy raw materials yield only a fraction of the returns from investments in metal processing. Hence, let us reap the maximum benefits afforded by the international division of labor on a scale exceeding the limits of regional groupings, such as CEMA. The profits will not be purely economic. A closer look at the prospects for trade between the socialist countries and the Third World would lead us to conclude that the import of crude oil from this region is one of the basic elements in strengthening the position of the socialist countries in the world market and the international arena. This, at least, is the author's conclusion after studying the material available to him.

"Fuels--Problem Number One"
Servis publicystyczno-informacyjny, 212 August 7, 1966, by Stanislaw Albinowski

When we read in a newspaper the words "fuel-power-industry-balance" they do not mean much for the majority of us. But when we switch on a washing machine which washes for us, when we take a streetcar or a bus instead of going on foot, when a crane working with a Diesel engine lifts walls or concrete buckets, then we can see that a machine does real work for us. Sometimes this work is so enormous that we would never manage to do it ourselves.

Economic progress depends on many factors. However, a fundamental technological-economic condition is an appropriate amount of energy which can be used for our purpose. Everything depends on this: from an increase in productivity to electrification of a household. It is here that the problem of fuel-power balance appears. There are always two sides of a balance. In this case the demand for energy on the part of the CEMA countries represents one side and the possibilities of satisfying this demand the other. Such a preliminary balance-sheet has already been prepared up to 1980.

Here a number of problems appear. We receive energy from the so-called primary raw materials. Not more than 10 percent of this energy is being processed into electric power, while the rest is used in a different way: by burning coal in boiler houses or by putting diesel engines into operation by means of petroleum or gasoline. So, what is the amount of these fuels? Are their resources sufficient to cover our future demands?

Considering the problem from the point of view of CEMA as a whole it can be stated that the already discovered deposits of raw materials for the power industry are inexhaustible. According to the estimate carried out at the beginning of the 1960s, the resources exceed 9,000 billion tons of standard fuel. One can hardly imagine such a great amount of fuel lying under the surface of the CEMA countries. It can be added for comparison that in 1960 the production of fuels in the CEMA countries did not reach one billion tons. At the present output rate these deposits would satisfy demands for 9,000 years! Moreover, every year brings new discoveries and new methods of extracting which allow for the exploitation of deposits previously considered unprofitable.

In spite of the existence of these resources the problem is by no means simple. On the contrary, it is extremely complicated. Serious disproportions can be noted in this field. The first of them refers to the geographical disposition of deposits. It is enough to say that the Soviet Union has over 90 percent of the power industry raw materials in CEMA while the participation of the USSR in the CEMA industrial production amounts to about 70 percent. The situation is the reverse in Czechoslovakia and East Germany which account for 6 to 8 percent of the industrial production but have only 0.5 percent of fuel resources. The second disproportion can be reduced to the fact that we have distinct predominance of hard fuels over liquid ones. Let us mention that the USSR possesses 8,700 billion tons of coal and "only" 28 billion tons of crude oil. Hard fuels are expensive. For example, in the

USSR the costs of mining 1 ton of coal are 20 times higher than those of drawing an adequate amount of natural gas of an equal calorific value, and 6 times higher than those of crude oil. The entire world turns its interest to oil and, unfortunately, we do not have much of it.

Another problem. Investment outlays for mining the power industry raw materials are extremely high. Let us take Poland: in the years 1961-65 every million zloties invested in the engineering industry brought an increase in production by 2.8 million zloties while in the fuel industry only 146,000 zloties! This difference is immense and it influences, to a considerable extent, the rate of growth of national income. This factor is of essential importance particularly when the power industry raw materials are so unevenly distributed. In fact, not the existing deposits of coal or oil are useful, but only those extracted. And this requires considerable expense.

Consequently, we are facing the following situation. In the present 20-year period the total industrial production of the CEMA countries is to increase five-fold, while the production of electric energy - nine-fold. On the other hand, in the same period the output of power industry raw materials within CEMA will increase from 970 to 3,400 million tons (in units of standard fuel), i.e. 3.5 times. The essence of this problem becomes obvious only if the situation in the USSR is considered separately from that in other CEMA countries.

Thus, in the Soviet Union the production of the primary energy - as calculated in standard fuel - will increase from 700 to 2,900 million tons, i.e. over fourfold and in the other CEMA countries from 270 to 500 million tons, i.e. by 85 percent only. While the Soviet Union will gain some surpluses which can be devoted to export, the other CEMA countries will have a shortage of over 100 million tons of fuels in 1980. Obviously, it would be impossible to liquidate this deficit through import from the Soviet Union (which would hamper considerably the growth of the Soviet national income) and, from the point of view of foreign trade such a solution would not be even purposeful.

Due to the shortage of fuels and the necessity of importing them, it lies in our interest (i.e. in the interest of Poland as well as the GDR, Czechoslovakia, etc.) to achieve the most economical import of fuels. In present technological conditions when atomic power industry is still a thing of the future, crude oil represents such a fuel. Although the possibilities of developing the output of crude oil in the Soviet Union are enormous, they have their technological-production and economic limits. As regards the latter, the question of profitability arises.

The deposits of crude oil in such countries as Algeria, Iran and others are much more efficient. This means that the investment and exploitation outlays per unit of output are lower. So, when developing cooperation within the CEMA it would be purposeful to consider the possibility of initiating and increasing the import of crude oil from the outside.

Of course, this problem is extremely complicated. An import of large amounts of oil (tens of millions of tons per year for the CEMA countries as a whole) from the "third" countries creates a number of political, economic and organizational-technological problems. However, appropriate solutions should be worked out because the economic development of our countries requires it.

"Petroleum: Policy of Equality of Rights and Policy of Plunder"
Neftyanik, 7, July 1966, pp. 34-37

In the modern era of transition from capitalism to socialism, when, speaking with Lenin's words, socialistic, bourgeoise and even sub-bourgeoise governments simultaneously exist on our planet, the problem about mutual relationships between them is one of the main problems in all world politics.

Two worlds, socialism and capitalism, objectively give rise to two world politics. The motto of socialism is the world and social progress equality and mutual benefits and international assistance in liberating nations from colonial domination. Capitalism is attempting to dictate over other countries, to enslave them and to envelop them with chains of neocolonialism.

The expansion of foreign trade relations and economic cooperation by the Soviet Union with young developing countries of Asia, Africa and Latin America is provided for by directives of the Twenty-third Congress of the CPSU in the Five Year Plan of development of the USSR national economy during 1966-1970.

A tendency toward trade between countries is determined by objective economic principles, since under conditions of international separation of labor foreign trade makes it possible for each country to get full and practical use of its resources, and in the political plan it creates a healthy basis in the world between governments with a different overall structure.

Although machines and equipment make up more than half of the USSR's export to developing countries, there is a large amount of fuel, crude oil and oil products, chemicals and other goods being exported. Trade relations with the Soviet Union help young governments secure their economic independence, proceed on the road of industrial progress, and increase the prosperity of nations.

Soviet petroleum and other commodities, feeding into fifty young governments, are paid for by their own traditional export commodities including cotton, wool, hides, nonferrous metals, rubber, rice, citrus fruits, tea, cocoa--beans, coffee and certain products of the national industry. For example, the trade agreement between the USSR and Cyprus for 1965-67 provided for a three-fold increase in turnover of merchandise with important items being crude oil and mazut. At one time the supplies of Soviet petroleum products gave the Ceylon Government an opportunity to create a national oil corporation and to put an end to the domination of foreign oil monopolies in the country. The trade agreement for 1965-67 provided for an increase in the volume turnover of goods by two times in comparison with 1964. In this agreement there was also provided a further increase in supplies of Soviet oil products.

In the new Five Year Plan, as noted in the directives for development of the national economy of our country, the Soviet Union will strive to expand foreign trade with industrially developed capitalistic countries, which will give rise to their own disposition.

SELECTED READINGS

Even before the Second World War the Soviet Union was one of the main exporters of oil. In the second half of this century it has resumed exportation of oil on a large scale. At the present time the large trade partners of the Soviet Union are Finland, England, Italy, Japan, France and West Germany. Crude oil and petroleum products, bituminous and anthracite coal, wood and other commodities, in which a shortage is being experienced in most of the Western countries, make up the major share of Soviet exports. We also export machines and give licenses for the use of Soviet technological processes. In return our country buys in these countries the most diverse commodities such as equipment for plants and installations for chemical, petroleum refining and other branches of industry.

In the 1965-69 trade agreement between the USSR and France, the turnover of goods, which is to increase by 70 percent compared with the corresponding previous period, it can be noted that Soviet exports of crude oil and petroleum products stand foremost. In trade agreements signed between the Soviet Union and Japan, Switzerland, Italy, Denmark, Australia, Finland and other countries, Soviet crude oil and oil product exports occupy a central position.

In recent years in the countries of Western Europe the demand for crude oil has undergone a rapid growth, but not having enough reserves they have had to resort to importing it from other countries. Some time ago the main supplier of crude oil consisted of the member countries of the International Oil Cartel in the Near East, North Africa and Venezuela, but in recent years independent firms, particularly in Italy and West Germany, have begun to actively compete with them. The cartel introduced the 50-50 principle where it equally divided the profit from extraction of oil with the country where it was extracted. But this was a predatory principle. Using it the American companies received, for example, 52 million Libyan pounds in 1962 of pure profit from the export of Libyan oil while Libya received only seven million pounds. Algeria receives 250 million francs under the same principle and the French companies earn 750 million francs in the transportation of crude oil and sale of oil products.

The ENI Italian company introduced a new concept -- 75-25 -- and with this opened doors into developing countries. In order to increase its competitiveness, this company, in an agreement with the USSR which was signed for a mutually beneficial base without any political stipulations, will receive more than 25 million tons of crude oil from the Soviet Union. All attempts from the hands of the International Cartel to destroy the Italian oil company failed. Final removal from the international market of this ominous intermediary, which is the International Cartel, would be economically advantageous for all nations, would facilitate approval of peaceful national ties between them and improve the entire international situation.

Trade and economic relationships between the USSR and developing capitalistic countries were based on mutual benefit and are an important element of establishing trust between countries. In developing her trade relations with other countries the Soviet Union proceeds from the directions of V.I. Lenin in that without commerce "for peaceful coexistence there would either be no foundation or it would be extremely difficult." (Vol 33, p. 348) Therefore it is natural that the USSR, successively leading world politics, gives all in order that international trade relationships were built on a base of mutual benefit and full equality of rights. In this connection the approval of reactionary circles of imperialism sounds more strangely as if the USSR, increasing the volume of oil export to other countries, will create some competition. More than this, they intimidate the Soviet "petroleum menace" by causing constant pressure on governments of both oil supplier and consumer countries.

As they say, lies have short legs. The third session of the Council on Commerce and Growth under the United Nations, meeting from January 25 to February 17, 1966, in its resolution noted that commerce and economic cooperation between socialistic governments and developing countries of Asia, Africa and Latin America are being substantially expanded. Simultaneously, a number of delegations have blamed discriminatory politics in commerce, carried out by the US, for the deep break of resolutions of the Geneva Conference.

Vast imperialistic monopolies with the aid of any manipulations artificially lower prices on stable products sold by young governments. According to UN data, as a result of similar operations, the developing governments, up to 1970, will annually lose seven billion dollars. The monopolies are continuously raising prices for their own products. This is pure robbery of the national riches of the nations of Asia, Africa and Latin America. A paradoxical situation is created: many countries have significantly increased production and export of their traditional commodities; meanwhile their revenue from foreign trade has remained low.

Policies carried out by imperialistic monopolies present much danger for small governments and nations throughout the entire world.

It is interesting how oil and gas, as the most advanced forms of fuel, rank so high in the economics of developed capitalistic countries in our time. This, of course, is good. But it is too bad that the imperialists do not strive to solve the oil problem on the basis of equality and mutual benefit.

For example, 73 percent of the total demand for energy in the US in 1962 was supplied by oil and gas, which is clearly understood, since out of the 23 industrial billion dollar corporations 11 are oil companies which had 60 percent of the total profit of the giants and the same portion of profit which was received by American capital abroad. Possessing the basic economic power, the oil monopolies of the US are the most aggressive force and the most violent enemy of progress and independence. It is known that the primary regions of oil and gas extraction in the capitalistic world, besides the US are the Near East and Latin America, and now North Africa. By this it can be explained that the imperialists are striving to establish, by any means, control over these regions. They carry bloodshed and poverty to the nations of these countries. The chief role in this policy belongs to the US as a world political policeman and exploiter of nations. If investments of United States oil monopolies amounted to 1.3 billion dollars in 1945, then in 1965 they exceeded 10 billion. Moreover, the US extracted hundreds of millions of tons of oil in other countries and because of this the multibillions of profits of the monopolies grow annually.

Venezuela, for example, they quite correctly call the rich land of peasants because here in the presence of fertile earth, a mild subtropical climate, vast reserves of iron ore, coal, nickel, gold, zinc, copper, bauxite, manganese and especially oil, give wealth to Venezuela while its 8.5 million inhabitants lead a half-starved existence.

The first oil well in Venezuela was brought in back in 1913 and now in this country 180 million tons of oil are extracted annually. With respect to the extraction of black gold, Venezuela ranks third in the world. Yet oil has not brought prosperity to the peoples of this country, but misfortune and suffering instead because the owners of oil riches were foreign monopolies: Standard Oil of New Jersey operating under the name of the Creole Petroleum Corporation, and the Anglo-Dutch Shell Oil Company. They changed Venezuela into a supplier of black gold, made her their victim, bound her up by one-sided agreements, and imparted a one-sided nature to Venezuelan economics.

According to calculations of the prominent Venezuelan economist Salvador de la Plaza, there has been more than 50 billion dollars worth of oil extracted in Venezuela during the first half of the century. But pennies are given to the lot of Venezuela while the bulk of profits rest in the safes of Wall Street and First National City Bank. Foreign oil monopolies annually receive 35 to 40 percent of the revenue from invested capital which amounts to almost five billion dollars. Monopolies of the US alone annually receive 450 million dollars from exploitation of oil deposits. Another line of economic theft is arbitrary "regulation" of prices. During the last five years the total losses, which Venezuela suffered from the lowering of prices, will exceed four billion dollars. Having received enormous profits, the oil monopolies do not shun those "trifles": at the beginning of 1966 the government of Venezuela charged American oil concerns with the fact that they avoided payment of taxes and demanded payment of 115 million dollars.

Having enslaved the country economically, Standard Oil of New Jersey has, for more than 50 years, determined the external and internal politics of Venezuela. The nations of Columbia, Peru, Bolivia, and a number of other countries have been in this same position. The oil monopolies of the US intervene in the internal business of the countries of Latin America. They not only maintain supremacy of the bourgeoise landowner oligarchy but also organize plots and state revolutions, plant strong-arm dictatorships, incite persecution and persecute patriots. And, therefore, it is fully a conformity that a movement of the masses is taking place in the countries of Latin America for economic independence of their countries. They do not want their countries to be an American "national forest."

SELECTED READINGS

Not far behind the US in the pillaging of other countries are imperialists of a lower rank. Let's take, for example, the richest oil region in the world. In 1964, the extraction of black gold in the Near and Middle East exceeded 400 million tons. The oil extracted here is one of the sources of vast profits for English imperialism. The Anglo-Dutch "Shell" monopoly and the English "British Petroleum" have a third of the extracted oil in this region. According to official figures, here alone pure annual-profits of the English companies exceed 250 million pounds sterling. And Shell Oil Company, at the start of 1966, reported to its stockholders that its revenues from petroleum sources in countries of the East amounted to 1.2 billion pounds between 1954 and 1964 while new capital expenditures amounted to only 150 million pounds.

British imperialism, prior to the war, strived to maintain its own position by all means in this part of the world. In connection with the fact that the neocolonialistic shield of the Southern Arabian Federation largely failed, British aircraft bombed villages in this region and soldiers carried out colonial military expenditures in Oman, in the Bahrain Islands, in Aden and other points. British imperialism has been attempting to undermine the government of Yemen for more than three years and the other Arab nations are not sitting still.

Having attempted to save the exhorbitant interests of oil colonialism the managing circles of England are now seeking the support of their old American partner. They are also striving to lean on feudal reactionary circles in these countries. At the end of 1965 emissaries of the king of Saudia Arabia conducted conversations in the Magrib countries for the purpose of creating an Islam Pact. There is a rumor about reactivation of the Eisenhower Doctrine to fill the vacuum in the Near East and about realization of the oldest Anglo-American dream about creating a union which would replace the Baghdad Pact and its heir--CENTO, for the struggle with Arabic nationalism and neutralism. President Nasser of the UAR called the Islam Pact an imperialistic pact. The forces of reactions will undoubtedly cooperate with imperialism because they fear the growth of an Arabic revolutionary struggle which presents a threat to their interests. Imperialism feels that an Arabic revolution is a danger to its oil monopolies. And this is understood: the securing of independence and development of Arabic governments and, to begin with, the UAR, going by a path of progress, will objectively signify the weakening of American and British positions in this region.

In our time the struggle for oil has become an important bond between nations in their struggle for national independence, opposed to the economic expansion of imperialism. In whose hands will the innumerable oil riches be found; in the very hands of the governments of Latin America, the Near and Middle East and their nations or in the hands of foreign monopolists? There is the key to the solution of many indigenous problems.

The experience of Latin America shows that in countries such as Mexico, Argentina, Brazil, and Chile, where state oil companies were formed, more favorable conditions for national development were created.

Pemex, as the government branch of Mexican industry, was formed in 1938 after nationalization of American and British oil companies. The oil, which is in complete possession of the government, annually yields approximately 12 percent of receipts in the national budget. The growth of oil extraction reached 129 million barrels in 1965. Almost all extracted oil is refined at domestic plants, primarily for internal consumption. Pemex daily yields a return of 2.5 million pesos to the state. Workers of the oil industry receive a higher wage than workers in other categories. Consequently they consider Pemex as their own creation.

The example of Mexico and other countries convinces the people of a country, where oil is extracted, of the necessity to nationalize the ownership of foreign monopolies in the petroleum industry and in other branches.

In 1965, the Peruvian Communist Party drafted all the progressive and democratic forces of the country to create a unified front to defend oil deposits. It was emphasized that the American monopoly, International Petroleum Company, having seized 90 percent of all oil production, ruthlessly robs the interior of the earth and profits by monstrous exploitation of Peruvian workers. President Illia of Argentina, in spite of political pressure, at the beginning of 1965 abolished oil contracts and began to take American enterprises under control.

Similar proceedings are taking place in other regions. At the start of 1965 the government of Syria issued a decree by which all petroleum and mineral resources were declared as government property. In January, 1966, twenty five oil companies, operating in Libya, were obliged to agree to increased deductions from their profits for use by the Libyan government. Resolutions of the OPEC conference deemed the end of their resistance. Member countries of this organization (Iraq, Kuwait, Qatar, Libya, Saudi Arabia, Iran, Indonesia and Venezuela) decided to not offer any concessions in their territories to those companies or their divisions which refused to obey the new law adopted in Libya.

This example shows that it is possible and necessary to oppose the greedy **appetites** of imperialists by a unified effort of nations.

In connection with the rising tendency toward nationalization of petroleum wealth, American bourgeoise propaganda, as usual, expounds theories which denounce the idea of nationalism. Simultaneously, government companies are continuously subjected to the attacks of American and other oil monopolies who see a large obstacle in them for accomplishing their own plans of neocolonialistic enslavement of peoples.

The Soviet Union has always stood and will stand on the side of liberating nations from colonial dependence. In the counter-balance to aggressive politics of imperialism, which has split the world into closed military, political and economic groups, the USSR and socialistic countries are carrying out a relentless struggle for peace and peaceful coexistence of all countries, based on strict adherence to economic and other interest of both large and small nations. It will oppose the imperialistic policy of neocolonialism for a policy of peace and social progress.

"The Sea, the Fleet, and Politics"
Izvestiya, April 20, 1966, p. 3

(An interview with the Minister of the Merchant Marine of the USSR, V. Bakayev.)

(A correspondent of the press agency, Novosti, put some questions to the Minister of the Merchant Marine USSR, Viktor Bakayev. The answers to these questions are presented for the readers' attention.)

<u>The 23rd Congress of the CPSU, of which you were a delegate, adopted directives for the Five Year Plan for development of the national economy of the USSR for the years 1966-1970. Comment, if you please, on the portion of the directives touching on the Marine Fleet.</u>

The increase in industrial production of the USSR by one and a half times, as contemplated in the Five Year Plan, and the further development of the external economic ties of the country will require a considerable expansion of sea transport. The freight turnover will increase within the five years from 208.8 billion ton-miles to 380 billion ton-miles.

It is our intention to increase the tonnage of the Soviet fleet by 50 percent. Each year in the course of the five year period our fleet will receive new ships from our country's ship-building plants and from the wharves of foreign lands with a total of tonnage of a million tons.

The qualitative makeup of the Soviet fleet will be improved considerably. In 1965, 59 percent of our ships had a speed of over 14 knots; in 1970 this will be over 71 percent.

Great changes will take place in Soviet ports; their productive capability will be increased 40 percent, the level of complex mechanization of work in loading and unloading will rise 90 percent. We can now **say** that in saturation level of mechanization lightening human labor and speeding the freight-transfer process, the Soviet ports take first place in the world.

<u>Can you compare the tempo of growth of the Soviet marine fleet with the tempo of growth of the world fleet</u>?

If you are speaking of growth in tonnage, then the following comparison is interesting: during the past fifteen years the tonnage of the world fleet increased 89 percent, that of the Soviet fleet - 4.2 times. An analysis of orders placed for ships enables one to predict that the world's fleet will increase their tonnage during the five year period by 20 to 25 percent, the Soviet by one-and-a-half times.

<u>What, in your opinion, are the prospects for shipping?</u>

Briefly speaking favorable. Judge for yourself the world population in 1965 reached 3,300,000,000 persons, and by 2000 it will have doubled. This will call for the inevitability of quick growth in vital resources and will greatly increase international trade. Consequently, marine traffic will also increase sharply. The prospects closer at hand are also favorable. Our estimates show that the total volume of world sea traffic, which in 1965 amounted to about 2 billion tons, may rise by 1970 to 2.5 billion tons, or possibly more. Figuring on this estimate and also on the prospects of the Five Year Plan for the development of the national economy of the USSR, we shall increase the tonnage of our own fleet.

<u>Will you impart the assumptions regarding the development of international cooperation in the sphere of navigation?</u>

In recent times it has seemed fitting to devote considerable attention to such prognostications. Analysis of the development of ships and ports, of the maritime policy of individual countries, of commercial relations between ship-owners and shippers shows that the consideration of these questions solely within the framework of existing navigational alliances, conferences, pools, and also technical intergovernmental maritime organizations, does not satisfy the growing demands of the majority of countries. Very many of them are maritime, already possessing or else planning to introduce their merchant fleets. The development of this trend without a doubt will necessitate an examination of the problems of maritime shipping on a broad and imposing international basis. The basis for such an examination of shipping problems may be the Committee on Shipping of the Council for Trade and Development of the UN.

If up to this time international shipping organizations have concentrated on the legal and technical aspects of shipping, then the statute of a new inter-governmental organ will make it possible to embrace much more widely the problems of world sea traffic, including the commercial interrelationships of ship-owners and shippers.

The development of inter-governmental relations itself shows that this is the only possible way suitable for most fully satisfying the interests of all countries which are members of the UN and at the same time of protecting the world against the discriminating actions of certain governments. Ten European nations and Japan have already joined in a committee at the government level for defense against the discriminatory measures in shipping established by the US Government in a unilateral manner.

How do you stand toward bilateral relations between countries in regard to shipping?

The highest form of good relations seems to be agreements between countries in merchant shipping. In them the broadest complex of questions can be solved. As is well known, the conclusion of such an agreement between Great Britain and the USSR was established at the time of the recent visit of Prime Minister Harold Wilson in Moscow. The maritime organs of the USSR plan to develop broadly the bilateral ties on a basis of equality and mutual advantage.

You mentioned the Committee of Ten and the discrepancies between American and West European shipping organs and companies. Please tell us about it in more detail.

Western maritime shipping today is a field of activity which to a great extent finds itself in the hands of private entrepreneurs. Here, accordingly, very many contradictions and discrepancies of every kind are arising. The chief discrepancy in recent years which evoked an aggravation of relations between the US and Western European ship-owners and reached the level of government relations, was the American ultimatum with a demand for raising the freight rates 15 percent on the passage of freight to the US. Such a rise in rates would inevitably lead to a lowering of competition by European shipping companies, an increase of prices for European goods in the American market, and would cause a noticeable loss of exports from these countries to the US.

Such action by the US has caused an outburst of indignation in European maritime countries. In order to defend the interests of the European ship-owners of England, Belgium, Holland, Greece, Denmark, Italy, Norway, the Federal Republic of Germany, France and Sweden, a special committee was formed, known in the world as the Committee of Ten. Later Japan also joined this committee.

SELECTED READINGS

The US lost the battle with the united forces of these countries and had to give up.

<u>Of what sort are the interrelations between the USSR and the US in regard to shipping?</u>

Such interrelations for practical purposes are non-existent. Soviet ships do not call at American ports because of the awkwardly complicated conditions established by government agencies of the US for visiting of its ports by ships under the flag of the USSR. American freight and passenger ships rarely visit Soviet ports; however, in the past year there were some tens of them among us. In the end, utilizing the right of reciprocity, we might have refused to admit the US ships to our ports. But, to speak mildly, why copy the unreasoned measures which were introduced, obviously, only for purely propagandistic purposes?

The United States, if it wished to trade with the USSR, and to judge from the report of the "Miller Committee" and the recent report of D. Rusk, that would be advantageous to them, willingly or unwillingly will have to abstain from special demands on Soviet ships on the occasion of visits to American ports.

<u>"Adventures of the Oil Kings"
Izvestiya</u>, April 19, 1966, p. 2

The numerous pleas of imperialism's advocates, as though there were not industrial groups who were interested in the construction of an international arrangement, can fool no one at present. Among the military conflicts of the last decade one can hardly mention one in which the oil empires of the US or England were not involved, directly or indirectly.

The Suez adventure in 1956, the intervention of the US and England in the Near East in 1962 and other war areas were provoked and to a considerable degree subsidized by the oil monopolies of the West for the purpose of protecting their imperialistic control over the Near East and North African sources of oil.

It is characteristic that the first call to organize an armed intervention against Cuba came from the lips of the uncrowned king of oil, Nelson Rockefeller, Governor of New York, and the first upon contribution into a fund for the subversive activity against Cuba was made by the oil monopoly, the Texas Oil Company. As a basis of such "primacy" is the fierce hatred of the American oil robbers for the island of Freedom which in 1960, upon rooting out the residues of imperialistic oppression, nationalized the refineries of the Rockefeller's ESSO and Texas Oil. Additionally, it appears very dangerous to them to have the close neighborness of that heroic island to the "second home of Rockefeller", Venezuela, where yearly the American monopolies extract more than 100 million tons of "black gold." Finally, the malicious seeds scattered by the oil robbers later erupted into the dangerous Caribbean crisis.

The quiet of the Arabian peninsula is periodically disrupted by the explosion of English bombs and shells, rifles and automatic weapons. Here again the smell of oil is mixed with the smell of powder. The Arabian peninsula is the richest storehouse of "black gold." That is why the English monopolists put forth their tentacles ever deeper here. In 1964, for example, about 160 million tons of petroleum were produced in the entire capitalist world under the control of English capital, of which 120 million tons were in the Middle East, including about 60 million tons in the Arabian peninsula.

But what explains the unhealthy increased interests of American imperialism in the southeast section of the Asiatic continent? Among other reasons there should be noted evidently the fact that in the entire vast expanse of the Eastern hemisphere to the east from the Persian Gulf only this region is relatively rich with oil. That is why the American monopolists produce about 60 percent of it here.

A close association exists between the western oil monopolists and the military establishments of the imperialistic states. It is based on the fact that the military assist the oil robbers to hold entire nations under their obedience, and these in turn generously supply the imperialist armies with fuel and lubricants. Yearly, for example, the American oil industry sells to the Pentagon more than 40 million tons of petroleum products (which, incidentally, is equivalent to their requirements in all of France) for 1.2 billion dollars. Also criticism is often heard in the Congress of the US on the action of the Pentagon which purchases petroleum products to please the monopolies, frequently at increased prices.

The press of the American oil industry, distinguished by unparalleled cynicism, has never hidden its rejoicing over the increased demands for petroleum products in connection with this or that military conflict. On the pages of the Texas journal "Oil and Gas" they are still singing praises in honor of the Korean adventure of the US, since it aided the oil magnates to obtain additional profits in supplies of fuel not only for the ground forces but also for the sea transport ploughing the endless space of the Pacific Ocean.

During the Caribbean crisis "Oil and Gas" stated with satisfaction that due to the activation of the sea forces in the region of the Gulf of Mexico, the purchases of petroleum products by the Pentagon, and consequently the profits of the oil kings, increased by 14 percent.

In July 1965 a Pentagon representative, General Senter, in a speech to the National Petroleum Council, gladdened the petroleum business with the report that in Viet Nam in the four previous months the American military forces' requirements for petroleum products had increased 300 percent, and the intensity of the sea shipments -- four times. Considering the extended nature of the war in Viet Nam the oil magnates have begun the construction of oil refineries in this area.

The western oil monopolies have extracted no small profits from the conflicts which the imperialists succeed in provoking between their former colonies and semi-colonies.

In 1918 V.I. Lenin, in "A Letter to American Workers", speaking about profits accumulated by the imperialists during the First World War, wrote, "For each dollar -- a lump of mud from the "lucrative" military supplies... For each dollar drops of blood..." Since then almost half a century has passed. But, as attest the facts presented above and many other similar facts, the nature of imperialistic monopolies has not changed. Just as before, they attempt to extract profit from the sufferings and misfortunes caused by war on mankind.

"Oil, Trade, and Politics"
International Affairs,
by B. Rachkov

Since the middle of the 1950s, the world has repeatedly found itself faced with the threat of world war, arising directly or indirectly and to a greater or lesser extent, from the machinations of powerful imperialist oil concerns.

In October 1956, following the nationalisation of the Suez Canal by Egypt, and in July 1958, after the overthrow of the monarchist regime in Iraq, the leading imperialist Powers mounted armed intervention in the Middle East for the immediate purpose of defending their oil interests in that area. In July 1960, it was the nationalisation in Cuba of plants belonging to the U.S. oil monopolies that caused Nelson Rockefeller, the uncrowned oil king, to issue his call for military intervention in Cuba.

The colonial war in Algeria lasted for almost eight years: it was waged against the Algerian people by the chief NATO countries, which were impelled by their grasping for the Sahara's natural resources, mainly oil. Conflicts over "black gold" brought about the tragedy of Bizerta in July 1962, the armed conflict between Algeria and Morocco in October 1963, the coups d'état in Iraq in February 1963 and in Brazil in April 1964, a series of terrorist acts in America, Europe and Asia, military operations in the south of the Arabian Peninsula, reactionary drives in India, Ceylon and Indonesia and various other events on different continents.

World politics smacks more and more of oil, and this is a reflection above all of the increased importance of that fuel in the world economy.

In the last decade, the consumption of oil products in the capitalist countries doubled and in 1965 stood at 1,300 million tons.[1] At the same time, there were marked changes in the

---

1. See World Petroleum, September 1965, p. 32.
Annual Refinery Review, 1965, p. 29.

SELECTED READINGS

capitalist energy balance. In the early 1950s, oil was increasingly displacing coal as a source of energy; by the early 1960s, it was definitely the chief source of energy. In 1955, the share of coal in the capitalist energy balance was 48 per cent, and that of oil 35 per cent; by 1965, the figures were 35 for coal and 45 for oil.

The petrochemical industry, whose importance for modern economy can hardly be overrated, has made its first large strides. It may be a young industry, but it already consumes up to 3 per cent of the oil extracted, that is, roughly as much as goes to agriculture, one of the oldest consumers. Between 1955 and 1965, the manufacture of semi-processed petrochemical products in the capitalist world increased from 12 to 40 million tons. In 1954, gross investment in the petrochemical industry of Western Europe totalled $240 million; 10 years later the figure had gone up to $3,000 million.

As oil becomes an ever more important factor in the capitalist economy, the capitalist countries become increasingly dependent on external markets for their supply of this vital commodity. An average of only one in ten countries has the necessary oil resources at home, while the others have to turn to external markets. In 1965, 56 per cent of the oil consumed in the capitalist countries was imported, as opposed to only 45 per cent ten years earlier. Since 1962, oil and oil products have accounted for more than half of all the international marine freight.

In 1955, gross investments in the oil industry in the capitalist world came to $63,000 million; by 1966, they were estimated at $150,000 million. In 1955, only two oil companies -- the American Standard Oil of New Jersey and the Anglo-Dutch Royal Dutch Shell, known as ESSO and Shell, were among the 10 industrial joint-stock companies with the largest gross income; in 1965, two other U.S. oil companies, Socony Mobil and Texaco, joined the group. The oil industry has no rival as far as concentration of capital is concerned. In the early 1960s, the assets of the seven biggest oil monopolies operating internationally were greater than the total assets of 24 automobile, steel and electrical engineering firms.

The oil tycoons, never noted for their modesty, have been shouting from the roof tops that their business is a "bulwark of free enterprise" and a "phenomenal stimulator of the modern economy". But one of the leading oil journals in the U.S.A. complains that "in the newly emerging countries of Asia and Africa as well as the well-established economies of Europe, the international majors are under pressure from political forces.... For a major oil company which is exploring, producing, refining, and marketing in perhaps 100 countries, political problems can assume gigantic proportions overshadowing all others."[2]

It was a sharp aggravation of the basic contradictions of capitalism, specifically of its petroleum market, that set in motion the political forces of which the oil tycoons complain.

Let us take, in the first place, the contradiction between the oil-rich developing countries in the East and the West and the International Oil Cartel. The members of the Cartel--five American monopolies, one British, one Anglo-Dutch and one French--have been exploiting the oil-rich countries under onerous concession agreements. For the peoples of Iran, Iraq, Saudi Arabia, Indonesia, Venezuela and other oil-rich countries, the struggle for their political and economic independence is largely becoming a matter of fighting the domination of the imperialist International Oil Cartel.

However, the nationalisation of the oil industry in Iran, undertaken in the early 1950s, is the only and unsuccessful attempt by an oil-rich country since the Second World War to eliminate imperialist exploitation by expropriating the foreign monopolies.

The reason for this does not lie only in the acute shortage of national personnel or, in some cases, the readiness of the ruling circles to compromise. The fact is that on its way from the oil well to the consumer, oil passes through a series of complex enterprises, the chief of which are the oilfields, the means of transportation, the refineries and the marketing organisations. But the monopolies have seen to it that the oil-rich countries do not possess the full set of these enterprises, but only, as a rule, enterprises for extracting crude oil. A country embarking on nationalisation would therefore hold

---

2. Oil and Gas International, April 1963, p. 31.

SELECTED READINGS

only one end of the production line--the oilfields, plus, at best, a few refineries.

Thus, in 1964, the oilfields of the developing countries in the Eastern Hemisphere yielded about 470 million tons of oil, while the capacity of oil refineries on their territory was scarcely above 60 million tons. Consequently, they could process only 13 per cent of the crude oil they extracted. In Venezuela, the proportion of refining facilities to volume of output is not over 30 per cent.

What is more, these countries do not have a tanker fleet to transport several million tons of oil or oil products. The tanker fleet, hundreds of oil refineries, tens of thousands of filling stations and in general the whole of the marketing network are beyond the reach of oil-extracting countries, because they are the property of foreign monopolies, most of whom are members of the International Oil Cartel. Thus, the Cartel is in a position to force a country nationalising the oilfields to curtail oil extraction for a long time by refusing to buy nationalised oil.

These measures are effective only because the oil-extracting countries have something of a "one-crop" economy, with oil either the only or the major source of national income. In Indonesia, Iran, Venezuela, Algeria, Iraq, Saudi Arabia, Libya and Kuwait, oil yields from 30 to 90 per cent or more of the national revenue and foreign exchange earnings. A stoppage of this source threatens the country with economic crisis, financial bankruptcy and political chaos. What is more, the oil monopolies, relying on domestic reactionaries and playing up demagogic anti-Communist ideas, have been trying to disorganise or even to destroy the patriotic forces who favour nationalisation.

Between 1951 and 1953, the International Oil Cartel not only organised the boycott of the nationalised Iranian oil on the world markets, which led to a drop in output from 33 million to 1 million tons and inflicted great damage on the Iranian economy, eventually paving the way for a return of foreign monopoly domination of Iran's oil industry. Besides, it mobilised the domestic reactionaries, who subsequently attacked not only the supporters of nationalisation among Iran's ruling circles, but also and chiefly the patriotic-minded masses of people.

It is noteworthy that when 12 death sentences were passed in the first trial of the patriotic leaders in Iran in September 1953, the shares of the British company exploiting Iranian oil registered a sharp rise on the London Stock Exchange.

In the last few years, the Cartel's positions have been attacked in Indonesia, where it is represented by two U.S. monopolies and the Anglo-Dutch Shell. In October 1960, the Indonesian Government obliged them to surrender to the state 60 per cent of their profits instead of the earlier 50 per cent. Three years later, it put forward a plan by which the monopolies had to transfer their marketing organisation to Indonesian state companies within 5 years, and their oil refineries within 10 to 15 years. In March 1965, the Government announced its intention to take over the management of the refineries in the very near future. A number of Left-wing organisations, including the Central Trade Union Federation of Indonesia, demanded outright nationalisation of the oil industry.

The U.S. oil monopolies responded by asking the White House to stop economic "aid to Indonesia and to take other sanctions against the recalcitrant country. In May and June 1963, a special U.S. mission tried to talk President Sukarno out of the offensive on the positions of U.S. capital. The reactionary U.S. press called for a simultaneous closure of markets for Indonesian oil and gave threatening reminders to Sukarno about the fate of Mussadiq, the Iranian Prime Minister overthrown by American and British monopolies in 1953 for his attempt to nationalise oil.

The recent persecution campaign against the Left-wing organisations in Indonesia, mass killings of progressive leaders, prohibition of the Communist Party, all these facts are very much in line with Western oil tycoons' wishes. Their ties with Indonesian reactionaries have a long history abounding in examples of close cooperation. Not so long ago--in 1958--they gave extensive financial assistance to the insurgent "provisional government" in Sumatra.

Upon the other hand, it is also highly characteristic that the stubborn struggle of the oil-producing countries against the International Oil Cartel is now chiefly assuming the form of establishing state oil companies, whose task is to make use of national manpower and material resources to work the country's oil deposits by organising the complete cycle of extraction, transportation, refining and export of oil products on a national basis.

Of course, the International Cartel of oil imperialism, has been doing everything it can to prevent the establishment or, at any rate, the normal development of state oil companies. In Venezuela, for example, the U.S. monopolies and domestic reactionaries have been blocking the establishment of such a company for two decades; even after its establishment they prevented it from operating for almost two years. It took a seven-year struggle to set up a national oil company in Iraq.

In December 1961, paving the way for the future state company, the Iraqi Government took away from the Iraq Petroleum Company--a branch of the principal members of the Cartel--the unworked sections of its concessions which constituted more than 99 per cent of all the concession territory. At the same time, the Iraqi Government demanded that the company sell it 20 per cent of its stock. This would have allowed Iraq not only to control the activity of the foreign monopolies in the country, but also to take part in production operations of the working concession enterprises, and to take a considerable part of their profits.

Iraq's determination to intrude into the holy of holies of the International Oil Cartel--the sphere of production and distribution of profits--threw the Western oil tycoons into confusion. They viewed the Iraqi Government's policy as a dangerous precedent. The neighbouring countries also began to demand the return of unworked oil-bearing territories under concession to Cartel members. This led to a reduction of the total concession territory in Kuwait by almost 50 per cent in 1962, and in Saudi Arabia, where the concessions covered almost the entire territory, to roughly 1 per cent in 1963.

To make the Iraqi example less attractive, the Cartel resorted to economic blackmail. It froze the extraction of oil and, consequently, Iraq's oil earnings, and increased the rate of output in countries whose ruling circles proved to be more amenable. The oil monopoly press made more and more threatening comparisons with the 1953 events in Iran.

The International Oil Cartel pressure on Iraq was supported by the ruling circles of the imperialist Powers. In late January 1963, the head of the Iraqi Government, Kassem, openly accused the United States of subversive activities on the territory of Iraq.[3] On February 7, Kassem gave his last interview, in which he announced his intention of promulgating within the next few days a law setting up a state oil company. The next day, the Kassem Government was overthrown and he himself was killed.

It soon became generally known that the American intelligence service took part in preparing the coup and exploited political conflicts in the country to ensure the interests of the oil monopolies. As usual, imperialism dealt its main blow at the Left-wing organisations. Thousands of patriots were killed and the oil monopoly press applauded the bloodshed. Three weeks later, a monopoly mouthpiece wrote: "What has happened so far in Baghdad has to be regarded as encouraging to the country's oil industry." Two years later, the same periodical cynically warned that the "IPC companies appear to be relying on this fact [i.e., the coup.--B.R.] to deter other governments from following Iraq's lead".[4]

Nevertheless, state companies are now operating in seven oil-rich developing countries. But in 1965, they extracted only about 7 million tons of oil, and refined less than a million tons. That is not a great deal, of course. The division within the patriotic forces opposing the Western monopolies had so far helped the monopoly tycoons to thwart the operation of the state companies, to blunt the edge of their anti-imperialist activity, and to prevent their use as an effective means of protecting national oil resources.

On the other hand, the working of oil deposits in developing countries by Western monopolies has been growing. Between 1955 and 1964, oil output--chiefly on foreign monopoly concession--in Asia, Africa and Latin America increased from 300 million to 680 million tons, that is, from 40 to 58 per cent of the total extraction of liquid fuel in the capitalist world. In consequence, it is highly important for the oil-extracting countries to use all practical means in their power to improve the terms of concessions and to establish at least indirect control over the activity of foreign monopolies. At present, concrete steps in this direction are being taken by the Organisation of the Petroleum Exporting Countries (OPEC).

3. See Washington Post and Times Herald, Feb. 3, 1963.
4. Oil and Gas Journal, March 4, 1963, p. 115; May 31, 1965, p. 53.

This was set up in September 1960, and now includes Venezuela, Kuwait, Saudi Arabia, Iran, Iraq, Qatar, Libya and Indonesia. In 1964, they accounted for almost 90 per cent of the oil extracted in the capitalist world for export. Through negotiation with monopolies and mutual consultations, OPEC members are working to secure higher concession payments and also to prevent overproduction and a drop in oil prices.

OPEC's tasks may be modest, but the members of the International Oil Cartel are alarmed at this first alliance of the countries they are exploiting. Accordingly they have made OPEC one of the main objects of their pressures and subversive activity. The monopolies try to avoid meeting all the members of the alliance together, and prefer to make wide use of the divide-and-rule principle, by meeting them separately to prevent them from taking joint action. Some Governments are given generous promises, others are intimidated, still others are given hand-outs at the expense of common OPEC interests. As a result, the members of the alliance often find themselves divided and their decisions are at times purely symbolic and of no practical significance.

The chief results of OPEC's activity are the so-called additional agreements with the principal concessionaires and also the agreements on output rates in member countries. The additional agreements, made public in December 1964, provide for slightly higher concession payments, which were to increase the oil earnings of the Middle East countries by 6 or 7 per cent. An increase of $140 million was expected in 1965.[5]

However, the lack of unity within OPEC is one of the reasons why negotiations with the monopolies on the additional agreements dragged on over more than two years and the concessions eventually made by the monopolies were only a third of those OPEC expected. Besides, the monopolies frankly warned OPEC that they would "jealously guard" their "right" to regulate oil output and "resist any attempt by governments to usurp it".[6]

In an effort to undermine OPEC's positions, the monopoly press has been describing its aims as running counter to the interests of the oil importing countries. For example, it has been said that the OPEC demand for higher royalties will

5. See Financial Times, Jan. 25, 1965; Middle East Economic Survey, Jan. 15, 1965.
6. Oil and Gas International, October 1965, p. 39.

inevitably result in higher prices for importers and consumers. But these assertions are quite groundless. In Western Europe, for instance, the consumer pays an average of $11 per barrel of oil products, of which only $0.74, that is, less than 7 per cent, goes to the oil-producing countries in the form of concession payments. The rest--93 per cent--constitutes the earnings of the monopolies and the taxes that go to the treasuries of the oil-importing countries.

Still, monopoly propaganda is making itself felt: until recently the influence of the OPEC has been very slight, although it has been in existence for more than 5 years. The U.N. Conference on Trade and Development held in Geneva from March to June 1964 was very indicative in this respect. Petroleum Times, a mouthpiece of British oil tycoons, wrote with some satisfaction: "Of 2,000 delegates, representing 120 nations, a majority represented oil consumers... Understandably, they kept quiet, when Algeria, Indonesia, Iraq, Kuwait, Saudi Arabia and Venezuela made general statements to promote their case for higher income from oil; contrary to expectations, no debate on the subject of oil prices followed."[7]

Contrary to the desire of the United States, Britain and some of their allies, the Geneva Conference recommended that the United Nations should recognise the international organisations set up by the principal exporters of primary products. OPEC is just such an organisation. In 1962, its efforts to obtain recognition from the United Nations were unsuccessful, but in June 1965, the U.N. Economic and Social Council unanimously approved the establishment of official ties with OPEC.

The period of what might be called OPEC's isolation is over. The struggle of the developing nations for a greater share in the use of their national resources, specifically oil, is now assuming international proportions.

Over the last decade, there has been a sharp aggravation of the contradiction between the Oil Cartel and the capitalist and developing countries lacking oil resources of their own. Consumption of oil in these countries between 1955 and 1965 increased from 200 million to 610 million tons a year. But the supply business remains, for all practical purposes, in the hands of the American and British members of the International Oil Cartel.

7. Petroleum Times, Aug. 7, 1964.

In 1963, the Cartel owned roughly 70 per cent of oil refining facilities and up to 72 per cent of the wholesale and retail marketing organisations in the capitalist countries outside the United States and Canada, and oil supplies to more than 100 countries were effected mainly by American and British refining and marketing firms. Importers have to pay in U.S. dollars and pounds sterling, and with the rise of energy consumption, payments for imports are becoming more and more burdensome.

Both the developed and the developing countries are eager to weaken or even to abolish the U.S. British oil dictatorship. In Italy, Japan and West Germany this urge is stimulated by the attempts of national capital to get into the highly profitable oil business. In India, Ceylon, the U.A.R., Brazil and other developing countries, which have to import oil, the struggle against the domination of British and U.S. monopolies has become one of the chief fronts in their anti-imperialist struggle. Many countries, left face to face with the International Oil Cartel, discard the prejudices of the cold war and establish business ties with Soyuznefteexport of the U.S.S.R.

From 1955 to 1964, the capitalist countries increased their purchases of Soviet oil and oil products from 3.8 million to 31 million tons. What they find most attractive in Soviet oil export is that the U.S.S.R. usually spends its oil earnings on their traditional export goods, a fact which has a favourable effect on the state of their payments balance and the economy in general. The practice of the Western monopolies is to remit the bulk of the oil earnings to banks in the United States and Britain, which has a detrimental effect on the payments balance of the importing country.

The members of the Cartel, especially the Americans, who hate to see any source of oil outside their control, have been viewing the development of Soviet oil exports with growing animosity. The more so as the mutually advantageous terms of the Soviet liquid fuel trade, and the help that the Soviet Union has given to a number of developing countries to set up their own oil industry, are in sharp contrast with the exploiting attitude of the imperialist oil monopolies. In an effort to reinforce its positions, the Cartel has been trying to scare the man in the street with the "Red menace" bogey and has been conducting a campaign of vilification, blackmail and terror against all those in the capitalist world who refuse to tolerate the dictatorship of the American and British concerns in the petroleum world.

The attacks of American and British monopolies on ENI, the Italian state oil and gas concern set up in 1953, and its president Enrico Mattei were without parallel in the history of "free enterprise". ENI angered the monopolies by daring to trespass on their "backyard", the Persian Gulf, and by importing Soviet oil.

By 1958, the attacks against the concern, in the form of press cuttings, the texts of parliamentary speeches and all manner of reports resulting from dubious "studies" ran to 23 volumes. Within the following 5 years, it is said to have at least doubled. Business and political circles accused Mattei and his supporters of Communism and fascism alternately. Members of the Italian Parliament connected with foreign monopolies, repeatedly demanded Mattei's dismissal and the liquidation of his concern. Many still remember Mattei's death late in 1962 under suspicious circumstances. Rumour had it that it was due to a secret death sentence passed on the ENI president by the enemies of his oil policy.

Many people believe, and not without reason, that oil monopolies also had a hand in the assassination in 1959 of Solomon Bandaranaike, a founder of the Sri Lanka Freedom Party in Ceylon, whose Government conducted a policy designed to eliminate the domination of foreign capital in the national economy, and chiefly American and British control of oil supplies. For many years, the ESSO, Shell and Caltex monopolies and their agents harassed the Indian Minister of State for Mines and Oil, K. D. Malaviya, who sponsored the state oil company in India and dared to investigate the activity of imperialist oil monopolies in the country.

To protect its interests in the oil-importing countries, the International Oil Cartel makes wide use of diplomatic and external state economic agencies of the United States and Britain. One circular issued by the U.S. oil chiefs openly calls on the United States and other interested countries to bring pressure to bear on the ruling circles of Italy and other Western countries whose oil policy has annoyed the monopolies. Diplomatic pressure in the interests of the British and American oil concerns--ranging from official notes and representations to calls by U.S. and British ambassadors on heads of government--has been applied in Italy, Japan, India, Ceylon, the U.A.R., Brazil, Uruguay and many other countries. Diplomatic action was frequently backed up by a suspension or threat of suspension of American "aid".

In December 1962, the U.S. troops stationed in the Far East made an attempt to boycott the products of the major Japanese national oil company Idemitsu Kosan. The Pentagon act aroused deep indignation in Japan, but on that very day was praised by an American Senator well known for his connections with oil circles.

Something of an international scandal flared up over NATO's intervention in the world oil market in early 1963. The leadership of this aggressive military bloc demanded that the Western countries stop exporting steel pipes to the Soviet Union in the hope of slowing down the development of the Soviet oil industry and the growing oil trade between the Soviet Union and the capitalist countries. The NATO embargo was a painful economic blow at the interests of importers of Soviet oil and was resolutely condemned in many countries, including even Britain.

The U.S. State Department hastened as usual to declare that neither Washington nor the U.S. monopolies had anything to do with NATO's action. However, World Petroleum, a mouthpiece of the oil monopolies, confirmed that "the first demand to use NATO and U.S.A. diplomatic channels to restrict trade in oil between the West and the U.S.S.R. has been made in November 1960 at an annual meeting of the American Petroleum Institute, by President Brockett of Gulf Oil and the then president of Standard Oil Co., N.J., Mr. Rathbone".[8]

In the importing countries, the monopolies and their International Cartel are working hand in hand with domestic reactionaries, frequently sponsoring military plots and coups d'état. One example is Brazil, which has had three presidents in four years.

Petrobrás, a state oil company set up in Brazil in 1954, has been since 1961 a genuine bulwark against the domination of the country's power system by ESSO, Texaco and Shell,

---
8. World Petroleum, May 1963, p.29.

owing to the vigorous support of the patriotically-minded public circles. With the help of internal reaction, the monopoly trio launched a frenzied campaign against Petrobras and this largely aggravated internal political conflicts in Brazil, as a result of which two presidents were forced to leave the country in quick succession.

The decree signed by President Goulart on March 13, 1964, nationalising private oil refineries whose owners had become factual agents of oil imperialism, was annulled after the April 1 coup. However, the oil monopolies failed to disrupt the business relations between Petrobras and Soyuznefteexport, which had been established in 1963 and which met Brazil's vital economic needs.

The conflict in the United States between the oil industry and other oil-consuming industries, which was sharply aggravated in 1961-1963, left its mark on world politics. The fact is, that U.S. oil industrialists, having secured artificial restrictions on the import of cheap oil into the United States, are keeping oil prices at home $7 a ton higher than abroad. Besides, 27.5 per cent of their net profits are exempted from taxation by virtue of the law on depletion allowance, which, according to many Americans, should have been abrogated in the 1930s.

Some capitalists object to these privileges since a considerable rise in the cost of power in the United States causes a rise in the cost of manufactured goods making them less competitive on the external market and creating additional difficulties for the U.S. payments balance. The Financial Times noted that, "In the long run American manufacturers may be unable to withstand international competition unless their fuel costs are kept as low as possible".[9]

The bulk of energy in the United States is being consumed in the country's north-east, with most of the losses from the high cost being borne by the states of New England, which have no fuel resources of their own. They have been sharply criticising the U.S. oil business for many decades. Year in year out, Senators and Congressmen from these states have accused Washington "of supporting the price of Texas crude oil at the expense of New England economy".

9. Financial Times, Dec. 12, 1961.

Prominent among them was John Kennedy, then Senator from the State of Massachusetts. Elected President in 1960, Kennedy put forward a programme in 1963 providing for higher taxes on industries enjoying "unfair or unnecessary preferences", that is, above all, the oil industry. It was to increase its tax payments to the Treasury by roughly $280 million a year. The Kennedy Administration had also prepared a programme for increasing imports of cheap oil into the United States, which was to have reduced the annual cost of liquid fuel to consumers, and, consequently, the profits of oil companies, by almost $3,500 million.

In 1963, the U.S. oil business mustered its forces and stiffened resistance to Washington. Texas Senator Tower promised to do everything in his power to prevent the Kennedy project from being implemented. The President of the American Petroleum Institute, the largest association of oil industrialists in the United States, said frankly: "The industry has no choice but to resist such proposals in every proper way."[10]

The American oil businessmen serried their ranks under the banner of outright anti-Communism and demagogic clamour about the "Soviet oil menace", thus infecting international economic relations at the beginning of the decade with the cold war virus.

Kennedy's assassination threw the supporters of his programme into disarray, at least for a short time. It had extremely important international repercussions and gave greater influence in Washington to the oil monopolies, the most aggressive force of imperialism.

No group of capitalist monopolies has given so many assurances of its non-involvement in politics as the spokesmen of the oil business. Operating simultaneously in dozens of countries with different state systems and frequent changes of government into the bargain, each of the oil empires endeavours to gain for itself a reputation of political neutrality. But the facts given above show that there is hardly a group of monopolies that has displayed as much destructive energy, craft, intrigue and special inclination to international piracy as the oil business in its efforts to secure control both of domestic political relations in various countries and of the principal factors of world politics.

---

10. Platt's Oilgram News Service, Jan. 25; July 2, 22; Nov. 15, 1963; March 2, April 7, 1965; World Petroleum, Dec. 1960, p. 4.

"From Seventh Place To Third"
Vneshnyaya torgovlya, 1, January 1966, p. 10-11
by E. Gurov, chairman of Soyuznefteexport

Concerning the tempos of expansion of Soviet oil exports it should be noted that not so long ago, in 1959, the first year of the Seven Year Plan, the export of oil reached 25 million tons; thus, in seven years oil exports have more than doubled. Our profits correspondingly have increased.

At the present time Soyuznefteexport sells more oil in the international market than was produced in the USSR in 1954. The annual increment in Soviet oil exports exceeds the volume of our oil sales in 1932 -- which was a record year for the export of crude oil and petroleum products from the USSR in the prewar period.

The past year was the first full year of operation of the Druzhba crude oil pipeline along its entire route - from the Volga steppes to Poland, the GDR, Czechoslovakia and Hungary. Certainly, not without the influence of this great TransEuropean trunk pipeline the increment achieved in the past year in Soviet oil exports to Socialist countries was double the usual.

In the past year Soyuznefteexport continued the search for new markets and contractors in capitalist countries. In 1965, for the first time, a long-term contract was concluded for the sale of a significant quantity of crude oil to the major government oil company of France - Compagnie Francaise Des Petroles.

Readers of the journal Vneshnyaya torgovlya also may be interested to learn that during the past seven years Soyuznefteexport, in terms of the volume of oil sold in the international market, has risen from seventh place to third place, behind only the American company "Standard Oil of New Jersey" (ESSO) and the Anglo-Dutch "Royal Dutch Shell" (SHELL).

Despite the unfavorable conditions of the oil market, in the past year our prices held, and in a number of instances an increase in prices was achieved in contracts with capitalist countries. We strengthened our efforts to find more rational transport of our goods.

In the new year Soyuznefteexport first of all will make considerable effort to fulfill the new major contracts, concluded in the past year, and second - to concentrate maximum effort on the further raising of the effectiveness of the Soviet export of crude oil and petroleum products. In 1966 we will strive for a further reduction in transport costs. We must give still more attention to the study of market conditions.

"Oil: Active And Passive"
Izvestiya, December 19 1965, p. 3
by E. Gurov, chairman of Soyuznefteexsport

(Statistics have been released on foreign trade of the USSR during 1964. For the Soviet export of crude oil and petroleum products, this was a jubilee year: 40 years ago the young Soviet Republic sold its first significant quantity -- at that time -- of liquid fuel -- 0.7 million tons. In the past year the USSR exported 57 million tons of crude oil and petroleum products valued at about 850 million rubles.)

The appearance of Soviet oil on the world market in the 1950's, after an extended absence, brought about by the war, was greeted by the large capitalist monopolies more with skepticism.

But these non-well-wishers quickly changed their tune. The wounds of the war having healed, the USSR proved to have the means to offer to other countries such a quantity of oil, from the proceeds of which it would be possible to buy not only coffee, but also whole industrial enterprises. From the underevaluation of our capabilities in the field of international oil trade, American and other western monopolists went to the other extreme. In their reports there appeared a note of panic. "For the American oil companies - informed the Associated Press - the appearance of the communists in the oil market is as unexpected and distasteful as was the first sputnik for the US Department of Defense."

Inasmuch as the mutually advantageous conditions of Soviet oil exports underscore the exploitator character of the activity of the Western and especially of the American oil monopolies, the latter began to regard the export of oil from the USSR as a menace to its own imperialistic position. Declaring these exports as "the main armament of the economic arsenal of the Kremlin", they carried out an energetic struggle against Soviet oil exports to capitalist countries. With this purpose in mind, a rich arsenal of resources was mobilized - from slander from the US Congress and pages of the bourgeoise press to economic and political pressure on those countries buying Soviet oil.

The propensity of various countries to mutually advantageous cooperation, however, proved to be stronger than the blackmail of the oil companies. In the ten-year period from 1955 through 1964, the export of crude oil and petroleum products from the USSR increased from 8 million tons to 57 million tons, including a growth in exports to capitalist countries from 3.8 million tons to 31 million tons. Among the more than 50 importers of Soviet crude oil and petroleum products, the largest is Italy, which in the past year imported from the USSR about 8 million tons of oil.

In terms of value, crude oil and petroleum products are second in the total volume of exports from the USSR only to machinery and equipment. In relation to the export to various capitalist countries, then crude oil and petroleum products are in first place.

We believe that prices must not be artificially raised or lowered. The discriminatory pricing policy carried on by the oil monopolies of the West is alien to us. Monopolies, in order to force out competition from a given market, often artificially reduce prices there, and strive to recover their losses by non-excusable raising of prices in those markets where their control is secure.

At the present time in the world oil market a number of unfavorable factors are exerting influence. Among these first of all there is noted the presence of a large surplus of producing capacity in the capitalist oil market. In oil production, this surplus exceeds 30 percent, and in refining it approaches 20 percent. (It is necessary of course to keep in mind that this excess is relative. It stipulates that imperialism is able to keep many countries backward. If the per capita consumption of oil in the developing countries were equated with the West European level, the whole capitalist oil industry could not meet the increase in demand.)

Excess capacity belongs chiefly to the major oil companies. Consequently they originated a bitter competitive struggle. But for some time past imperialism increasingly has enlarged surplus producing capacity because it regards this as its unique antidotal against the national independence movement in the Arabian countries, in Iran, and in Venezuela. The chairman of the Mellon oil monopoly "Gulf" observed not long ago, that "rumors of the nationalization of the oil industry are disturbing to the stockholders", attempting to soothe the latter by the fact that "Gulf" is establishing in various regions new capacity, which is an "excellent guarantee in the event of loss of any source of crude oil".

The monopolies of the US and England have applied much effort in order to prepare for themselves "reserve positions" in Alaska and in other scarcely populated corners of the world, where a people's movement for independence is not impending. Despite the vast production possibilities in Iran, Iraq, and Saudia Arabia, the monopolists increasingly force oil extraction, for example, in the small Arabian kingdom of Abu-Dhabi, where not more than 15,000 people live. Production having begun here only three years ago, the monopolists already have raised output to 15 million tons per year.

With regard to the excess capacity in the refining of crude oil, its rapid growth, for example, in a number of West European countries, to a large degree was dictated by the interests of the aggressive NATO military bloc.

But excess capacities not only hang as a sword of Damocles over the national liberation movement in the oil-rich regions. They are also the principal cause of the sharply worsened conditions in the world oil market. With surplus production capacities at their disposal, the oil monopolies could not refrain from using them as a tool of competition in recent years. These capacities enable them to produce additional quantities of crude oil and petroleum products, which they sell at reduced prices in order to force out their competitors. As a result, from the end of the past ten-year period the capitalist oil market suffers from overproduction, prices have a tendency to decline. The income of the Arabian and other oil-rich countries from each ton of "black gold" declines from year to year.

Five to seven years ago the chairmen of the major oil monopolies themselves acknowledged the negative influence of the extremely large excesses of capacity on the conditions of trade in crude oil and petroleum products. However, since that time they have changed their opinion. During the past year the leaders of Esso, Shell, BP and other monopolies praised the excess capacities, and the past president of Esso characterized them as "necessary and even good".

It should be remembered that the creation of this "good" already has cost the capitalist world a minimum of 25 billion dollars. With this sum all of the Middle and Near East could have been transformed into a flourishing garden. In its own glory the unproductiveness of expenditures on the establishment of reserve capacities in the capitalist oil market is comparable only with military expenditures. They are being paid for by the consumer masses. The cost of the oil fields and refineries in conservation invisibly is present as an integral element in the approximately 11 dollars which West European consumers pay for a barrel of crude oil through the system of prices for gasoline, heating, illumination.

Ninety percent of all consumers of liquid fuel live in countries not having their own crude oil. In light of this it becomes clear that surplus production capacities harm not only the exporting countries but also the oil importing countries, placing an additional financial burden on them. There is no doubt that western oil monopolies sacrifice the economic interests, both of the countries rich in oil, and the countries not having oil, for their own political ends.

It is in the interests of all countries to strive for normal conditions in the trade of crude oil and petroleum products, eliminating the artificial competition in the world market.

The world oil market is an important part of contemporary international economic life. Oil in many respects is the leading commodity in world trade. Not less than 8 percent of the value of international trade turnover, and more than one-half of all of the freight moved by sea, is oil. The incomes from oil to a significant degree determine the level of prosperity of the peoples of the Near and Middle East, North Africa, and also Venezuela, Columbia, and a number of other countries. There can be no doubt that regulation of relationships in this market would favorably affect the economic development of many countries and the expansion of mutually advantageous economic contacts between peoples.

"On The Determination Of The Economic Effect Of Foreign Trade"
<u>Planovoye khozyaystvo,</u> 11, November 1965, pp. 39-44

Let us examine the economic conditions for the export and import of raw materials and fuel. Even within the limits of the country mining enterprises have dissimilar natural (mining-geological) conditions and technical-economic indicators of extraction of one and the same product. Thus, the cost of extraction of crude oil at the "very best" deposits of the USSR, characterized by the highest yield (per well) and the greatest concentration of reserves, is ten times lower, than at "the worst". The cost of extraction of crude oil at "the best" is several times lower, and at "the worst" several times higher, than for the country as a whole. The cost of extraction of coal in the USSR at the most economic basins was 15 and more times lower, than at coal basins and deposits with poorer mining-developmental conditions.

Inasmuch as in every country the number of deposits of mineral resources, characterized by the most amenable mining-developmental conditions of extraction, are limited and their productive possibilities cannot meet the national economy demands of a given type of raw material or fuel, then as a rule, deposits are developed which are characterized by less amenable economic indicators.

Deposits of useful minerals differ one from the other not only by mining-developmental conditions of extraction but by their economic-geographic situation. Unequal distances of sources of supply of raw material and fuel from the basic areas of consumption, differing economic conditions of transportation (for example, differing amenability for transport of raw material and fuel in dependence on their quality), stipulate a varying level of effectiveness of utilizing the deposit.

It follows to take into account that economic indices of extraction at even "the worst" parts, deposits or basins must confirm with rational conditions of development and the wide use of all progressive achievements of contemporary techniques. "The worst" and "the best" deposits differ only by their natural and economic-geographic conditions, but not by level of techniques and organization of production. In general it is more correct to say not "the worst" parts, deposits, or so forth, but additionally drawn into exploitation in connection with exports or removed from exploitation as a result of imports.

The differentiation of natural conditions and economic-geographical situation of deposits, limiting the possibility of exploitation of only "the best" under capitalism leads, as is known, to the formation of mining differential rent. In the determination of the effect of export it is necessary under socialism to consider the differentiation of the natural and mining-geological conditions of the basins and deposits, and also the variance in their economic-geographical situation. A socialist country, expanding domestic extraction of raw material or fuel for export, cannot consider the worsening of the economic indicators (in part, an increase in the cost of extraction) owing to the necessity of additional drawing into exploitation of basins and deposits with less amenable mining-exploitational and economic-geographic conditions. With an increase in the scale of extraction the economic indices of extraction may be lowered even in the previously developed deposits in view of the working of the deeper - lying strata and of their less productive capacity and so forth. Of course, that drawn into exploitation in connection with these losses is at the expense of exports. In practice this means that the determination of the effectiveness of export of raw material or fuel should be based not on the average economic indicators (cost of production, specific capital expenditures) for the country as a whole but on indices, corresponding to conditions of extraction at deposits and parts, additionally drawn into exploitation in connection with exports and in the majority of instances having worse mining-developmental conditions or characterized by less amenable economic-geographic situation.

Ignoring this condition in the determination of the effect of foreign trade with specific types of products may lead to incorrect conclusions. Thus, there is created an exaggerated idea about the high effectiveness (and profit) of the export of crude oil from the USSR, in particular, Urals-Volga, owing to the fact that, basically, calculations are taken from average branch-of-industry indices of cost of production or indices of cost of production of crude oil at that deposit which has the best economic indices in the Soviet Union. In our opinion, there is no basis for determination of the effectiveness of the export of crude oil according to its cost of production in the Urals-Volga, inasmuch as for satisfying the internal requirements of the country oil reserves of other regions are used, the cost of production at which are several times higher.

The effect of export of energy coals from the Donets Basin also is being determined incorrectly. The possible scales of extraction of these coals are comparatively limited and may not meet the requirements of the European regions of the USSR for energy coals and fuel in general. As a result of the export of Donets coals, for the compensation of the additional deficit of energy fuel in the European part of the country, it is necessary either to expand the extraction of significantly less economic coals of the Moscow Basin, or to increase the import of Kuznets coal. Under such conditions the effectiveness of export of Donets coal must be determined not by their cost of extraction, but by the cost of production of the replacement coals (as a result of the export) with additional expenditures resulting from transport costs.

Determination of the effectiveness of export of goods, considered as one of a group of interchangeable products, also is being done incorrectly. For example, mazut and coal are interchangeable types of energy fuel. In the export of an interchangeable product the economic indicators should be examined not of each type of this product (group) but of its replacement. For example, in the export of mazut one should not proceed only from its cost of production. Exporting mazut from the USSR (as a rule, from the European regions) we must to meet consumption requirements replace it with other types of fuel: more expensive Donets and Kuznets coals or natural gas from Central Asia. In this, the effectiveness of export of mazut must be determined on the basis of economic indicators (cost of production, specific capital expenditure) of extraction and transport of either Donets or Kuznets coal, or of Central Asian natural gas. In practice foreign trade organizations determine the effect of export of mazut according to the average branch-of-industry cost of production, which leads to a mistaken conclusion about the great advantage of its export.

In the determination of the effectiveness of import of raw material and fuel, calculations should be based not on average branch-of-industry economic indicators, but on indicators of cost of production and capital expenditures, corresponding to those deposits or basins, the development of which the importer of a given or interchangeable type of raw material or fuel is freed from.

The economic effect of the import, for example, of Polish coal into the Northwest or Central regions of the USSR must be determined on the basis of the economic indicators of extraction and transport into these regions of Pechora or Moscow Basin coal (or Donets), which may be replaced by the imported coal, whereupon it follows to use not the average economic indicators of extraction of Pechora or Moscow Basin coals, but indicators of the cost of production and capital expenditures, which correspond to extraction displaced by imports.

"Western Oil Monopolies Try To Blackmail India"
Radio Moscow, September 2, 1965;
Summary of Pravda article by Yuriy Davydov

(Text) The Western oil firms say they have been insulted. They are trying to pretend that they are indignant. While they robbed India they did not complain. When they were told to stop robbing, however, they began to speak of alleged injustice. The American Esso Oil Company recently published a booklet entitled "Profits from the Sales of Oil and Its Products." The booklet's authors pretend they are benefactors who are not understood. They complain that the Indian Government has restricted their activities in the country and has, as they say, deprived them of profit legally made.

For many years the oil monopolies have lined their pockets at the expense of the Indian people. By using the monopolies for the marketing of oil products in the country, the British and American monopolies are making fantastic profits. The Indian Press Agency declared that the profits that foreign oil firms were getting in India were among the highest they got anywhere in the world. Until recently the oil companies themselves fixed the oil prices in India. They demanded payment only in foreign currency and did everything in their power to

SELECTED READINGS 295

obstruct the home oil industry from development. This policy deprived India of the necessary resources for financing important economic projects, and it hindered its industrial progress.

The Indian Government set up a state oil corporation to overcome the country's dependence on the foreign monopolies. With the aid of the Soviet Union, India is now producing its own oil. Several oil refineries have been built in (word indistinct) India. To combat the constant sabotage of the monopolies, the government reached agreement on Soviet oil deliveries that are paid for in rupees. Last May the Indian Government told the oil firms that their products would be paid for in rupees. Realizing that their robbing activities would end, the Western firms launched open blackmail. In June they stopped selling oil products, including gasoline, kerosene, and diesel fuel. By articially creating a shortage of fuel, the firms brought transport to a standstill in some parts of the country. This frank sabotage by the oil companies aroused a storm of indignation in the country and the government introduced steps to restrict the activities of the firms.

Times have become hard for the oil robbers. An end is coming to their many years of ruling the roost in India. That is why the firms are pretending to be offended and they are trying to blackmail the Indian Government.

"France Buys Soviet Oil"
Izvestiya, June 26, 1965, p. 2

(The long-term contract that was signed in June 1965 for delivery of Soviet oil to the French government concern UGP has caused wide comment. An Izvestiya correspondent asked Ye. P. Gurov, chairman of Soyuznefteexport to discuss the new contract. His statement follows.)

"Our contract with the government concern UGP (Union Generale de Petrole, General Oil Union) is the largest in the entire history of Soviet-French oil trade. UGP and ANTAR, a firm working with it, will buy more than 6 million tons of Volga and Ural oil in the upcoming Five-Year Plan.

"Not long ago Standard Oil, Socony Mobil and other US oil companies completely controlled the French oil market. However, the influence of purely French firms has now increased considerably. UGP, which was established in 1960, together with the Compagnie Francaise de Petrole (French Oil Company) has become one of the leading firms in the French oil market.

"As recently as this spring, Albert Nickerson, chairman of the board of directors of the huge US oil company Socony Mobil, went before the US Congress to call on Western countries to limit Soviet oil imports sharply. He declared that 'only Russia' would benefit from oil trade with the Soviet Union, and the Western countries, he said, would fall into a 'dangerous dependence on Soviet deliveries.' Our contract with UGP is a new proof of how groundless this conjecture was. It is a mutually advantageous trade agreement.

"Here is evidence from the American weekly periodical Petroleum Intelligence: 'The French government is killing two birds with one stone by this contract: it is diversifying the sources of supplying the country with oil and at the same time finds one more market for the goods of its manufacturing industry.'

"What does this weekly have in mind? Algerian oil is playing an ever greater part in French energy. France annually imports more than 20 million tons of oil. However, Algerian oil is light, whereas France also needs heavy and medium oils. We have such types. Mixing them with the lighter Algerian oil will increase the output of petroleum products, which France needs, and at the same time will promote a further increase of oil imports from Algeria.

"Moreover, as a rule, Soviet foreign trade organizations will buy French goods with the money from oil exports to France. For example, in 1963, Soyuznefteeksport sold the French firms more than 16 million rubles worth of oil and petroleum products. This money was used to buy equipment for the chemical industry, staple fiber, etc. Of course, such a trade basis actually promotes French exports.

"As far as our profits are concerned, the Soviet Union has vast reserves of 'black gold' and is interested in profitably selling part of it to other countries. Thanks to the new contract, our oil and petroleum product deliveries to France are practically double.

SELECTED READINGS 297

Our receipts from these exports will exceed tens of millions of rubles. At the same time, goods will be bought in France that will more fully meet some Soviet needs and demands of the people"

Finally, Ye. P. Gurov took up the prospects of Soviet oil exports to France. "There are still possibilities of further growth in oil trade between the USSR and France," he declared. "Total French import of oil and petroleum products is more than 50 million tons per year and is growing by 4.6 million tons annually. With such a large import, France obviously could import more Soviet oil. On the other hand, the USSR could buy considerably more machines, equipment and other goods from France."

SOVIET ATTITUDE TOWARD WESTERN OIL HOLDINGS IN THE MIDDLE EAST AND NORTH AFRICA

"The Boom in the Oil Business"
<u>Ekonomicheskaya gazeta</u>, 9, February 1969, p.43

    This year the Suez Canal - the greatest hydrotechnical construction project of the past century - will celebrate its 100th anniversary. Ships under the flags of 60 countries have carried through the Canal in years past one-seventh of all freight moving in world trade. It could today serve to further strengthen world economic relations, but for the past two years the Canal has been closed because of Israeli aggression. 15 ships are trapped in the Canal, and the accumulating silt symbolizes in itself the inability of capitalism to use for the good of all people the capability of the Canal.

    The closure of the Canal has brought about losses not only for the Arabian governments but also has strongly impacted upon the economies of those West European countries which together with the United States actively supported the Israeli aggressors. Out of 242 million tons of freight, moved through the Canal in 1966, almost 200 million tons were destined for West European countries or were exported by them to Asia, Africa and Australia.

    The transportation of these goods around the Cape of Good Hope has extended their lengths of haul by several thousand kilometers, and the duration of movement by 6 to 10 days. Transport expenditures have risen accordingly and, correspondingly, so have prices. The very greatest loss to the countries of Western Europe has been in the transport of more than 150 million tons of Arabian and Iranian crude oil, which earlier was moved through the Suez Canal. During the first six months alone of closure of the Canal England had to pay 200 million dollars more for its oil. West Germany had to pay 90 million dollars more and Italy - 80 million dollars. The total sum of overpayment by West European countries during the 1.5 years of closure of the Canal has reached 1.4 billion dollars, of which England has had to pay about 0.5 billion dollars.

SELECTED READINGS 299

These numbers were used in a speech by the English Minister of Foreign Affairs G. Roberts. He also mentioned the tardy re-opening of "this important trading route." The declaration of the English minister testifies to the bankruptcy of the policy of toleration of the Israeli aggressors, a policy which is followed by English imperialists despite the interests of the majority of the countries of the world and the national interests of England itself.

In many countries of Europe, Asia and Africa a growing persistence is heard, demanding the opening of the Suez Canal. Under these conditions even the American oil business, the role of which in the unleashing of the Israeli adventurists is generally known, is attempting to rehabilitate itself in the eyes of the world community.

The American publication Oil and Gas Journal has written the following: "The closure of the Canal has placed an expensive burden on oil trade. This conflict economists consider one of the reasons for the devaluation of the pound sterling, the recurring exchange crisis and internal difficulties in France. Look at the facts. The closure of the Canal has placed a financial burden on England and Italy just as it has on Egypt, and has raised the cost of oil all around the world. The Canal must be reopened as soon as possible and, if necessary, with Western aid."

The implications of this statement are clear. In the very same article the following is admitted: "Inasmuch as the oil industry not only overcame the crisis, but adjusted to long-term difficulties, brought about by the continued inactivity of the Canal, inasmuch as the war will not break out again on a large scale, it is easier to regard the current situation as normal. In the end, the majority opinion is that (the closure of the Canal) was an 'unfortunate happiness,' for the domestic oil producers and other sources of oil to the West of the Suez create for us a greater independence from Middle East oil."

To reckon the Middle East situation as "normal", but the misfortune of hundreds of thousands of Arabs, and tremendous losses to (their countries) to be looked upon as "unfortunate happiness" is possible only through the great profits gathered in by the American oil producers as an aftermath of the Israeli aggression.

Economists usually divide the US oil companies into two camps. The first includes the hundreds of small and average-sized firms - "domestic" grouping, which as a rule, produces crude oil by itself. The other camp is made up of 5 giants of the oil business, the Rockefeller "Standard Oil of New Jersey," "Standard Oil of California,"the Mellon "Gulf Oil" and "Texaco," controlled by Chicago and New York banks.

Out of the 440 million tons of crude oil, which was produced abroad by the "big five" in 1967, almost 240 million tons were produced in the Middle East. American oil empires each year bring back to the United States a billion dollars of pure profit, which they obtain as a result of exploitation of the Middle East oil fields. The evil inclinations of the imperialists value the preservation of their control over these unique sources of "black gold" in the capitalist world - the basic reason behind all the Middle East adventurists of the past three centuries, including as well the Israeli aggression against the Arab nations in June of 1967.

From the moment of closure of the Suez Canal interested circles in the West spread the version that the inactivity of the Canal was in the interests of only the small and average American oil firms. Actually, the profit of a number of "domestic" companies in the last quarter of 1967 rose on the average by 20 percent as a result of enlarged demand and higher prices.

Much less has been said about the vast profits taken by the "big five," although they exceed by many times the profits of the small and average "fish" of the American oil industry.

"If you look for indicators that the Middle East crisis was profitable for the American oil companies, then you need look no further than their incomes during the third quarter," wrote the October 1967 issue of the Petroleum Intelligence Weekly. "It is possible that the best example is that of the largest oil company in the world (Standard Oil Company of New Jersey). Its profits were almost 16 percent higher than in July-September of the previous year. The jump in profits came about not only because they increased their extraction of oil in the US, but increased deliveries of Venezuelan crude oil by 36,000 tons daily."

The import into Western Europe in the second half of 1967 of additional volumes of American and Venezuelan oil, which is 7 to 10 dollars more expensive per ton than Middle East oil, placed a heavy burden on the balance of payments of these countries, and the additional dollar flows into the United States enabled the government of L. Johnson to battle against the threatened de-evaluation of the dollar. The English newspaper "Times" not without envy and annoyance then turned attention to "the important supplements to the American balance of payments through the export of expensive, unprofitable crude oils, which in other times America would not be able to sell."

Upon the movement of Middle East oil around Africa the demand for American oil in Western Europe was reduced. The rates of growth of profits of the "domestic" oil producers subsided. The profits of the largest oil companies continue at present to rise at high rates. Thus, in 1968 the profits of the average and small companies increased by only 3 percent, but those of the "big five" and the Anglo-Dutch monopoly (Royal Dutch-Shell) rose by 10 to 15 percent.

The major reasons for the growth in profits of the oil imperialists of the US are the high prices dictated to West European importers, the forced delivery of more expensive oil from the American concessions in North Africa and raising chartering rates in connection with the movement of oil around the Cape of Good Hope.

In the 1.5 years after the beginning of the Israeli aggression the profits of the "big five", in comparison with a corresponding period prior to the aggression, rose from 4.6 to 5.5 billion dollars. In the history of the capitalist economy such high rates of enrichment of a group of companies within one branch of industry is without precedent. The major factor behind the prosperity of the oil industry is named as the "Suez crisis", unambiguously by the English publication Petroleum Press Service.

Pumping out of Western Europe since June 1967 about 1.5 billion dollars, of which almost one billion they put into their own pockets, and the remainder was divided among the "domestic" oil producers and the South African racists, profiting from the need for a thousand ships, the oil companies of the US began to think about how to convert the competitive factor into direct income. Certain unrest caused the exit of L. Johnson and his advisers from Texas, called the "Texas mafia", from the White House.

In the end of 1968 the bulletin of the oil producers "Platt's Oilgram", exposed its point of view on political relationships toward the Near and Middle East to the new President. It recommends "avoiding the stratification of continuous antagonism toward Arab governments, inasmuch as the sole reason for the American presence in the Middle East is oil."

These authors frankly admitted that the US "in no way is interested in reopening the Canal." Moreover, they did not object to keeping the Canal closed in order to bring about an unpleasant situation for the Soviet Union."

The undesirability of the oil monopolies to dispose of their advantages connected with the closure of the Canal, are so great that they, in essence, are undertaking a new reckless attempt to protect profits in turn of anti-Communist forces, by pointing out that in the end the opening of the Canal will be more to the advantage of the Soviet Union than to the West.

This fact speaks to the absurdity of this approach. In 1966 a total of 21,350 ships passed through the Canal, of which only 1,469 flew the Soviet flag and 832 the flag of other socialist countries. It is not difficult to determine that the blockade of the Suez Canal nine times out of ten does not impact on Socialist countries.

On the liberal relationship of the United States toward the Israeli aggressions, without a doubt the fact that, using the closure of the Suez Canal, the American oil industry extracts additional profits from other countries, exerts influence, as does the United States in its economic and military blocs.

SELECTED READINGS

"A Soviet View of Western Oil Interests in the Near East"
<u>Vyshka</u>, July 29, 1966, p. 3

All the woes of the Arab East are derived from its petroleum riches -- a tempting bait for imperialist expansion in this corner of the world. The odor of petroleum excites the appetites of all the imperialist monopolies, from the USA to Japan. Twenty-five foreign petroleum companies hold interests in the Libyan petroleum alone. As might be expected, the Anglo-American companies have captured the overwhelming majority of the oil fields in the Arab countries.

The petroleum magnates hold firmly to the Arabian petroleum, which is the cheapest in the world, especially in that period when the demand for petroleum products is increasing. Thus, while in 1940 all the countries of the world produced a little more than 250,000,000 tons of petroleum, in 1966 this index rose to 1,505,000,000 tons. The Arab countries have colossal petroleum reserves. In the capitalist world, Kuwait and Saudi Arabia occupy the leading place in petroleum reserves. And in the entire Arab East petroleum reserves by 1964 amounted to more than 56 percent of all the reserves available in the capitalist world.

Western petroleum companies are rapaciously exploiting the Arab petroleum. Petroleum production in the Arab East in the period from 1940 to 1964 increased by a factor of 66, while the average world index in this same period increased only four-fold. While in 1957, the fraction of Arab petroleum on a worldwide scale amounted to 16 percent, in 1961 it was 21.8 percent and in 1964, 26 percent.

These data testify eloquently to the fear of the colonialists of the ever growing rise of the national-liberation struggle of the peoples of the Arab East. President Nasser recently commented that "Imperialism feels that the Arab revolution is a great danger to its petroleum monopolies." Considering this circumstance, the western companies, especially in recent years, have been manifesting a great activity in those countries which up to this time were subordinate to them. For example, the Iraq Petroleum Company is operating actively in Oman. Next year it is proposed to export Oman petroleum via a petroleum pipeline which will connect the interior regions of the country with the shore of the gulf.

At the same time, the West has paid great attention to the petroleum riches located on the bottom of the Persian Gulf. Disputes between Saudi Arabia, Kuwait, and Iran concerning the distribution of the petroleum regions between them have somewhat hampered this. Negotiations between these countries began eight months ago, and they have not yet ended.

The enormous appetite of the western monopolies for Arab petroleum is increasing even more because of the cheapness with which it is produced. For each ton of liquid fuel, western companies pay 5 to 7 dollars, and they sell fuel oil, gas oil, and gasoline for 10-30 dollars per ton, lubricating oils for 50-150 dollars, and petrochemical products for several hundred or even several thousand dollars per ton. Besides this, while the monopolies pay 46 dollars per hectare of a petroleum field in Venezuela, in the Near East the cost is only 14 cents.

The productivity of wells in the Arab East is unusually high. According to data from Arab sources, the average productivity of one of the wells in Qatar is 154 times greater, and in Kuwait 179 times greater, than the productivity of the same sort of wells in the USA.

The petroleum magnates pitilessly exploit the labor force in the petroleum-producing Arab countries. Thus, a worker in Iraq receives only an eighth or a tenth the wages that a petroleum worker in the USA does. And in the Persian Gulf region things are even worse. In this region, wages are only a twelfth to a tenth what they would be in the USA. Incidentally, here the gates have been thrown open to immigrants from other countries, chiefly non-Arab lands. The petroleum companies really need cheap labor.

The facts indicated explain adequately convincingly why the production of a barrel of Arab petroleum costs 0.5 cent on the average, while the cost in Venezuela is 21 cents, and in the USA one dollar and twelve cents. Thus, the enormous profits flowing into the banks of New York and London from the rapacious production of petroleum in the Arab East every year amounts to 2.5 billion dollars.

In order to guarantee this lion's share in the future also, the colonizers are assigning part of it to the struggle with the national-liberation movement. All the facts testify that the western petroleum companies are financing subversive activities against the republican regime in Yemen. The aspiration to suppress the struggle of the peoples of South Arabia, Oman, and Bahrain for their rights clearly shows how much the West fears a revolutionary war, which might spread to neighboring Saudi Arabia and Kuwait, which are as good as the estates of the petroleum monopolies. As the Arab press reports, the petroleum magnates have allotted millions for the realization of the planned Islamic Pact. Bills of indictment in a case of a group of plotters, published in February of this year in Cairo, sheds light on their secret connections with employees of the British Shell petroleum company. Petroleum also plays a great part in the degree of interference of the west in the internal political life in the Near East. For example, before World War II, the imperialist power in the Arab East was in the hands of the "British lion", which now chiefly plays the part of an assistant to the CIA, which determines and translates into reality the policy of the West with relationship to the peoples of the Arab countries. The following fact reflects this to a considerable degree: while before the war Britain controlled 79 percent of the petroleum production in the countries of the Near and Middle East, at the present time American capital controls about two-thirds of the production of these countries, and British capital less than one-third.

What do the Arabs get for their "black gold"? These riches bring them only misfortune. With the exception of the petroleum of the UAR, it is all in the hands of bitter enemies of the Arab people. We will take Kuwait, for example, whose "prosperity" is very frequently mentioned in the West. This country, with a population of less than half a million, occupies the third place in the capitalist world in petroleum production, but it suffers from frequent epidemics, and 80 percent of the population is absolutely illiterate. Even children are forced to work instead of studying. According to an admission by the American journal Foreign Affairs, in Kuwait thousands of people are forced to live in miserable hovels.

In the sheikhdom of Bahrain, things are even worse. At the will of the American petroleum companies and British authorities, in Bahrain, which has a population of 150,000, where the eternal flame of oil wells has long been burning, 90 percent of the population is illiterate, 40 percent of the children have never crossed the threshold of a school, and there is only one hospital.

The fact that the state of the economy of the majority of the Arab countries, because of the interference of western businessmen, is associated primarily with income from petroleum, which amounts to from 1/3 to 9/10 of the total national income in the petroleum-producing Arab countries, also causes apprehension.

Thus, the Arabs get little good from their uncounted petroleum riches, while, according to a comment by the newspaper Al-Akbar, the income from the Arab petroleum could "transform the Arabian deserts into a garden of paradise" in a few years.

The Arabs have no other way out except liquidating the positions of the western petroleum companies. Only such a way will give a guarantee for the progress of the Arab countries along the path of independence and prosperity.

The west cannot help but feel that the dawn of a new life has appeared on the horizon of the Arab East. Since 1960 Saudi Arabia, Iraq, Kuwait, Libya, and Qatar have been members of an organization of countries exporting petroleum, created for the joint struggle against coercion by the western petroleum monopolies. Two years later in Iraq the National Petroleum Company was created. Important amendments, considerably increasing the income of the Libyan government from the exploitation of the petroleum resources of the country, were added to the Libyan petroleum law that has existed since 1955. The progressive forces in the Arab East are persistently demanding that the Arab petroleum be transformed into a real political weapon against the plots of the colonialists.

The time for the western pirates to be sent back where they came from has long been overdue, and the petroleum of the Arab East should be left to its real owners -- the Arabs.

"Oil Without Myths"
Izvestiya, May 23, 1966;
by B. Rachkov

The emblem of Royal Dutch Shell, one of the biggest oil companies of the capitalist world, is a seashell, which has long symbolised the desire for travelling, for seeking places and things new and unexplored. Sometimes this company puts out prospectuses with reproductions of "The Birth of Venus," by Botticelli, an Italian painter of the Renaissance. It depicts the goddess of love and fertility emerging from a seashell.

The operations carried out by the US and British monopolies with the "black gold" are, indeed, yielding abundant fruit. Hardly any other corporation can compare in wealth, financial and economic might and political influence with Rockefeller's Standard Oil, Mellon's Gulf Oil, British Petroleum, and other oil empires.

But there is another pole in the oil business, the pole of economic backwardness and poverty. It can easily be observed in the adobe huts of Iraq, Iran and Venezuela and among the Bedouins of the Arabian Peninsula.

The Middle East countries account for about 70 per cent of the known oil reserves in the capitalist world. They are providing two out of every three tons of liquid fuel consumed in Western Europe and practically all the oil needed by the Asian countries. This wealth is in the hands of foreign monopolies. Last year, for instance, 210 million tons of "black gold" were taken out by US companies, 130 million by British, and 50 million by French and other companies. In the last three years alone they obtained 6,000 million dollars net profit here, while investing less than 800 million dollars.

US and British imperialism have been greatly concerned in the last decade over ways of consolidating their financial and economic positions in the general system of capitalist economy, and of saving the dollar and the pound sterling from devaluation. Particularly fond hopes are pinned on the foreign expansion of the oil companies.

US and British economy is considerably supported by oil monopolies, active in practically all the capitalist countries, placing big orders at home for equipment for their overseas branches. Such orders do have a stabilising influence on employment, export and balance of payment, but such encouragement of the foreign currency and economic position of the United States and Britain harms the Arab countries, Iran and Venezuela. Suffice it to say, that while the oil companies in the period 1964-1967 are implementing nearly 50 projects in the construction of new and expansion of old refineries outside of the United States and Britain, they will not put up any new or modernise any old refinery in the oil-rich countries of the Middle East.

Imperialists of every shade show their full solidarity with the oil monopolies preferring to see this area a prominent supplier of crude oil. For almost a third of a century Iran in vain pleaded with the Western concerns as to the need for building a metallurgical plant in that country. West German firms demonstratively refused to build a hydropower station on the Euphrates in the same way as American banks and concerns earlier categorically refused to finance the construction of the hydrocomplex on the Nile. When studying such facts the entire sinister idea behind the recommendations drawn up by an authoritative American commission becomes absolutely clear. This commission without a trace of humour advised that only shops for manufacturing combs, pots and flash-lights be built in the countries of the "Afro-Asian Arc."

The implementation of these recommendations would help the neocolonialists not only in preserving their profits in the Near and Middle East but also to preserve the as yet strong survivals of semi-feudal relations in these countries and from which imperialism as well naturally draws advantages. In particular, under the present system a considerable part of the income received from oil designated for the Arab countries and Iran falls into the hands of the sheikhs and other representatives of the local nobility which spend them on articles of luxury or deposit them in the armoured cellars of the City and Wall Street. The exact figure on Near and Middle Eastern deposits in US and British banks is unknown. However, according to the most modest estimates they have already exceeded the 3,000 million dollar mark. The currency system of American and London bankers without this "stand-by" would perhaps have crumbled or at least experienced an extremely acute crisis. The interest alone received by the banks from loans issued using the accounts of the sheikhs runs into tens of millions of dollars.

Every time the peoples of the Near and Middle East make a new step on the road to their liberation, be it the nationalization of the Suez Canal, the overthrow of the monarchy in Iraq, or the liberation war in the south of the Arabian Peninsula--imperialism strives to punish them by sword and fire. The smoke over the ruins of cities and villages, the blazing crops force one to recall not Botticelli's painting of Venus but the cruel "exploits" of the medieval conquerors.

Adventurists, out for the wealth of the Arab East, have appeared in every possible toga. The tragedy of the Middle Ages which was explained by the desire "to defend the holy places of Christianity from the Moslems" was recently repeated as a farce in the hypocritical concern of the western powers for the unity of the Moslem world, in the persistent attempts to set up a so-called Islam Pact. However, the new "civilisers" consider anti-communism their best pretext for aggression in the Near and Middle East.

Representatives of the oil monopolies are most profuse on the need to protect by all means this area from a "communist threat" and increase allocations for military bases in the Gulf of Persia and in the Red Sea. Others call for a more intensive exploitation of Arab and Iranian natural wealth so that it should not fall into the hands of Communists, etc. However, Morse, the prominent American Senator, once gave a fitting reply to this. He pointed out: "The American tax-payers are being deceived for the sake of the privileges of American companies who receive vast profits from foreign oil. We are constantly lulled by a theory according to which we should drain all oil deposits before they fall into the hands of Communists. I declare to the American people that this is utter nonsense. What the companies actually want is to spread their control over a maximum possible number of oil sources."

The USSR builds its relations with the countries of the Near and Middle East on the basis of full equality and mutual benefit. The imperialist states give clear preference to those countries of this region where big oil deposits have been discovered, whereas the Soviet Union does not differentiate between the countries which have oil and those which do not have it. The West tries to retain this region as its raw-material appendage, whereas the USSR gives aid to the Arab countries and Iran in developing their multi-branch national industry.

For example, the United Arab Republic and the Syrian Arab Republic which have more developed economies were regarded by Western oil empires as a profitable market, as a region less suitable for oil extraction. They hindered geological surveying there and used their deliveries of liquid fuel for the economic and political blackmailing of these countries. However, the searches for oil started there by Soviet specialists several years ago have already resulted in the discovery of a number of oil deposits which will allow the UAR and the SAR to discontinue the import of liquid fuel. This prospect becomes all the more feasible since oil refineries have been built in the republics with the aid of socialist countries to ensure the home manufacture of oil products.

The socialist countries help these republics, as well as Iraq and Iran where the monopolies developed oil extraction to the detriment of other industries, in building the foundation of metallurgy, machine-building, electrical engineering, the textile industry, etc. A hydrotechnical complex being built on the Nile with Soviet aid will bring the UAR an annual income of more than 250 million pounds sterling. At present oil does not bring that much money even to Iran where foreign monopolies have been extracting oil for decades now and from where they annually export more than 70 million tons of oil.

As a rule, the countries of the Near and Middle East pay for the Soviet Union's technical assistance and other aid with the deliveries of cotton, wool, citrus fruit and other farming produce. With the development of the national branches of mining and other industries there the basis for economic cooperation will be greatly extended. A portent of this is the Soviet-Iranian agreement signed in January 1966. In accordance with the agreement, Iran will pay for the deliveries of Soviet equipment for an Iranian metallurgical plant largely with the supplies of natural gas needed by the Soviet Central-Asian Republics.

The recent Soviet-Syrian communique will play an important part in the development of mutually beneficial economic ties between the USSR and the Arab countries. It cites, among other measures, the construction of a hydrotechnical scheme on the Euphrates with Soviet aid.

In the past the bountiful banks of the ancient Nile and the fertile interfluve of the Tigris and the Euphrates supported over 60 million people. In the past few decades they have not been able to feed half that number. The domination of the oil monopolies in the Near and Middle East has not, clearly, promoted the economic regeneration of this cradle of human civilisation. The way to genuine economic and cultural progress leads not through imperialism's "boons" sweetened by boastful publicity and the servile press, but through labour feats of the peoples of the developing countries leaning on the mutually advantageous cooperation with the USSR and other socialist countries.

"Oil Kings Call the Tune in Western States"
Radio Moscow, March 29, 1965

The United States is considered a vivid example of the power exercised in the Western state by oil businessmen. We do not have to mention dozens of names of U.S. Congressmen who have become members of Congress by virtue of the support of oil kings. These "Kings of Oil," as they call themselves, consider their main task to be that of defending oil company interests by any means. Not only are the interests of oil businessmen defended by members of Congress, but also by members of the U.S. Government. It is known that the law firm of (Elwan and Cromwell?) works for the interests of the oil empire of Standard Oil. Simultaneously, this company is considered, in many cases a springboard for entering the U.S. Foreign Service. This is not a coincidence.

In recent years, a large number of Standard Oil employees have entered the U.S. Foreign Service. The late John Foster Dulles, who acted as one of the directors of this firm, occupied the position of U.S. Secretary of State for sometime. Several members of the U.S. Government are, simultaneously, stockholders in oil companies. Among these, for instance, is John McCone, chief of the CIA, a position of considerable importance in the United States. John McCone is the second largest shareholder in the Standard Oil Company of California, one of the biggest U.S. oil monopolies, which carries out its operations abroad jointly with the monopolistic Texas Oil Company. These two companies work in foreign countries under the name Caltex. Douglas Dillon, Secretary of the Treasury under the late President Kennedy and presently under President Johnson, is also closely connected with oil companies. His bank--Dillon, Read and Coy, Inc.--finances many operations of U.S. oil companies abroad. It has become commonplace to appoint an oil company owner as a deputy U.S. interior secretary.

There are international connections, established by U.S. law, between the government and oil owners. The White House has a so-called "national council" on oil affairs, which includes the magnates of the biggest U.S. oil companies. This council officially constitutes a consultative body for the government, but in fact it determines U.S. strategic-technical plans on oil affairs. The attempt by the late President Kennedy to reduce the big privileges of the U.S. oil monopolies was the first U.S. attempt in this field, and minimized the duties and rights of the national council for oil.

These measures greatly angered U.S. oilmen. It is believed that this restriction was among the chief factors which caused the assassination of President Kennedy in Texas, the center of the U.S. oil industry. It is known that two weeks after the assassination of the U.S. President, the previous rights and duties of the national council for oil affairs were restored.

The governments of other Western states are also closely connected with oil companies. The British Government, for instance, owns 52 percent of the stocks of the British Petroleum Company. The French Government owns about 35 percent of the stocks of Companies Francaise de Petrole. This amount of stock is sufficient to induce any government to pursue a policy satisfactory to the oil kings.

"The Dollar in Africa"
<u>Aziya i Afrika Segodnya,</u> July 7, 1964;
by M. Kogan, an economist

In the last few years the US monopolistic circles, with the Rockefellers playing the leading role, are trying to take the European colonial powers' place under the African sun.

The might of the Rockefellers, the American oil kings, is mainly based on six oil corporations: Standard Oil Co. (New Jersey), Standard Oil Co. (Indiana), Standard Oil Company of

California, Marathon Oil Co. (the former Ohio Oil Co.), Socony Mobile Oil Co. Inc. (the former Socony Vacuum Oil Co. Inc.), and the Atlantic Refining Co.

At present about 500 million dollars, i.e., almost half of the direct US monopoly investments in Africa are spent on prospecting for and producing oil.

Libya, where the first large oil deposits have been discovered in the last few years, is the main objective in the Rockefellers' expansion on the African continent.

Nineteen big companies and corporations are taking part in the bitter economic battle being waged for oil in the waterless expanses of Libya: 12 American, one British, one Anglo-Dutch (Royal Dutch Shell group), two West German, two Italian and one French. Foreign capital investments in Libya for prospecting and oil production, rose from 210 million dollars in 1960 to approximately 800 million dollars at the end of 1962. According to the British Economist, the Standard Oil of New Jersey, a Rockefeller monopoly, alone invested about 150 million dollars in Libyan oil deposits at the end of 1961.

In Libya, beside the Standard Oil Co. (New Jersey) and its branches, there are also a number of other American firms, the majority of which are linked directly or indirectly with the Rockefellers. A large concession for the prospecting for oil and for transport planning covering 248,000 sq. kilometres was received by the Oasis Oil Co. of Libya Inc. in 1960. The Rockefellers' Marathon Petroleum Libya Ltd. owns large shares in the Oasis Oil Co. And the Mobile Oil Libya Ltd., another Rockefellers' concern, is actively participating in oil prospecting in the extensive Libyan desert.

The other monopolistic groups competing with the Rockefellers' oil companies are forced to accept a secondary role.

Assessing the US monopolists' "stake" in Libya The New York Times pointed out recently: the capital invested in prospecting for oil is no longer a magnificent mirage in the Libyan desert--it is gushing. The four years' search is beginning to pay. ... The investments of 19 companies are bringing real results. 165,000 barrels of oil are sent to the world market daily via the oil pipes and exit ports.

In 1963 the export of crude oil from Libya amounted approximately to 21 million tons, an impressive figure. The Esso Standard Libya Inc., controlled by the Rockefellers, was the first to supply the capitalist world market with Libyan oil.

The rich deposits of "black gold" in the Algerian Sahara have also attracted the companies and corporations belonging to the Rockefellers' empire. The omnipresent Standard Oil Co. (New Jersey) has received a sizable concession there as well. After the oil fields discovered in Hassi-Messaurd and Edgel began to produce, the Standard Oil Co. (New Jersey) tendered a strange "ultimatum" to the French Government--it demanded for itself a 50 per cent quota of all capital invested.

The "ultimatum" was accepted and in June 1959, the French Government gave the mixed Franco-American group exclusive rights for prospecting for oil on 20,000 sq. kilometres of territory bordering on Tunisia. The French press noted with unconcealed satisfaction that the Standard Oil Co.--Companie Francaise de Petrol (Algerie)--agreement was a capital act. That was a recognition by the powerful American oil company of the Sahara's significance as a large world centre for producing oil.

In their feverish search for "black gold" the monopolies are penetrating farther south into the deserts of the "Spanish" Sahara.

The American capital invested there adds up to tens of millions of dollars (28.7 million dollars in 1960), the greater portion of which belongs to the Rockefeller monopolies, principally to the Atlantic Refining and Marathon Oil companies. The concession received by the Atlantic Refining Co. alone back in 1960 was over 2.4 thousand sq. kilometres.

The Rockefeller oil companies are energetically penetrating into the farthest regions of the vast African continent. The Socony Mobile Oil Co., for instance, is taking part through its sister companies in oil prospecting, and in the production of oil already found, not only in Libya and the Algerian Sahara, but in Gabon, both the Congo Republics, Tunisia and Nigeria. A branch of the Socony Mobile Oil Co.--the Mobile Oil Egypt--is participating (50 per cent) in the exploitation of the UAR oil fields. Apart from Libya and the "Spanish" Sahara, the Marathon Oil Co. is prospecting for oil in the Somali Republic and Tunisia.

The US imperialist expansion is inevitably accompanied by "Esso", the trade mark of the Standard Oil Co. (New Jersey), the largest oil monopoly in the capitalist world, controlled by the Rockefeller family. In Libya it is Esso Standard (Libya); in Algeria Esso Saariene. As the Rockefeller companies forcefully penetrate into Africa they come up against the stubborn resistance of the British, French, Belgian, Italian and other capitalists.

The "oil business" is the Rockefellers' basic interest in Africa. Nevertheless in their expansion a big role is also played, along with the oil monopolies, by other numerous corporations, companies and banks belonging to the Rockefeller industrial empire.

In West Africa, for instance, the Republic of Guinea with its rich mineral resources, is an important objective of the expansion. The key positions in Guinea's economy are held by FRIA, an international consortium created for the production of alumina from bauxite found in extremely rich local mines. French, American, British, West German, and Swiss capitals have been invested in FRIA. However, American interests dominate because 48.5 per cent of FRIA's joint stock capital belongs to Olin Mathieson Chemical Corporation, a big ammunition industry, headed by Laurence Rockefeller. Of the 140 million dollars invested in FRIA approximately 70 million belong to the Rockefeller family.

The B.F. Goodrich Company, one of the "big five" in the US rubber industry, is working in Liberia. It is headed by David Rockefeller.

Rockefeller banks--Chase Manhattan and The First National City Bank of New York--have their branches in South Africa (The First National City Bank of New York (South Africa) Ltd.), Liberia (The Bank of Monrovia) and other parts of the continent. The Rockefeller monopolies play an important, and on occasion a decisive, role in the development of the as yet unstabilized national economy of the young African states.

Official US representatives accredited to the governments of some young African countries are directly connected with the Rockefeller "oil business." A vivid example, in this connection, is Charles Darlington, a New York democrat, appointed on August 22, 1961, as the first American Ambassador to the Republic of Gabon. For a number of years Darlington was the representative of the Rockefeller Socony Mobile Oil Co. in London and later in the Middle East. In Gabon important oil deposits, which produced about a million tons in 1962, are being exploited.

The fact that their henchmen occupy the key posts in the state apparatus of the USA helps a great deal in the intense expansion of the Rockefellers' interests in Africa. The banker Douglas Clarence Dillon has been appointed US Secretary of Finance; not long ago he was the president of one of the largest investment banks in New York--Dillon, Read and Co.--which owns, together with the Rockefellers, extensive concessions of rich oil lands in North and East Africa. The present head of the US Central Intelligence Agency, John McCone began his career in the Standard Oil Co.

The financial magnates of the United States--Rockefellers, Morgans, Mellons, Du Ponts and others--are fraudulently and hypocritically using NATO, the military-political imperialist trust, to present their imperialistic expansionist plans to the developing countries under the guise of the "collective policy" of the so-called free world. However, the African peoples' independence is seriously threatened by the usurpation and rapacious plunder of Africa's natural wealth by the Rockefellers' monopolies and banks.

MAJOR PIPELINE DEVELOPMENTS

"For the Study of the Operation of Gas Pipelines Under Northern Conditions"
Stroitelstvo truboprovodov, 3, March 1968, p. 14

A number of problems have arisen in connection with the long-range plans for construction of large-capacity gas pipelines in Tyumen Oblast and the Komi ASSR. Broadly postulated experimental work is needed for the resolution of these problems.

How large must the separation be between the parallel links of the gas pipeline? What should the distances be between the trunk pipelines and other installations? How will the designs of increased-diameter pipelines work? What effect will the new methods of laying them have? What will the heat exchange between the pipes and the soil in sectors where there is permafrost be like?

Answers to these and other questions can be obtained only through study from all angles of the actual operation of northern gas pipelines under actual conditions.

As far back as several years ago scientific workers of the VNIIST (All-Union Scientific Research Institute for the Construction of Trunk Pipelines) began full-scale investigations. In particular, an extensive experiment was conducted in the region of Ukhta in order to study the zig-zag system of aboveground pipelines.

This experiment resulted in the disclosure of a number of shortcomings in the supporting structure. It was established that the design of the fixed supports did not assure the complete immobility of the pipeline. "A"-shaped supports using suspension mechanisms, for practical purposes, do not permit the correct regulation of the entire gas pipeline system, and this results in overstraining some individual support mechanisms. It was also determined that a system of suspension-type supports is feasible only for gas pipelines with diameters no larger than 426 millimeters.

The results of these investigations were taken into account during the planning of the Taas-Tumus -- Yakutsk -- Pokrovsk gas pipeline, which was laid using sliding supports for the most part.

Further investigations are already in progress on the Taas-Tumus -- Yakutsk gas pipeline.

It was established that a number of the theoretical computations approved in SNiP (Construction Norms and Regulations) were not consistent with the actual operation of the gas pipeline. A new system - straight-line laying with slightly curved sectors - was brought into being on the basis of the study of the zig-zag system which uses sliding supports and an analysis of its defects.

The new system possesses a number of advantages over existing ones. In particular, it makes possible a decrease in the expenditure of pipe by from 1.5 percent to 2 percent as compared to the zig-zag system, also a decrease in the number of curved inserts, and a lowering of hydraulic resistance, etc.

112 kilometers of the Taas-Tumus -- Yakutsk gas pipeline have already been laid using the new system. In the near future this system will be used on gas pipelines to the settlement of Tazovskoye and to Norilsk (an over-all distance of around 200 kilometers.)

However, just as is the case with other systems, straight-line laying with slightly curved sectors needs further study. This will be done on the experimental gas pipeline which it has been decided to build on the route Ukhta - Torzhok.

A substantial part of the gas pipeline from the northern part of Tyumen Oblast - Ukhta - Torzhok will be above-ground or laid above-ground with a dirt covering.

The basic problems connected with these methods of laying pipelines must be solved by the time of the beginning of construction of subsequent sections of the trunk lines Northern Tyumen Oblast - Ukhta - Torzhok and Northern Tyumen Oblast - Urals.

The experimental pipeline must be situated beyond the compressor station located in the immediate vicinity of Ukhta. It will go into the zone of the so-called hot sector, which will make it possible to observe its operation not only under pressure, but also under circumstances where thermal conditions are changeable. Soils in this sector are mainly loams and sandy loams, and bogs are sometimes encountered.

It is planned that the test pipeline will be in the form of a parallel line having a diameter of 1220 millimeters to the main underground sector of the gas pipeline, which will also have a diameter of 1220 millimeters.

"Distribution Of The Oil Industry And The Movement Of Oil Freight"
Transport i khraneniye nefti i nefteproduktov, 3, March 1968, pp. 26-28

In connection with the growth in extraction of crude oil and the production of oil products the share of oil freight, in all freight movement, has increased, especially for the railroads. During 1950-65 the share of oil freight in total freight-turnover (measured in ton-kilometers--editor) accomplished by railroads has increased from 8.5 percent to 14.4 percent. At the present time, in terms of ton-kilometers performed by rail, oil freight holds second place behind coal.

Table 1
Dynamics in the Relationship of Volumes of Production and Movement of Oil Freight
(Million tons)

Indicator	1955	1960	1965	1966
Crude oil extraction	70.7	147.9	242.9	265.0
Volume of movement of crude and products by all types of transport (including for export)	166.6	331.8	526.4	574.7
Coefficient of movement (volume of movement related to crude production)	2.36	2.26	2.15	2.17

The higher the coefficient of movement the higher are the expenditures in the movement of 1 ton of produced crude oil. For this reason a reduction in this indicator speaks to an improvement in the location of enterprises connected with the extraction and refining of crude oil.

By 1965 the relationship of the volumes of crude oil output in the Urals-Volga and the Caucasus in relation to the national total had changed sharply. The share of the Caucasus declined from 31 percent in 1955 to 17 percent in 1965, and at the same time the share of output in the Urals and the PoVolga increased from 58 percent to 72 percent.

A still more significant shift took place in the location of oil refineries, bringing refineries closer to major areas of consumption (the Center, the Ukraine and Siberia). The share of the regions of the European and Asiatic parts of the USSR (excluding the Urals-Volga and the Caucasus) increased during 1956-65) by more than 3 times. If in the total volume of movement of oil freight to the Northwest, west, central and eastern regions the share of the Transcaucasus and North Caucasus reached 12 percent in 1955, then by 1965 it had declined to 5.7 percent. Consequently the number of movements by mixed water-rail in the delivery of oil to these regions declined. In the total movement of oil freight the share of the Caucasus regions declined during the 10 years by 3.4 percent.

Table 2
Share of the Urals-Volga and the Caucasus in the
Total Movement of Oil Freight

	1955	1960	1965
Oil Freight movement by all means			
Million tons	166.6	331.8	526.4
Percent	100.0	100.0	100.0
Of which, from the Urals-Volga & Caucasus			
Million tons	133.0	270.2	404.3
Percent	80.0	81.5	76.8
From Urals-Volga			
Million tons	90.9	194.6	288.7
Percent	54.6	58.6	54.8

The construction of oil refineries in Omsk, Irkutsk, Yaroslavl, Ryazan and Moscow permitted the more rational supply of oil products to the Far East and Northwest.

SELECTED READINGS 321

Table 3
Oil Freight, Moved by All Types of Transport in 1965
Urals-Volga and

Transport	USSR Million Tons	%	Caucasus Million Tons	%	Urals-Volga Million Tons	%
Total	526.4	100.0	404.3	76.8	288.7	54.8
Railroad	222.2	100.0	129.1	58.1	91.9	41.4
Pipeline	225.7	100.0	208.2	92.2	176.1	78.0
Maritime	53.5	100.0	47.6	88.9	1.4	2.5
River	25.0	100.0	19.4	77.6	19.3	77.2

In terms of length of haul oil freight occupies one of the leading places among the types of freight moved in large amounts. Thus, for example in 1965 the average length of haul for railroads in general was 807 kilometers, for hard coal it was 681 kilometers, for ore and heavy metals 813 kilometers, but at the same time the average length of haul for oil freight reached 1,262 kilometers. This reflects the inadequate movement of oil refining enterprises toward major consuming areas.

Aviation gasoline produced in the Far East is carried to the PoVolga, gasoline is moved from the Transcaucasus to everywhere, and in Siberia flows of aviation gasoline from the Guryev refinery meet aviation gasoline coming from the Far East. For lube oils, produced in the Transcaucasus, the average lengths of haul are the greatest for any of the oil products.

Further complications are to be found in irrational movements of crude oil. Crude oil is moved from Kazakhstan for refining in the Center and the Urals, with the simultaneous transport to Kazakhstan of crude oil from the PoVolga. Such irrational flows - a consequence of the incorrect profile of the Guryev oil refinery in terms of the local crude oil. The development of capacity at the Guryev refinery was carried out without considering expansion of the Orsk refinery, created earlier and designed for the refining of lube-base crude oils from Kazakhstan. For this reason from year to year the demand for Guryev crude oil increases at the Orsk refinery which, with the existing level of crude oil output in Western

Kazakhstan and expansion of the Guryev refinery, has enlarged these cross-flows of crude oil. According to preliminary calculations such irrational flows cost more than 3 million rubles annually, an amount equal to the cost of construction of an oil refinery with an annual productivity of 2 million tons.

The establishment of new refining centers in West Siberia (Omsk) and Eastern Siberia (Irkutsk) far removed from oil deposits of the Urals-Volga but still based on the refining of these crudes, with an inadequate study of adjacent regions in terms of supply of products and in terms of specialized production of individual products, has led to the creation of two types of irrational flows: ready products encounter crude oil and other products. Surplus production in Western and Eastern Siberia of various types of mazut, transported close to Murmansk Oblast, illuminating kerosine and special gasoline (never saw this term used before--editor) transported even to the Center -- all have led to excessive transport. In 1965 in the irrational movement of fleet mazut alone from these regions a sum of more than 2 million rubles was spent. Incorrect specialization of the above refineries also has resulted in these surplus petroleum products moving counter to the basic flow of light petroleum products (automobile gasoline and diesel fuel), which are transported by product pipeline and moved by railroad to Siberia and the Far East from the PoVolga.

Counter flows of the same-type oil products are also encountered in the movement of automobile gasoline from Saratov in an easterly direction (to Kazakhstan) with the simultaneous delivery of Ufa automobile gasoline to western regions (to the Ukraine, Belorussia, and the Pri-Baltics). The production of automobile gasoline at each of these refineries of selected types corresponding to the structure of demand in adjacent zones would permit the reduction in volumes of work of the railroad system of about 600 million ton-kilometers and would save about 1.5 million rubles annually.

Table 4
Freight Turnover and Expenditures in the Transport
of Oil Freight in 1965*

	Freight Turnover			Transport Expenditures (Million Rubles)		
	Actual	Irrational Flows				As %
	(Billion	Billion	As % of		Irrational	of
Oil Freight	TKM**)	TKM**	Actual	Total	Flows	Total
Crude oil	219.2	6.1	2.8	319	15	4.7
Oil products	258.2	65.3	25.3	625	163	26.1
Total	477.4	71.4	15.0	944	178	18.9

*All calculations exclude movements for export
**Ton-kilometers.

"Capital Investment in Crude Oil and Product Pipelines in the
USSR and Their Factual Effectiveness"
Ekonomika neftedobyvayushchey promyshlennosti,
3, March 1968, pp. 45-47

The total volume of capital investment in pipeline transport in the USSR during the 10-year period 1955-65 reached about 900 million rubles, of which almost 60 percent was made in the second five years. The dynamics of capital investment by years (as a percent of 1955) are given in the following table, from which it can be seen that the greatest investment was made in 1963 (253 percent of the basic year).

Table

Year	USSR Total	Crude pipelines	Product pipelines
1955	100	100	100
1956	96	87	112
1957	105	120	76
1958	126	155	75
1959	113	125	92
1960	145	184	76
1961	219	290	195
1962	230	339	36
1963	253	381	27
1964	166	252	15
1965	75	111	11

This sum was almost completely determined by investment in crude oil pipelines, inasmuch as investment in product pipelines that year totaled only 27 percent (of 1955 level). Such a "peak" in dynamics of capital investment was brought about by construction on the "Friendship" pipeline, and also continuation of construction in 1963 on major crude oil pipelines (Omsk-Irkutsk, Gorkiy-Yaroslavl), and expansion of Tuymazy-Omsk-Novosibirsk No. 1 and Tuymazy-Omsk-Novosibirsk No. 2 and others, as a result of which 95.6 percent of the capital investment was directed to the construction of productive facilities.

A large part of the capital investment in product pipelines was made in the first five years of the period under study, and during the course of 1961-65 it systematically declined, for in these years only completion and expansion of construction of projects begun earlier had been undertaken.

The share of Glavneftesnab RSFSR (The Main Administration for Oil Supply, RSFSR) of capital investment in pipeline transport of the USSR is 68 percent (presumably for the years under study - editor). Of the total sum of capital investment in pipelines for this Main Administration 76.6 percent was directed to the construction of crude oil lines and 23.4 percent to the construction of product pipelines.

The "peak" of capital investment in pipelines in the RSFSR was reached in 1961, at 177 percent of the base year of 1955, and separately for crude oil pipelines - 223 percent. In the subsequent years the volume of capital investment was less than it was in 1961. Despite the fact that the tempos of pipeline construction in the USSR did not decline, the volume of capital investment by Glavneftesnab RSFSR during 1962-65 was less than in 1955-61. The reduction in the tempos of construction of crude and product pipelines derived from the practical absence of construction of product pipelines following completion of the Kuybyshev-Bryansk product pipeline in 1962.

The structure of capital investment in crude and product pipelines in the USSR is shown in the following:

	(Percent)	
	Crude pipelines	Product pipelines
By structure:	100	100
Construction-installation	88.6	89.0
Equipment	6.3	5.9
Other	6.1	5.1
By direction:	100	100
Linear	73.1	60.4
Site	12.6	28.6
Other	9.3	11.0
By purpose:	100	100
Non-productive	3.0	6.0
Productive	97.0	94.0

"Prospects for the Development of Pipeline Transport of Crude Oil and Petroleum Products"
Transport i khraneniye nefti i nefteproduktov,
11, November 1969, pp. 6-8

Trunk pipelines are the most effective means of transport of crude and products; for this reason the system of trunk crude and product pipelines in the USSR has been expanded at rapid rates. The development of this pipeline network during 1959-66 is illustrated in Table 1.

Table 1

Development of Trunk Crude Oil and Product Pipelines During 1959-66

Indicator	1959	1966
Length of system (thousand kilometers)	13.8	27.9
Including:		
Crude oil pipelines	10.5	22.7
Product pipelines	3.3	5.2
Movement of crude and products by pipeline (million tons)	94.9	247.7
Including:		
By crude oil pipelines	81.6	225.6
By product pipelines	13.3	22.1
Average length of haul (kilometers)	356	666
Ton-kilometers performed (billion)	33.7	164.9
Including:		
By crude oil pipelines	27.5	146.6
By product pipelines	6.2	18.3
Share of pipelines in total oil freight turnover (ton-kilometers) (%)	15.5	31.6
Cost of movement of crude and products (million rubles/billion ton-kilometers)	1.36	0.97

Table 2

Characteristics of Trunk Pipelines, by Diameter, as of 1 January 1967

Diameter (Millimeters)	Pipe Capacity (cubic meters/ kilometer)	Pipe Weight (Tons/ kilometer)	Length (Kilometers)	Capacity of all Pipe (cubic meters)	Total Pipe Weight (thousand tons)	Million Ton-kilometers/year	Percent of Total
219	32.67	37	1,251	40,870	46.3	1,049	0.5
273	51.45	52	2,275	117,048	118.3	3,320	1.5
325	73.99	64	3,139	232,255	200.9	4,830	2.2
377	100.66	84	2,982	300,019	250.5	8,994	4.0
426	129.40	102	70	9,085	7.2	371	0.2
529	207.39	108	9,337	1,936,400	1,008.4	59,166	26.6
630	295.94	123	441	130,509	54.2	3,307	1.5
720	402.43	146	4,990	2,008,128	728.5	59,720	26.8
820	504.92	187	2,108	1,064,371	394.2	40,588	18.3
1,020	780.30	274	1,274	994,102	349.1	40,810	19.4
Total	-	-	27,867	6,832,760	3,157.6	222,155	100.0

In the system of trunk crude and product pipelines the length of pipelines having a diameter of 529 millimeters or higher reaches 65 percent, and their capacity - 92 percent of the total. As of 1 January 1967 the system of trunk crude and product pipelines by diameter was characterized by the data presented in Table 2. The average diameter of crude oil pipelines, which have a total length of 22,690 kilometers is 582 millimeters, and for product pipelines, which have a total length of 5,177 kilometers, the average diameter is 441 millimeters. For the entire system of crude and product pipelines, 27,867 kilometers, the average diameter is 560 millimeters.

According to Directives of the 23rd Party Congress, during the Five Year Plan the volume of transport of crude and products by pipeline (ton-kilometers) is to increase by about 2 times, and pipeline construction is to total about 12,000 kilometers.

According to preliminary data, the planned development of pipeline transport of crude and products during 1966-70, in comparison with the previous five-year period, is given in Table 3.

Table 3

Characteristics of Pipeline Transport of Crude Oil and
Petroleum Products in 1966-70, As Compared with 1960-65

Indicator	1960 Actual	1965 Actual	1970 Plan	Increments During: 1961-65 Absolute	1961-65 Percent	1966-70 Absolute	1966-70 Percent
Length of pipelines at end of year (kilometers)	17,322	27,217	39,818	11,257	64.9	11,239	39.2
Including: Crude pipelines	13,042	22,217	31,615	9,175	70.3	9,398	42.3
Product pipelines	4,280	5,000	8,203	2,082	48.6	1,841	28.9
Movement by pipelines (million tons)	129.9	225.9	346.3	95.8	73.7	120.6	53.7
Average length of haul (kilometers)	394	650	815	256	65.0	165	25.4
Share of pipelines in movement of oil freight (percent)	40.3	45.3	47.7	5.0	—	2.4	—
Share of pipelines in ton-kilometers of oil movement (percent)	17.8	30.6	40.4	12.8	—	9.8	—

From these data it is clear that in 1966-70, compared with 1961-65, the movement of crude and products by pipeline is to significantly increase. This is connected with the completion of powerful pipelines of diameters 529 to 1,020 millimeters and the further increase in the loading per kilometer. Construction of crude oil pipelines will retain its priority. It should be noted that it would be economically expedient to build product pipelines in the Ukraine, in the new Lands, the PriBaltics and in other directions, along which a great mass of light products is moved by railroad. However, the allocation of resources is toward crude oil pipeline construction, without which it would not be possible to increase oil output in individual regions.

In the current Five Year Plan the western and northwestern system of pipelines will be further expanded as a result of completion of the crude oil pipelines Polotsk-Ventspils, Yaroslavl-Kirishi, Almetyevsk-Kuybyshev No. 2, Brody-Uzhgorod No. 2, Almetyevsk-Gorkiy No. 3, and also the product pipelines Ryazan-Moscow, Kirishi-Leningrad and parallel pipelines along other sectors (of the system). There will be completed pipelines for the movement of high-sulfur crudes of the Central Volga and the Urals.

However, particular attention in the Five Year Plan is being given to the construction of new powerful crude oil pipelines from the prospective oil extracting regions of Tyumen Oblast, the Mangyshlak Peninsula, Northern Caucasus, and Turkmen. During 1966-70 there will be completed in these regions about 5,500 kilometers of crude oil pipelines, including Shaim-Tyumen, Ust-Balyk--Omsk, Mangyshlak-Guryev-Kuybyshev, Ozek-Suat--Malgobek, Malgobek-Tikhoretsk, Baku-Batumi No. 2.

The particulars of pipeline transport over a long distance of Mangyshlak, Stavropol, and Turkmen high-paraffin crudes are related to the congealing of this crude at ordinary temperatures; for this reason it will be necessary to build "heated" pipelines or the transport of the crude with water, with preliminary thermal (heat) treatment, with the addition of surface-active agents, and so forth. Additional difficulties in the development of pipeline transport during 1966-70 will be encountered in the laying of pipelines in Tyumen Oblast across flooded taiga, and also in the movement of Turkmen crude to the Ukraine with multiple transfers from one type of transport to another.

In the subsequent Five Year Plan (1971-75) there will be completed second sections along the Friendship crude oil pipeline, and the expansion to planned capacity of Almetyevsk-Gorkiy, Yaroslavl-Kirishi, Mangyshlak-Guryev-Kuybyshev, Ozek-Suat--Malgobek-Tikhoretsk, Baku-Batumi No. 2, and Ukhta-Yaroslavl crude oil lines. In addition, it is planned to build these new crude pipelines: Ust-Balyk--Krasnoyarsk, Omsk-Pavlodar, and also crude oil lines for the movement of crude from new producing regions in Belorussia, the Ukraine, Eastern Siberia, and so forth. With the availability of corresponding resources in 1971-75, the product pipeline system will also be expanded.

"A Chronology of Pipeline Construction in the Soviet Union"
Stroitelstvo truboprovodov, 11, November 1967, pp. 2-29

Year	Accomplishment
1928	In November, on the eve of the 11th anniversary of the Great October Socialist Revolution, the first welded crude oil pipeline in the USSR was completed. This was the Groznyy-Tuapse crude oil line, about 250 millimeters in diameter, and 618 kilometers in length, with a design working pressure of 50 kilograms/square centimeter. In the construction of this pipeline, the progressive method of electrode welding of pipe was used for the first time in the world.
1930	Completion of the second crude oil pipeline between Baku and Batumi.
1936	Completion of the Guryev-Orsk crude oil pipeline, 325 millimeters in diameter and 709 kilometers in length, to be used for the transport of crude oil from fields of Western Kazakhstan to the Orsk refinery.
1941	Laying of a 200-millimeter gas pipeline, 70 kilometers long, from the Dashava deposits to the city of Lvov.

1942	First gas flow at the Elshanka deposit in Saratov Oblast. The first production of natural gas in the USSR for industrial useage. In the course of 20 days the gas pipeline Elshanka-Saratov was built, to deliver gas to industrial enterprises and electric powerplants in Saratov.
1943	From the gas deposits in the region of Buguruslan, a gas pipeline 325 millimeters in diameter and 160 kilometers in length, was laid to the city of Kuybyshev. The gas pipeline Pokhvinstnevo-Kuybyshev, diameter of 325 kilometers and length of 135 kilometers, was also built. Laying of the 127-kilometer gas pipeline between Voy-Vozh and Ukhta, 325 millimeters in diameter, was completed. And in 8 months the crude oil pipeline Astrakhan-Saratov was built.
1944	A governmental decision was made to lay the first pipeline in the USSR for the transport of gas over a long distance. This was the Saratov-Moscow line.
1945	Construction begun on the Saratov-Moscow gas pipeline, 325 millimeters in diameter and length of 788 kilometers, with a design working pressure of 55 kilograms/square centimeter (until then working pressure in gas pipelines did not exceed 35 kilograms/square centimeters).
1946	Construction of the Saratov-Moscow gas pipeline completed.
1947	Delivery of crude oil from the Tuymazy fields to the Ufa refinery began.
1948	The gas pipeline Dashava-Kiev was built, diameter of 529 millimeters and length of 512 kilometers; and also the Kokhtla-Yarve--Leningrad line, 529 millimeters in diameter and 202 kilometers in length.
1949	Discovery of the Shebelinka gas-condensate deposit in Kharkov Oblast. Automatic welding under flux first used on construction of the Dashava-Kiev-Bryansk-Moscow gas line.

1950	Discovery of the North Stavropol gas deposit.
1953	First commercial flow of gas in Central Asia.
1955	The length of transmission gas pipeline reached 4,861 kilometers (in 1945 the length was only 620 kilometers).
1956	Discovery of the Gazli gas deposit in Central Asia. The gas pipeline Stavropol-Moscow, 1,254 kilometers in length, was completed a year ahead of schedule. 720-millimeter pipe was used. The crude oil pipeline Tuymazy-Omsk was completed, the first oil pipeline to be built of 529-millimeter pipe.
1957	Construction of the TransSiberian crude oil pipeline Ufa-Omsk-Irkutsk begun. During the course of 2.5 months the 720-millimeter gas pipeline between Shebelinka and Dnepropetrovsk (176 kilometers) was laid.
1958	By November installation of the Dzharkak-Bukhara sector of the Dznarkak-Bukhara-Samarkand-Tashkent gas pipeline. Development of the Bukhara-Khorezm gas-bearing region, the largest in the USSR, was begun. Construction of the 376-kilometer gas pipeline Shkapovo-Ishimbay-Magnitogorsk was completed, the first of a series of transmission gas pipelines for the delivery of gas to the Urals. The crude oil pipeline Almetyevsk-Gorkiy was placed into operation.
1959	The Serpukhov-Leningrad gas pipeline was completed; its diameter was 720 millimeters and the length - more than 800 kilometers. The major gas fields Zeagli-Darvaza were discovered in the Karakum desert. Construction of pipelines 1,020 millimeters in diameter was mastered. The first gas pipeline of 1,020 millimeter pipe was Krasnodar Kray - Serpukhov.
1960	The high-mountain gas pipeline Karadag-Akstafa-Tbilisi-Yerevan was completed, supplying gas to the republics of the TransCaucasus. The diameter of the pipe was 529 to 720 millimeters, and the length (with branches) 856 kilometers. The Central Asia

gas pipeline Dzharkak-Bukhara-Samarkand-Tashkent was completed in its entirety. It was decided to build the "Friendship" crude oil pipeline, which would have no equal in Europe in terms of capacity, distance over which the crude would be moved, and in number of pumping stations. The gas pipeline Dashava-Minsk, 622 kilometers long, was built (443 kilometers was built of 820-millimeter pipe, and 219 kilometers of 720-millimeter pipe).

1961   Laying of the Saratov-Gorkiy-Ivanovo-Cherepovets gas pipeline, 529, 720, and 820 millimeters in diameter was completed. The first kilometers of pipe on the Bukhara-Ural gas line were installed. Crude oil was obtained from a well near the village of Megion, which laid the beginning of the oil extracting industry in Western Siberia. The first commercial flow of crude oil was obtained on the Mangyshlak Peninsula. The crude oil pipeline Gorkiy-Ryazan, 720 millimeters in diameter and 394 kilometers in length, went into operation.

1962   Laying of the gas pipeline from Krasnodar Kray to Serpukhov, a distance of 1,000 kilometers, with a diameter of 1,020 millimeters which, together with the Stavropol-Moscow gas pipeline, linked the southern gas-producing regions with the Center of the country. The pumping of crude oil from the USSR to Czechoslovakia through the Friendship pipeline was begun.
And pumping of oil to Moscow through the crude pipeline system Almetyevsk-Gorkiy-Ryazan-Moscow was also begun. The product pipeline Penza-Bryansk was completed in its entirety. The crude oil pipeline Tikhoretskaya-Tuapse was completed. The Bayram-Ali gas deposit in the Karakum desert was discovered. At the Novo Moskovsk metallurgical plant the manufacture of 1,020-millimeter pipe with one longitudinal weld was mastered.

1963   The first section of one of the most powerful gas pipelines in the world, Bukhara-Urals, diameter of 1,020 millimeters and length of 2,200 kilometers, was completed. Construction of the Moscow Gas Ring, a pipeline designed to supply gas to tens of cities, population points and industrial enterprises in Moscow Oblast, was finished. The gas pipeline Ordzhonikidzhe-Tbilisi, diameter of 529 millimeters and 729 millimeters, with a length of 200 kilometers (357 kilometers including the Maykop-Nevinnomyssk branch line) was placed into use. Pumping of crude oil to Poland through the Friendship pipeline begun. Crude oil moved from the city of Mozyr to the city of Schwedt in East Germany by means of the Friendship pipeline. Operation of the crude oil pipeline Gorkiy-Yaroslavl begun. The Minnibayevo-Kazan gas pipeline No. 2, 529 millimeters in diameter and 282 kilometers in length, completed. This pipe-handles stripped associated gas coming from the Minnibayevo natural gasoline plant. Construction of the gas pipelines Kzyl-Tumshuk--Dushanbe and Leninabad-Fergana completed. At the Chelyabinsk pipe mill the manufacture of 1,020-millimeter pipe with 2 longitudinal welds was initiated. A test batch of pipe, of new grade steel, with a yield strength of up to 55 kilograms/square centimeter, was produced.

1964 Movement of crude through the entire Friendship pipeline system begun. The length of the pipeline within the USSR is 3,004 kilometers, in Poland - 675 kilometers, in East Germany - 27 kilometers, in Czechoslovakia - 836 kilometers, and in Hungary - 123 kilometers. The diameter of pipe used (millimeters): 426, 529, 720, 820 and 1,020. Pumping units with a productivity of 7,000 cubic meters per hour were first used on this pipeline. In the 5.5 years of operation of the pipeline, it has moved about 76 million tons of crude oil. Of this quantity, more than 53 million tons were delivered to Eastern Europe. The last stage of the Trans-Siberian crude oil pipeline Ufa-Omsk-Novosibirsk-Krasnoyarsk-Irkutsk was completed. The diameter of this pipeline is 720 millimeters and its total length is 3,682 kilometers. The cost of movement of crude through this pipeline is 4 times less than the cost of movement by rail. Construction of the gas pipeline Taas-Tumus--Yakutsk-Pokrovsk, the first transmission gas pipeline to be built in the Arctic, was begun. A portion of the pipeline is being built above-ground.

1965   The first gas pipeline in Western Siberia, Igrim-Serov, 1,020 millimeters in diameter and length of 497 kilometers, was completed. Laying of the first oil pipeline in Western Siberia, Shaim-Tyumen, 529 millimeters in diameter and length of 410 kilometers, also completed. The oil pipeline itself was built in 18 months. All of the planning-investigation work, construction of the line, equipping of the fields and construction of a pumping station was accomplished in 22 months, despite the great difficulties related to climate and the geological-hydrological characteristics of the route. The crude oil pipeline Uzen-Zhetybay-Shevchenko, on the Mangyshlak Peninsula, was completed for use. Work on the gas pipeline Belousovo-Leningrad, 1,020 millimeters in diameter and 875 kilometers in length, has unfolded. Construction initiated on the gas pipeline Mubarek-Tashkent-Chimkent-Dzhambul-Frunze-Alma Ata, designed to supply gas to the republics of Central Asia and Kazakhstan. The length of the pipeline is 1,317 kilometers, and the diameters will be 529, 720, 820 and 1,020 millimeters. By the end of 1965, the length of oil and gas pipelines of the Soviet Union reached 75,000 kilometers. The length of gas pipelines during the Seven Year Plan increased by more than 3 times and by the end of 1965 totalled more than 42,000 kilometers. The cost of transport of gas declined by 25 percent during 1959-65.

1966   On the base of gas deposits discovered in Central Asia construction of the Central Asia-Center pipeline was begun, one of the most important projects of the Five Year Plan. Construction of the second Bukhara-Urals gas pipeline was finished. The total length of both lines, including loopings, is more than 4,500 kilometers. The gas pipeline from Serov to Nizhnyy Tagil, the end point of the Bukhara-Ural gas pipeline, was placed into operation. Thus, gas from Tyumen Oblast joined in a single flow with gas from Uzbek. The Magnitogorsk-Ishimbay gas pipeline No. 2 was completed, as was the gas pipeline Nizhnyaya Tura-Pashiya-Lysva-Chusovaya-Perm, with a diameter of 1,020 millimeters and a length, including branch lines, of about 364 kilometers. In a record period (in 8 months) the Achak gas deposit was explored, planned, equipped and linked with an existing pipeline.

1967 Construction of the Central Asia-Center gas pipeline No. 1, with a diameter of 1,020 millimeters and a length of about 3,000 kilometers, was completed one year ahead of schedule. The second crude oil pipeline in Western Siberia, Ust-Balyk--Omsk, with a diameter of 1,020 millimeters and a length of about 1,000 kilometers, was finished. The gas pipeline "Bratstvo" (Brotherhood) by means of which gas from the deposits in the Western Ukraine will be delivered to Czechoslovakia, was put into operation. Construction of the Ostrogozhsk-Belousovo gas pipeline, 1,020 millimeters in diameter and 533 kilometers in length, was completed. Work was initiated on the construction of a powerful system of transmission gas pipelines from the North of the Tyumen Oblast to regions of the Center, West and the Urals. The length of this system will be 7,430 kilometers. Of this, the Northern branch will account for 5,394 kilometers (Nadym-Salekhard-Ukhta-Kotlas-Cherepovets-Torzhok-Minsk; from Cherepovets two pipelines will be built, to Leningrad and Gorkiy); the Urals branch will account for the remaining 2,036 kilometers (Nadym-Serginy-Komsomolskiy-Ivdel-Serov-Nizhnyaya Tura - Sverdlovsk-Chelyabinsk; from Nizhnyaya Tura gas pipelines will be built to Perm, Berezniki, Izhevsk). The productivity of the system will be 130 billion cubic meters per year. The technology of construction of pipelines 1,220 and 1,420 millimeters in diameter is being mastered. Scientific, construction, and technological problems, connected with the future building of ultra-powerful gas pipelines 2 to 2.5 meters in diameter, are being resolved.

"Technical Progress in the Transport and Storage of Crude Oil and Petroleum Products"
Neftyanoye khozyaystvo, 10, October 1967, pp. 48-57

In June 1918 the whole oil industry of Russia was nationalized and administration of this important branch of the economy was transferred to the government. At that time there were only 91 oil bases in operation in the country, and the length of trunk pipelines for the movement of oil reached 1,120 kilometers with an annual productivity of 2.5 million tons. During the past 50 years pipeline transport has taken gigantic strides forward, as illustrated in Table 1.

Table 1

Development of Pipeline Transport in the USSR

Year	Total Length of Oil Pipelines (Thousand kilometers)	Tons-Originated (Million)	Ton-Kilometers (Billion)
1917	1.12	2.5	0.50
1928	1.60	1.1	0.70
1940	4.10	7.9	3.80
1945	4.40	5.6	2.70
1950	5.40	15.3	4.9
1960	17.3	129.9	51.2
1965	28.2	226.0	146.6
1966	29.65	247.7	164.9
1967 Plan	33.15	271.5	180.0
1970 Plan	41.1	452	300.0

The participation of pipelines in the total volume of movement of oil freight has increased from year to year, as Table 2 indicated.

Table 2

Share (Percent) of Various Types of Transport in the
Movement of Oil Freight

Means of Transport	1940 Tons-Originated	1940 Ton Kilometers	1965 Tons-Originated	1965 Ton Kilometers
Pipeline	11.8	5.7	45.3	30.7
Railroads	44.4	54.7	44.4	58.6
River	14.5	18.2	7.5	6.0
Sea	29.3	21.4	5.3	4.7

The volume of work performed by pipelines in 1966 reached 165 billion ton-kilometers. Of the 217.9 million tons of oil produced in the RSFSR in 1966, a total of 204.6 million tons or 93.8 percent was moved by pipelines under Glavneftesnab (Main Administration for Oil Supply) RSFSR. Product pipelines handled 21.1 million tons.

In the future it is planned to connect the oil pipeline Ust-Balyk--Omsk by pipelines with the oil deposits of Tomsk Oblast. From these deposits it is also planned to build a powerful oil pipeline to Anzhero-Sudzhensk, forming a ring of pipelines.

35 to 40 years ago the maximum diameter of trunk oil pipelines did not exceed 300 millimeters. At the present time more than 65 percent, by length, of trunk pipelines have diameter of 500 to 1,000 millimeters. It is planned to build in the near future oil pipelines having a diameter of 1,220 millimeters, and later on - 1,420 millimeters and greater.

For cutting steel consumption and reducing construction costs a new type of electrically-welded oil pipeline pipe was developed in 1962. From this type the manufacture of pipe ranging in diameter from 219 to 1,220 millimeters, with wall thickness of 4 to 20 millimeters, is scheduled. Pipe of straight or spiral weld must have a length of 18 to 21 meters, and for their production the use of low-alloy steel.

Table 3

Cost of Movement of Crude Oil and Petroleum Products
by Pipeline, in Dependence on Diameter

Pipeline Diameter (Millimeters)	Pipeline Productivity (Million tons/year)	Cost of Movement of Ton-kilometer (Kopecks)
219	0.9	0.220
273	1.5	0.160
325	2.3	0.140
377	3.3	0.130
426	5.0	0.120
529	7.5	0.100
630	11.7	0.075
720	17.0	0.065
820	26.0	0.063
1,020	47.5	0.060

In the past 25 to 30 years, for the pumping of crude oil and petroleum products by pipeline use of highly productive centrifugal pumps have taken the place of cumbersome and low-productive piston pumps. The productivity of the best piston pump, NT-45, is about 150 cubic meters/hour, and contemporary trunk pipeline centrifugal pumps reach 7,000 cubic meters/hour.

Characteristics of centrifugal pumps, used at the present time for the movement of crude oil and petroleum products by pipeline are given in Table 4.

Table 4

Characteristics of Centrifugal Pumps, Used by Trunk Oil Pipelines

Type of Pump	Productivity (Cubic Meters per Hour)	Static Head (Meters of Liquid)	Revolutions per Minute	Capacity (Kilowatts)	Pump Coefficient in Pumping of Water
8MB-9X2	180-500	350-250	2,950	280-460	0.70 (about)
8ND-10X5	320	425	2,950	690	0.73
10N-8X4	500	740	3,000	1,500	0.73
12N-10X4	750	740	3,000	2,000	0.73
14N-12X2	1,100	370	3,000	1,600	0.80
16ND-10X1	2,200	230	3,000	1,600	0.86
20ND-12X1	3,000	220	3,000	2,000	0.84
24ND-14X1	4,000	210	3,000	2,750	0.86
24ND-17X1	5,000	210	3,000	3,200	--
28ND-19X1	7,000	220	3,000	5,000	0.90

"A Portfolio on the Central Asia-Center Gas Pipeline"
(Sources and pages as cited)

Gas from the Gazli and Achak fields is being moved through the pipeline to the Saratov underground storage facility as well as to the city of Gorkiy. The daily delivery to Saratov has reached 12 million cubic meters. Uzbek and Turkmen gas is already being burned at enterprises in Gorkiy, Cherepovets, and others. (Turkmenskaya iskra, September 15, 1967, p. 1)

The Achak deposit is located 30 kilometers from the transmission gas pipeline. By the end of the year 14 wells will be producing. Reserves at Achak are estimated at 175 billion cubic meters. Within a month Moscow will be receiving 10 million cubic meters of Central Asian gas daily. This is equal to about one-quarter of Moscow's requirements at present. By November construction workers will have completed the laying of a second, 40-inch gas pipeline to Leningrad (from Moscow). Thus, the route of the Central Asian gas will be increased by still another 700 kilometers. (Pravda, October 5, 1967, p. 2)

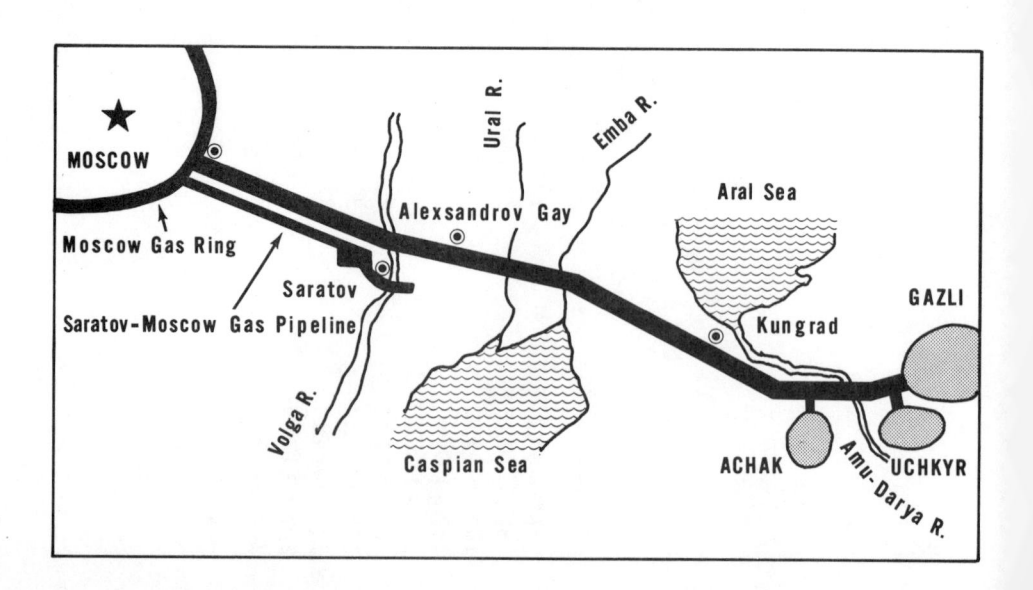

The total length of the pipeline is 2,751 kilometers, of which 1,500 kilometers were laid in arid regions. Upon completion of the second section of the system, the annual carrying capacity will reach 25 billion cubic meters. (Sovetskaya Rossiya, October 5, 1967, p. 1)

The magnitude of construction of the 2,750-kilometer gas pipeline, of pipe 1,020 millimeters in diameter, with a productivity of 10.5 billion cubic meters per year, is evident when compared with the first gas pipeline built in the USSR. This was the Saratov-Moscow gas pipeline, 800 kilometers in length, with a diameter of 300 millimeters and a productivity of 0.5 billion cubic meters per year. With the aim of stepping up the completion of the pipeline, there was a change in the established practice of exploration and completion for production of gas fields. Already, in the initial stage of commercial exploration of the deposit according to the results of testing the first exploratory well yielding gas flows, according to geological analogy with other deposits of the given gas-bearing structure (region), and with geophysical methods, the configuration of geological structure was determined and preliminary evaluation of the gas reserves made. Then a technological scheme was established concerning the test-commercial development of the deposit as well as planning documentation for equipping the first section of the deposit. After this, several exploratory-development wells were drilled, the deposit was equipped in terms of the minimum output scheme and was linked with the transmission gas pipeline. All this took place in one year. Under normal practice, this would have required 4 to 5 years. Laying of the second, parallel pipeline is moving ahead at full speed. It is being built of 1,220-millimeter pipe and will have a productivity of 15 billion cubic meters per year. (Pravda, October 6, 1967, p. 2)

The first section of the pipeline will deliver to Moscow, industrial cities of the Center and the PoVolga 10 billion cubic meters of gas per year. Calculations indicate that cheap Gazli gas will permit the savings in capital investments in the fuel industry of 450 million rubles and more than 170 million rubles in exploitation expenditures. In the winter time Moscow consumes 40 to 42 million cubic meters of gas daily. With the linking of the pipeline to the Moscow gas ring, an additional 9 to 10 million cubic meters may be obtained. (Sovetskaya Rossiya, October 6, 1967, p. 1)

Since the discovery of the Achak gas deposit, it has produced 522 million cubic meters. The cost of production of 1,000 cubic meters of Achak gas on the average is 60 kopecks. The cost of preparing 1,000 cubic meters of Achak gas is 14 kopecks. (Turkmenskaya iskra, October 7, 1967, p. 1)

In 1968 the Achak gas deposit will produce not less than 3.5 billion cubic meters, and by 1970 - 8 to 9 billion cubic meters. (Turkmenskaya iskra, October 6, 1967, p. 3)

The Central Asian gas pipeline links up the Moscow gas ring at the city of Voskresensk. The Central Asian pipeline was built in less than two years, and the first section of the Achak gas field was equipped in about one year. The Elshano-Kurdyum (Saratov Oblast) underground gas storage facility, with compressor station, was completed in 6 months. The total cost of the work, carried out during 1966-67, exceeds 200 million rubles. From July 15th the pipeline has delivered about 10 million cubic meters of gas to the regions of the Center. In 1968 delivery will increase to 16 million cubic meters per day, and upon completion of the second pipeline, 48 inches in diameter, annual deliveries will increase to 25 billion cubic meters, or 68 million cubic meters per day. This is double the daily consumption of gas in Moscow in the summer at the present time.

"An Example of Fraternal Cooperation"
Neftyanik, 7, July 1967, pp. 16-18

Socialism has established a new type of international relations. The socialist countries have accumulated valuable experience in using various forms of mutual economic ties. They have established international organizations, and, foremost, the Council for Mutual Economic Aid (COMECON) which provides for the coordination of their planning activities. During the 18 years of its existence COMECON has done much for the organization and development of cooperation. On its recommendation about 1,800 types of products have been included in specialization, and large enterprises adapted to intergovernmental specialization and cooperation have been built or are under construction.

History has fully confirmed Lenin's thesis that the problem of fuel and raw materials is actually an international problem. As a result of the combined efforts of the COMECON countries it became possible to speed up the development of their fuel-raw material base, and to create a net of powerful oil and gas pipelines together with extremely large oil refineries, not only in the USSR but in the other socialist countries as well. An immense complex in this net is the oil pipeline "Friendship," which began operating in January 1962.

During the current Five Year Plan Soviet oil workers will produce a total of about 1.5 billion tons of crude oil, of which tens of millions of tons will be pumped along the "Friendship" pipeline alone to Poland, GDR, Czechoslovakia and Hungary. This oil pipeline permits them not only to increase fuel consumption from year to year, but also to make more progressive the structure of the fuel-power balance of the countries which do not have their own oil in sufficient quantities.

About 1,200 enterprises, including 12 petroleum refineries and 48 chemical plants, have been constructed in the socialist countries with the economic and scientific-technical assistance of the USSR. Similar relations exist also between other socialist nations. The German Democratic Republic, for example, for the past 10 years has supplied the fraternal countries with more than 900 sets of equipment for entire enterprises or installations, including many in the USSR. This confirms the existence of authentic fraternal international relations and mutual aid between the peoples of the socialist nations.

In only three years, from December 1963 to December 1966, of operation of the "Friendship" pipeline, the GDR has received 10 million tons of Soviet oil. The right flank of the German industry is the petroleum refining and the petrochemical combine in Schwedt, whose annual productivity since the operation of the second stage in the beginning of 1966 has reached 4 million tons. By the beginning of 1967 it was already covering half the needs of the country for fuel. The construction of a new center of the petrochemical industry in the GDR is being completed. Through a 300-kilometer pipeline from Schwedt Soviet oil will reach the "Leuna-2" combine this year. Specialists from a number of socialist countries are participating in the construction. At the end of last year the construction of still another oil pipeline was begun in the GDR. It is being laid from Schwedt to the Berlin region. It is planned to extend it to Dresden in the future.

The scientific-technical cooperation between the Soviet Union and the German Democratic Republic received broad development in the field of petrochemistry, which was accomplished by means of the mutual transfer of scientific-technical documents, joint research, and the exchange of experts and scientific delegations. In recent years alone many delegations of specialists from the GDR visited a number of Soviet oil refineries and petrochemical enterprises. Our experience proved quite useful to the German friends during their creation and development of a new branch of industry in the GDR -- petrochemistry. At the petrochemical combine "Leuna-2" Soviet specialists helped their German comrades make optimum use of the facilities producing ophelines, and at the plant in Luzendorf, helped put into operation the new deparaffination units. The German friends for their part share with us valuable experience in the production of cables, acetylene, chemical fibers, and domestic chemical products.

Close fraternal relations have been established between the USSR and Poland. At the present time Poland is satisfying its petroleum needs 100 percent from Soviet deliveries. The Polish people have called the Plock oil refinery and petrochemical combine a giant of chemistry. Following the start of operations of the first stage alone it is refining annually no less than 2 million tons of oil, making 15 kinds of products. The USSR is building a powerful atmospheric-vacuum tower for the second stage of the enterprise; when it is operative the combine will refine more than 5 million tons of oil annually.

A direct economic effect on Poland from the "Friendship" oil pipeline is evident today in that oil transport costs alone have decreased three-fold. With the Plock combine in operation great and important changes are occurring in the life of the Polish state. As is well known, coal has been the basis for chemistry in Poland for a long time; but now coal is giving way to oil and gas. With the changeover to oil the Polish chemical industry enters a new, higher stage of development.

Construction in Plock is remarkable in that a new cadre of petrochemists has risen and is growing here. A broad net of schools and courses for training highly qualified specialists has been established at this combine. Many Polish engineers and technicians have already received specialized training at petrochemical enterprises in the USSR, Czechoslovakia, Hungary, the GDR, and Rumania. The combine director, P. Novak, stated in a conversation with Soviet journalists, "Alone Poland would not have been able to erect an enterprise of such size, of such exceptional complexity. The fraternal socialist countries came to our aid. We received from them the basic equipment, which was perfected from a technical point of view. This is why we call the combine a "little international."

The Pulavy nitrogen fertilizer combine is called a Hercules of chemistry by the Poles. Its first stage will give agriculture almost a million tons of fertilizer, which is as much as all the country's enterprises produce now. The capacity of the second stage will be more than a million tons. And here the "international" can again be observed in action. The USSR, Czechoslovakia and the GDR will supply the major part of the new equipment. The combine will operate on natural gas, coming by pipeline from the Soviet Union.

For the past two years prospecting operations for new deposits of oil and gas have been conducted in Poland. Soviet specialists are assisting Polish engineers and workers. The USSR has supplied the newest equipment which allows the drilling of wells 4,500 meters deep, geophysical apparatus, transport means, and technical documents. The prospecting has yielded the first successes. Commercial deposits of oil have been found in the region of Mekhuva and Bokhni near Cracow. Oil was also found in the Lyublin, Lodzinsk and Zelenogursk provinces. In the Lyubachub area there were found large reserves of natural gas. It will be used for the gasification of the industrial enterprises of Silesia. In 1966-70 the volume of drilling operations in Poland will increase not less than 50 percent. As it has been in the past, chief attention will be given to the search for new oil and gas deposits. There will be an especially large range of operations on the Polish lowlands and in the Carpathian foothills. In accordance with the concluded agreement the Soviet Union in the future will assist Poland in oil and gas exploration.

The Socialist Republic of Czechoslovakia, as is well known, is a developed industrial country. Machine-building and metallurgy are its key branches of industry. But in the last few years chemistry has also become such a branch. The first stage of the Slovakian petroleum refinery "Slavnaft", which produces a variety of products, is already in operation. Four more stages will be constructed. The Czech people call the gas pipeline, which stretches from the Soviet border to Bratislava, the construction of friendship. By the beginning of 1967 its length on Czech territory already has reached 360 kilometers. It has been tested. The first thousands of cubic meters of the "blue fuel" will pass through the line this year. By 1970 the total annual delivery of gas from the USSR will reach 1 billion cubic meters. Several chemical enterprises are being constructed along the pipe line, the biggest of which is the Strazhsk chemical fertilizers combine. After the first stage of the gas pipeline is put in operation, the extension of the line for another several hundred kilometers is planned.

Soviet - Czech economic ties are being systematically expanded. In 1966 an agreement between our countries concerning the technical cooperation in the construction of oil refinery and chemical industry projects was signed in Moscow. The Soviet Union will supply Czechoslovakia with the complex equipment for them.

Planning and prospecting operations for the expansion of the Soviet-Czech section of the "Friendship" oil pipeline area already underway. A second line will be laid parallel to the first one, which was built in 1962. Putting into operation the second stage of the trans-European oil pipeline will permit a significant increase in the delivery of liquid fuel to the fraternal socialist countries. A supplier of pipe for the "Friendship" oil pipeline and the Dashava-Bratislava gas pipeline is the Khomutovsk tube mill of Czechoslovakia, which is increasing its productive capacity. It produced 265 thousand seamless pipes in 1966 alone.

Antonin Novotny, appraising Soviet-Czech economic cooperation, said: "The development of cooperation with the USSR has created a new firm foundation for the rebuilding of the structure of our industry, for providing our industry with oil and natural gas, for the construction of atomic power stations, for mutual productive specialization and cooperation."

The delivery of Soviet oil to Hungary is increasing. While in 1965 it amounted to 1.8 million tons, and in 1966 about 2.5 million tons, in 1970 the annual volume will reach 4 million tons. Scientific and technical cooperation between the Soviet Union and Hungary is expanding and growing stronger. The newspaper "Nepsabadshag" noted that in 1965 600 Hungarian specialists visited the USSR. During the same period 270 Soviet specialists visited Hungary. Hungarian specialists have studied Soviet experience first of all in the fields of mining, oil refining, oil production, machine-building and communications. The Soviet specialists, in turn, are studying the manufacture of a number of products of light industry and pharmacology. Soviet-Bulgarian economic relations are successfully developing. The Soviet Union supplies Bulgaria with various types of machinery and equipment, and also important industrial raw materials, of which oil is the most important. The symbol of the inviolable Bulgarian-Soviet friendship is the many industrial construction projects in Bulgaria, including the petrochemical combine in Burgas. Oil derricks have been raised near Pleven and in Dobrudzhe through the strength of this friendship. Having received their education and skills in the USSR, Bulgarian specialists, with the aid of their Soviet friends, are discovering yet new subterranean riches in the country.

Trade is an important form in the broad and varied economic ties of the socialist nations. Its volume is continually increasing. While in 1961-65 commodity circulation in the COMECON states amounted to about 99 billion rubles, in 1966-70 it will be approximately 140 billion rubles.

Foreign trade and socialist division of labor advance like two interdependent aspects of international economic relations, whose goal is the maximum economy of social labor in all spheres of the national economy.

Every historical stage of development of collaboration among the socialist countries has its own inherent specific forms and methods which are being constantly perfected. In recent years they have been assuming more and more a new character qualitatively. Side by side with regular trade such forms are being developed as multilateral agreement of plans, specialization and cooperation in productivity, joint capital investment, joint directorates and enterprises, established on the basis of agreements between two or several countries. Economic reforms, effected in many COMECON countries, are creating the conditions for shifting the center of gravity in the organization of economic activities to economic key factors and methods.

A world system of socialist countries should have a good, stable economic base. The Soviet Union, fulfilling its international obligation, satisfies a considerable part of the import needs of the other countries for oil, ore and other raw material commodities. The question of the close coordination of economic development on the basis of the material interests of both import-nations and export-nations was raised by life itself.

Here is one of many examples. In September 1966 an agreement on cooperation in producing oil in the Soviet Union was concluded between the USSR and Czechoslovakia. The agreement provided for an increase in deliveries of oil to Czechoslovakia after 1970. Our country will receive from Czechoslovakia in 1966-74 machinery, equipment, pipe, and other commodities necessary for oil extraction. Their payment will be realized by deliveries of oil to Czechoslovakia in quantities additional to those stipulated by the long-term trade agreement between both countries for 1970. The conclusion of analogous agreements with other countries is also proposed.

The purpose of international long-term credit in the relations between the socialist countries is changing: previously it was accepted as a form of aid from one country to another to carry out industrialization and overcoming these and those economic difficulties; now, however, it is used to stimulate the rational cooperation of production and increase the effectiveness of collaboration.

SELECTED READINGS 351

"Ultra-Powerful System of Gas Pipelines From North Tyumen
Oblast to Regions of the Center, West, and Urals"
Stroitel stvo truboprovodov, 3, March 1967, pp. 9-13

In the 50th year of Soviet rule, organizations of the Ministry of the Gas Industry are beginning operations in the North of our Motherland which are unique in scope and character. It involves utilization of the northern regions of the West Siberian oil and gas province, which in the next 5 to 10 years should become one of the main gas producing regions of the Soviet Union.

In northern Tyumen Oblast alone, 110 to 120 billion cubic meters of gas will be extracted per year.

The main gas bearing regions of West Siberia are Berezovo, Novoport, Tazovskoye, Pur, and Urengoy. From here, gas will be supplied to regions of the Center, West, and Urals through a powerful system of gas pipelines, the construction of which is to begin this year.

The system of gas pipelines from North Tyumen Oblast to regions of the Center, West, and Urals is divided into two branches -- Northern and Urals.

The northern branch passes along the Nadym-Salekhard-Ukhta-Cherepovets (or Rybinsk) - Torzhok - Minsk route. From Cherepovets it is planned to lay two independent gas pipelines to Leningrad and Gorkiy.

The Urals branch will be laid along the Nadym-Serginy-Komsomol'skiy-Ivdel-Serov-Nizhnyaya Tura-Sverdlovsk-Chelyabinsk route. From Nizhnyaya Tura it is planned to lay gas pipelines to Perm, Bereznikiy, and Izhevsk.

Gas lines from the Novyy Port (Novoport) and Vuktyl gas deposits respectively will be linked with the Northern branch in the region of Labytnangi and Ukhta. The Urals gas pipeline in the region of Komsomol'skiy connects with the Igrim-Serov gas pipeline and goes further almost to Chelyabinsk following along the route of existing gas pipelines.

Structural Decisions

The organization of transport of 130 billion cubic meters of gas per year to the European part of the country (85 billion cubic meters) and to the Urals (45 billion cubic meters) requires qualitatively new engineering decisions.

First, it is necessary to become oriented to the use of pipes of only large diameters -- 1,220 and 1,420 millimeters. The replacement of pipe having a diameter of 1,020 millimeters with pipe having a diameter of 1,420 millimeters reduces the steel input by 25 to 30 percent, and the capital investments (including the construction of compressor stations and auxiliary sites) by 20 percent.

Second, it is necessary to become oriented to the installation of gas-pumping aggregates with a capacity of 10,000 to 16,000 kilowatts. Only under this condition can the compressor stations be compact and easily controlled.

Third, technological communication at the head sections of the gas pipelines must be only by radio relay.

This article examines the structure of the line section of the gas pipeline. However, technical decisions on other structures associated with the main line are of great engineering interest and give rise to the necessity of the accelerated working-out of many serious problems.

Line section. According to climactic and engineering-geological conditions, the route of the gas pipelines can be divided into two characteristic regions. The first is widespread perpetually frozen ground. Located in this region are portions of the Northern branch between the head installations at the gas fields and the Abez' (railroad) Station, and portions of the Urals branch between the head installations and the Nadym River.

The second region consists of the remaining sectors of the route with a total length of about 5,000 kilometers.

In the region of perpetually frozen ground, the most expedient and reliable system of construction is the laying of gas pipeline above-ground on supports.

However, other decisions might arise in the process of a detailed study of ground conditions along the route. Not excluded, in particular, is the laying of gas pipeline of the surface in embankments, which from the technological point of view (retention of a minimum range of temperature fluctuations) is the most acceptable.

Above-ground laying is recommended according to the system developed by the All-Union Scientific-Research Institute of Hard Alloys. Under this alternative, the pipe are laid on stationary and sliding supports. In view of the weather conditions and the operational requirements, the lower portion of the pipe is located 0.7 meters from the surface of the ground.

Preliminary calculations have indicated that the distance between the sliding supports can be 50 meters for gas pipelines 1,420 millimeters in diameter, and 45 meters for pipelines 1,220 millimeters in diameter. Between the stationary supports the distance is 700 and 630 meters, respectively.

Piles of metal pipes are recommended as supports. They are more economical than reinforced concrete piles, and they have less weight and are more easily transported. In comparison with the "snake" system proposed by the All-Union Scientific-Research Institute of Hard Alloys, this system is distinguished by simplicity and economy and it ensures the best operation of the gas pipeline.

The line section outside the zone of perpetually frozen ground should be completed with the pipe laid on the surface or underground (depending on the hydrogeological conditions).

River crossings. In the first region, the route of the gas pipeline crosses a number of large rivers -- the Pur, Nadym, Ob, and Usu.

The Pur and Nadym have similar hydrogeological conditions. At crossings, they are characterized by broad bottom-lands with ox-bow and channels. The bottom-lands are submerged to a depth of about one meter, and the depth of the bed in the low-water period fluctuates from 0.5 to 1.5 meters. The width of the Pur and Nadym rivers in the low-water period is 1,300 and 1,000 meters, respectively. The width of the flooded bottom-lands during the high-water period is 12 and 24 kilometers, respectively. The spring elevation of water begins simultaneously with the ice flow in the second half of May and continues for about 3 weeks. The fall ice flow lasts for 3 to 7 days, and the period of stable ice on open water begins, as a rule, in the second half of October.

According to preliminary study, these rivers can be crossed by an above-water bridge structure. Sag pipe crossings here are less economical because of the perpetually frozen ground which make up the bottom lands. However, the final selection of the structure of crossings will depend on the results of detailed studies and the technological system of transport of gas on the whole.

Characteristics of the Gas Pipeline Routes From North Tyumen Oblast to Regions of the Center, West, and Urals

7,430 kilometers -- total length of system.

5,394 kilometers -- length of Northern branch.

2,036 kilometers -- length of Urals branch.

1,748-number of water-barrier crossings.

118 - number of railway crossings.

236 - number of highway crossings.

2,410 kilometers -- perpetually frozen sections.

4,734 kilometers -- forested sections.

1,835 kilometers -- swampy sections.

2,810 kilometers -- surface flooding (including swamps).

582 kilometers -- rocky ground sections.

Note: The length of sections includes feeder pipelines from the fields, and also the gas pipeline from Novyy Port to Kharp.

SELECTED READINGS

The route crosses the Usu River twice. At the first crossing, the width of the river is about 100 meters, and at the second crossing, it is about 580 meters. Recommended for the first crossing is an above-water bridge structure, and for the second crossing, a sag-pipe.

Of special interest is the crossing of the Ob River. The gas pipeline crosses the Ob in the region of Salekhard. Here the width of the bottom-land in places reaches 30 kilometers, and at the crossing, 8 kilometers. The width of the bed is 2.5 kilometers, and the depth is 14 to 20 meters. The duration of flooding of the bottom-land during the spring high water is an average of 80 days, but it sometimes reaches 122 days. Flooding of the bottom-land continues from late May to early September. The period and nature of the ice appearances on the Ob are the same as on the Pur and Nadym rivers.

Three alternatives for crossing the Ob were examined: above-water bridge, under-ground tunnel, and under-water sag pipe.

According to detailed calculations, the cost of one linear meter of bridge crossing is 9,000 to 10,000 rubles; tunnel, 5,000 to 6,000 rubles; and sag pipe, 2,500 to 3,000 rubles. In this connection, the State Institute for Design and Planning in the Gas Industry has recommended the designing of a sag-pipe, seven-thread crossing with pipe 1,020 millimeters in diameter.

On the remaining section of the route, where there is no perpetually frozen ground and the gas pipeline on the whole is laid underground or on the surface, considering the hydrogeological characteristics of the large rivers, crossings are recommended in the form of sag pipes using pipe 1,020 millimeters in diameter.

Crossings of medium and small rivers can be either sag pipe, beam, or suspension, depending on the specific conditions.

<u>Crossings through swamps.</u> The gas pipeline can cross innumerable numbers of large and small swamps along the route at the head section above ground, and at the sections west of the Abez Station and south of the Kazym River,

above ground or at ground level. These recommendations are based on the experience of designing and building the Igrim-Serov, Gorkiy-Cherepovets, Belousovo-Leningrad, and a number of other gas pipelines.

Pipes and fittings. According to preliminary pre-design studies, the Northern branch must be four-thread (one thread with a diameter of 1,220 millimeters; and three threads with a diameter of 1,420 millimeters), and the Urals branch must be three-thread (one thread with a diameter of 1,020 millimeters, the second with a diameter of 1,220 millimeters, and the third with a diameter of 1,420 millimeters). Thus, the first thread of the Northern branch and the second thread of the Urals gas pipeline will be built of pipe with a diameter of 1,220 millimeters (Igrim-Serov-Nizhniy Tagil gas pipeline, built of pipe having a diameter of 1,020 millimeters, is the first thread of the Urals branch). In the future, when the mass production of pipe with a diameter of 1,420 millimeters is underway, the second, third, and fourth threads of the Northern branch and the third thread of the Urals branch will be laid with these more economical pipes.

It is recommended to install on the gas pipelines equal-flow shut-off fittings of a spherical design which can operate normally at temperatures of up to minus 60 degrees Centigrade in the Northern regions and up to minus 40 degrees Centigrade in the other regions.

In this connection, the design bureaus and the fittings plants have the task of developing a design and initiating serial production of stopcocks with diameters of 1,200 and 1,400 millimeters which are adapted to operation under conditions of the Far North. If this task is not resolved in a timely manner and unequal-flow stopcocks must be installed on the gas pipelines, the capacity of the transmission lines will decline by 5 to 8 percent or by 6 to 10 billion cubic meters of gas per year.

<u>Main Technical-Economic Indicators of Gas Pipelines from North Tyumen Oblast to Regions of the Center, West, and Urals</u>

130 billion cubic meters of gas - productivity of system.

9 million tons - total steel used.

18,767 kilometers - total amount of pipe.

including:

    11,487 kilometers - 1,420 millimeters in diameter.

    6,169 kilometers - 1,220 millimeters in diameter.

    1,111 kilometers - 1,020 millimeters in diameter.

    50 - approximate number of compressor stations.

1,443 million rubles - capital investment in gas extraction.

4,619 million rubles - capital investment in transmission pipeline transport and underground storage of gas.

1.38 rubles - cost of extraction of 1,000 cubic meters of gas.

3.16 rubles - average cost of transport of 1,000 cubic meters of gas on the Northern branch.

1.62 rubles - average cost of transport of 1,000 cubic meters of gas on the Urals branch.

4.47 rubles - cost of 1,000 cubic meters of gas to Leningrad.

3.14 rubles - cost of 1,000 cubic meters of gas to Chelyabinsk.

    Ballasting. Of great importance in construction of the Northern and Urals gas pipelines is ballasting. Because of the numerous swamps and water currents, it will be necessary to install about 3 million cubic meters of reinforced concrete weights or about 60,000 meters of screw anchors.

    The replacement of the reinforced concrete weights with anchors will permit a reduction in expenditures for bracing pipe by a minimum of 150 million rubles. Therefore, the mass use of screw anchors is recommended which, moreover, will reduce the volume of transport operations in the roadless regions of the North.

Organization of Construction

The equipping of unique gas fields and the laying of an ultra-powerful system of transmission lines using large diameter pipe under the severe conditions of the Far North are operations on a scale and complexity unequalled either domestically or in world practice.

Manpower and material-technical resources. The work on the routes of the gas pipelines must be organized in such a way that in perpetually frozen areas and in taiga-swamp areas construction will be carried out only through the complex method. The complex must be carried through to the lower production unit.

Evidently it will be necessary in the system of the Ministry of the Gas Industry to create specialized administrations for the completion of pile supports in regions having perpetually frozen ground. In the other regions, we can retain the current standard principles of subordinate contractual-specialized construction.

According to preliminary calculations, proceeding from the assigned periods of construction and considering the number of working days per year (in Northern Tyumen Oblast not more than 200, in the central portion of Tyumen Oblast and in the Komi ASSR, 260, and in the other sections of the route, 286 days), an approximate annual need for manpower and material-technical resources has been established.

Item	Total	Including Northern gas pipeline	Urals gas pipeline
Number of workers	42,000	30,000	12,000
Number of construction machines	5,000	3,500	1,500
Number of motor vehicles	5,500	4,000	1,500
Including cross-country vehicles	3,500	2,500	1,000
Living quarters for builders stationary (thousand cubic meters)	200	150	50
mobile (number of trailers)	3,500	2,500	1,000

Construction materials and products prefabricated reinforced concrete (thousand cubic meters)	320	230	90
panels for buildings (thousand square meters)	140	100	40
crushed stone and gravel (thousand cubic meters)	550	400	150
steel pipe (thousand tons)	1,300	1,000	300
timber materials (thousand cubic meters)	550	400	150
screw anchors (thousand tons)	50	35	15

Bases of construction. In accordance with the conditions of laying the route, zones have been planned for which specific proposals are made to provide for the needs of construction.

The route of the Northern branch is divided into four zones: 1 -- gas fields, gas collectors to Nadym, and the Nadym-Salekhard sector; 2 -- the Salekhard-Kotlas sector; 3 -- the Kotlas-Cherepovets-Leningrad sector; 4 -- the Cherepovets-Minsk and Cherepovets-Gorkiy sectors.

For construction in Zones 1 and 2, it is necessary to create a support base of construction materials in Ukhta (plants for the manufacture of prefabricated reinforced concrete, large-panel home building, and ceramic gravel). It is also necessary to build mechanized stone and sand pits in the region of Kharp and a stone and gravel pit in the region of Ukhta. We must create two lumber enterprises in the regions of Tarko-Sale and Salekhard, and also shops for carpenter products in Ukhta. Moreover, in the region of Ukhta it is necessary to build enterprises for the capital repair of motor vehicles and construction equipment, a metal-structures plant, bases of specialized organizations, and bases for mechanization. In the region of Salekhard (Labytnangi), it is proposed to organize bases of the Administration of Under Water-Technical Operations, as well as bases of specialized organizations.

Needed besides the above sites are a transshipping base with moorage and a pioneer production base in Labytnangi, a receiving-transshipping base with moorage in Salekhard, a water-railway receiving-transshipping base and a pioneer production base in Nadym, a transshipping base with moorage and a pioneer production base in Urengoy and Tazovskoye, and a pioneer construction base in Ukhta. In the future, to link the Novyy Port deposit to the Northern branch it will be necessary to create a railway transshipping base in the region of Kharp-Obskaya and a transshipping base with moorage in the region of Novyy Port.

The construction of the Northern branch in Zone 2 will, moreover, be ensured by the available enterprises in the Vorkuta, Syktyvkar, and Kotlas industrial centers.

Construction in Zone 3 must basically be supported by the enterprises of Leningrad and Cherepovets, and in Zone 4, by plants in Gorkiy, Yaroslavl, Kalinin, and Minsk, and by enterprises of the Ministry of the Gas Industry.

The Urals branch is divided into two zones: 1 -- from Nadym to the Kazym River; 2 -- from the Kazym River to Sverdlovsk and Perm.

Construction in Zone 1 will be supported by the same enterprises as those in Zones 1 and 2 of the Northern branch, and the sector from the Kazym River, by enterprises of the Ministry of the Gas Industry located in Chelyabinsk Oblast.

For the Urals branch it is planned, moreover, to create a transshipment base with moorage in the region of Sergina, and receiving-transshipment and pioneer bases on the Kazym River and on the right bank of the Ob River.

In the areas of all the compressor stations along the Northern and Urals gas pipelines, according to pre-design recommendations, section production bases are being created.

Preparatory work. It is expected to take 7 to 8 years to complete construction of the entire gas pipeline system, but it is planned to complete in 1967 a large amount of work, mainly preparatory.

The main work in the preparatory period includes construction of transshipping and support bases; organization of section construction bases at compressor stations and gas fields; creation of stationary and mobile housing settlements; restoration and completion of a railway to open workers' movement during construction along the Salekhard-Nadym-Pur (Urengoy) route; establishment of winter quarters for fields; organization of bases for workers supply section, aerial transport and medical services; and clearing of routes and areas of timber, establishment of approaches and shelves at crossings of mountain regions of the Urals, and provision of electric power and water supply, communications, etc.

Each of the above measures is linked with broad engineering and economic research. Take the creation of stationary and mobile housing settlements. Here it is necessary not only to determine the amount of stationary and mobile housing and the planning of settlements and their dislocation on the route, but also to consider the organization of feeding the builders in the remote regions of the Far North, the creation of medical services, methods of combatting blood-sucking flies, etc.

Even the short list of basic work during the preparatory period indicates the complexity of the tasks which planning and production organizations must resolve this year.

Planning and scientific-research institutes of the Ministry of the Gas Industry are now working out plans for gas pipelines and fields, and the builders are beginning construction of the ultra-powerful gas pipelines.

"Technical-economic Indicators of the Development of Transmission Pipeline Transport of Gas During 1959-65"
Gazovaya promyshlennost, 12, December 1966, pp. 20-21

During the Seven Year Plan the transmission pipeline transport of gas underwent tremendous development. At the beginning of 1959 the system of transmission gas pipelines measured only 12.2 thousand kilometers. During 1959-65 in our country about 30 thousand kilometers of transmission gas pipeline were built, including branch pipelines. The growth in the transmission gas pipeline system in the USSR during 1959-65 (end of year) is illustrated in Table 1.

Table 1

Year	Total Length of Gas Pipe line (km)	Transmission Pipelines km	Percent Total	Branches km	Percent Total	Local Pipelines km	Percent Total
1959	16,494	11,102	67.3	1,547	9.4	3,845	23.3
1960	20,983	14,392	68.6	2,181	10.4	4,410	21.0
1961	25,329	16,678	65.8	3,504	13.9	5,151	20.3
1962	28,492	19,748	69.3	4,695	16.5	4,051	14.2
1963	33,033	23,888	72.3	5,962	18.1	3,183	9.6
1964	36,908	26,724	72.3	7,125	19.3	3,059	8.4
1965	42,273	30,269	71.6	7,986	18.9	3,018	9.5

The average annual increment in the length of gas pipelines reached 4.3 thousand kilometers, compared with a growth of only 0.2 thousand kilometers during 1940-50; 0.5 thousand kilometers during 1951-55; and 2.8 thousand kilometers during 1956-58.

During the Seven Year Plan the volume of gas transported (by pipeline) increased from 20.2 billion cubic meters in 1959 to 103.3 billion cubic meters in 1965, that is, by almost 5 times.

Table 2 presents the share of gas moved by transmission gas pipeline as a percent of total extraction of gas during 1959-65.

Table 2

Indicator	1959	1960	1961	1962	1963	1964	1965
Extraction of natural and associated gas (billion cubic meters)	35.3	45.3	59.0	73.5	89.8	108.6	127.6
Share of gas, moved by pipeline, as percent of volume produced	51.5	57.6	63.4	69.4	79.8	80.6	81.1

An important indicator, characterizing the development of transmission pipeline transport of gas is the volume of work performed (Table 3).

Table 3

Year	Volume of Transport Work Billion Cubic Meters-Kilometers	As Percent of 1959	Average Distance Gas Moved Kilometers	As Percent of 1959
1959	10,374	100	570	100.0
1960	15,782	152	607	108.2
1961	23,200	220	620	108.2
1962	31,950	310	629	110.5
1963	45,270	430	650	114.1
1964	58,024	560	660	115.8
1965	70,229	680	680	119.2

The basic factor, permitting (along with the growth in average distance moved) the increase in the volume of transport work, is the growth in the use of large diameter pipe (Table 4).

Table 4

Diameter (millimeters)	1959	1960	1961	1962	1963	1964	1965
	Length of Pipelines (Percent)						
100-273	15	12	11	11	11	10	10
325-529	48	45	42	41	39	39	38
720-1,020	37	43	47	48	50	51	52
of which, 1,020	0.5	3.2	5.0	7.2	11.2	13.8	17.8
Average diameter at end of year (millimeters)	514	553	574	581	605	614	628

During the Seven Year Plan the number and capacity of compressor stations increased significantly (Table 5).

Table 5

Year	Number of Stations	Index	Capacity (Thousand KVT)	Index
1959	18	1.00	130.2	1.00
1960	21	1.17	256.7	1.97
1961	28	1.55	564.7	4.41
1962	37	2.05	910.2	6.99
1963	52	2.89	1,190.0	9.14
1964	71	3.95	1,538.8	11.82
1965	81	4.50	1,868.8	14.36

The given data indicate that during the Seven Year Plan the number of compressor stations increased by 4.5 times, and their capacity increased by 14 times. It follows that the average capacity of one compressor station (KS) increased from 7.2 thousand kvt in 1959 to 23.1 thousand kvt in 1965, or by 3.2 times. The growth in the capacity of compressor stations increased as a result of raising the capacity of the gas-transmission aggregates. The growth in the average capacity of compressor stations during 1959-65 is shown in Table 6.

Table 6

Year	Average Capacity of One Compressor Station Absolute (Thousand KVT)	As Percent of 1959	Average Number of Aggregates per Station Absolute	As Percent of 1959	Average Capacity Per Aggregate Absolute (Thousand KVT)	As Percent of 1959
1959	7.2	100	6.8	100	1.05	100
1960	12.2	169	8.1	119	1.51	144
1961	20.2	280	9.2	135	2.18	208
1962	24.6	342	9.1	134	2.70	257
1963	22.9	318	8.5	125	2.70	257
1964	23.0	319	7.8	115	2.82	269
1965	23.1	321	8.1	119	2.81	268

Also during the Seven Year Plan the cost of transport of gas was reduced by 25 percent, and the cost of transport work - by more than one-third.

The number of workers, occupied in the transport of gas, increased by 3.4 times, and the volume of transport work by 6.8 times. Thus, labor productivity increased by 1.6 times or by 3.1 times (if considering the volume of transport work per one worker). The growth in labor productivity is illustrated in Table 7.

Table 7

Year	Number of Workers	Labor Productivity Million Cubic Meters/Worker	As Percent of 1959	Volume of Transport Work per Worker Billion Cubic Meters Kilometer per Worker	As Percent of 1959
1959	5,894	3.10	100.0	1.09	100.0
1960	8,565	3.04	98.0	1.46	134.0
1961	10,876	3.46	111.6	1.74	159.6
1962	12,644	4.02	129.8	2.27	208.3
1963	14,899	4.66	150.3	2.66	244.0
1964	18,556	4.71	152.0	3.12	286.3
1965	20,453	5.05	162.9	3.44	315.6

Table 8 presents data characterizing the relationship of payments to the budget of net income realized and capital investment in the construction of transmission gas pipelines.

Payment of the budget of net income realized from the sale of gas exceeded by 1.5 times the capital investment in transmission gas pipeline construction during the Seven Year Plan.

Table 8

Indicator	1959	1960	1961	1962	1963	1964	1965	Total
Total Capital Investment according to the Government Plan and from special sources (thousand rubles)	214,401	238,007	213,549	229,317	291,682	315,540	373,772	1,876,268
Total Payments to State Budget and Contribution of Profits to Capital Construction (thousand rubles)	110,716	146,008	230,140	341,097	496,532	648,264	809,153	2,781,910

"Certain Questions on the Economics of the Operation and Development of Oil Pipeline Transportation of the USSR, 1958-64"
Ekonomika neftedobyvayushchey promyshlennosti, 2, 1966, 35-39

In accordance with the growth in extraction and refining of oil and the consumption of oil products in the USSR during 1958-64, there has been a considerable increase in the volume of transport of crude oil and products.

In 1964, the total volume of movement of oil freight by all types of transport was 484 million tons. Of the total amount of freight turnover (measured in ton-kilometers) of rail transport, 15.5 percent was oil; of river transport, about 25 percent was oil; and of maritime, about 48.9 percent for oil (including export).

Tables 1 and 2 give the volume of transport of oil (tons originated) according to type of transport.

Table 1

Movement* of Oil and Oil Products by Various Types of Transport

Transport	1958 Million tons	%	1959 Million tons	%	1960 Million tons	%	1961 Million tons	%	1962 Million tons	%	1963 Million tons	%	1964 Million tons	%
Rail	112.5	45.4	132.0	46.2	151.0	46.6	168.4	47.3	190.5	47.3	206.4	46.8	219.2	45.3
River	16.1	6.5	17.5	6.1	18.4	5.7	20.5	5.7	21.2	5.3	22.8	5.2	23.9	4.9
Pipeline	94.9	38.3	111.3	39.1	129.9	40.2	144.0	40.3	165.2	41.1	185.4	42.1	212.6	44.0
Coastal	24.4	9.8	24.5	8.6	24.1	7.5	24.2	6.8	25.5	6.3	26.1	5.9	28.0	5.8
Total	247.9	100	285.3	100	323.4	100	357.4	100	402.4	100	440.7	100	483.7	100

* Tons - originated

Table 2

Freight Turnover (Ton-Kilometers) of Oil and Oil Products

Transport	1958 Billion ton-kms	1958 %	1959 Billion ton-kms	1959 %	1960 Billion ton-kms	1960 %	1961 Billion ton-kms	1961 %	1962 Billion ton-kms	1962 %	1963 Billion ton-kms	1963 %	1964 Billion ton-kms	1964 %
Rail	154.0	70.6	182.1	71.0	205.4	70.5	230.6	70.2	252.5	68.7	269.6	66.8	287.0	64.9
Pipeline	33.8	15.5	41.6	16.2	51.2	17.6	60.0	18.2	74.5	20.2	91.0	22.6	112.1	25.3
River	15.7	7.2	17.4	6.7	18.5	6.3	20.6	6.2	21.1	5.8	22.4	5.5	25.1	5.7
Coastal	14.7	6.7	15.4	6.1	16.3	5.6	18.1	5.4	19.6	5.3	20.5	5.1	18.2	4.1
Total	218.2	100	256.5	100	291.4	100	329.3	100	367.7	100	403.5	100	442.4	100

Table 3

Development of Network of Crude Oil and Products Pipelines

	Pipelines		Including			
			Crude Pipelines		Product Pipelines	
Years	Thousand kilometers	Percent of 1958	Thousand kilometers	Percent of 1958	Thousand kilometers	Percent of 1958
1958	14.5	100.0	11.0	100.0	3.5	100.0
1959	16.7	115.2	12.3	111.9	4.4	125.8
1960	17.3	119.4	13.0	118.2	4.3	123.0
1961	20.5	141.4	15.9	144.6	4.6	131.5
1962	21.7	149.7	16.7	151.9	5.0	143.0
1963	23.9	164.8	18.9	171.9	5.0	143.0
1964	26.9	185.6	21.9	199.1	5.0	144.0

As is evident from the data in Tables 1 and 2, the share of tons-originated and ton-kilometers of crude oil and oil products by pipeline transport is steadily growing, largely as the result of a decline in the freight turnover of these commodities by rail. Quantitatively, tons-originated and ton-kilometers increased considerably during the 7 years examined. With a total increase in ton-kilometers for all types of transport of more than twofold, the increase for rail transport was almost 1.9 times, and for pipeline transport, 3.3 times.

The role of pipelines in the transport of crude oil and products during the period examined increased as the result of an increase in the length of their network.

As is evident from Table 3, the length of the crude oil and products pipeline network increased from 1958 to 1964 by 85.6 percent, particularly in trunk crude oil pipelines. The fastest growth of the crude oil pipeline network is explained by the increase in the volume of refining and the necessity of the continuous supply of the oil refineries with raw material. The volume of tons-originated of oil pumping during 1958-64 increased by 134.7 percent, and ton-kilometers by 243.3 percent. In the total volume of ton-originated in 1964, the share of crude oil was 90 percent, and in ton-kilometers, it was 84.3 percent.

Built and put into operation during these years were such large trunk crude oil pipelines as Al'met'yevsk-Gor'kiy II, Gor'kiy-Ryazan', Gor'kiy-Yaroslavl', and Tuymazy-Omsk-Irkutsk. In 1964, work was completed on the last sections of the trans-European trunk oil pipeline "Druzhba," the length of which is more than 3,500 kilometers in the USSR alone.

Now 90 percent of all the oil refineries of the USSR are supplied with oil by pipelines, and only several relatively small refineries receive raw material by sea or rail. Thus, pipeline transport has a dominant role in supplying raw material.

Things are different with the transport and supply of oil products to the main consumer regions of the USSR. The amount of oil products handled by product pipelines from 1958 to 1964 increased by only 59 percent, and ton-kilometers by 181 percent. Here the role of pipeline transport is still low as the result of the small length of the product pipeline network and the limited technical facilities for handling only light products, mainly gasoline and diesel fuel. Until recently about 75 percent of all oil products were hauled by other means of transport, which considerably increased transport expenditures to the consumers. Moreover, because of the insignificance of the product pipeline network and the lack of branch lines from the main lines, there were sometimes seasonal difficulties, such as during the period of early completion of navigation on Siberian rivers and during the periods of intense consumption of light oil products in certain large agricultural regions (New Lands, the southern regions of the USSR, etc.). It must be noted that a large volume of light products have been transported until recently from Ufa refineries to the west by rail. This, besides increasing their cost, created an additional burden on the heavily loaded Ufa-Kuybyshev sector (of the railroad). According to data of the Institute on Complex Transportation Problems, the hauling of oil products for short distances is much more expensive than their handling by pipelines even if the latter were built of small diameter pipe.

The length of the product pipeline network in the USSR in 1963-1964 was only 18.5 percent of the total network of trunk crude oil and oil product pipelines, and in the US, it was almost 45 percent. Such attention in the US to the development of product pipelines in the presence of a dense network of railways and highways again confirms the great economic effectiveness and advantages of pipelines over other transportation. According to data of the State Institute for Designing Special Structures in the Gas Industry, total transport expenditures for oil movement and transfer by all types of transport were about 740 million rubles in 1960. For individual types of transportation, these expenses (including trans-shipping operations) were allocated as follows (in percent):

Rail	81.4
Pipeline	7.6
River	4.1
Maritime	3.4
General expenses for these types of transport	3.5

Thus, even with the consideration of possible inaccuracies in these computations, it can be stated that with a relatively large amount of crude oil and oil products handled by pipelines, transport expenses were very small in view of a lower cost in comparison with rail, as well as a relatively low share of expenditures for trans-shipping.

Current distribution of the crude oil and products pipelines network is determined to a large extent by the location of the oil extracting and refining industry. A large share of the trunk crude oil pipelines is located in the Urals-Volga region, in the North Caucasus, and in the Transcaucasus, i.e., where the major fields and refineries are located. Routes of trunk products pipelines link individual oil refineries with the largest consumers of light oil products: the central oblasts of the RSFSK, the Ukraine, Siberia, and North Kazakhstan.

In early 1965, the network of trunk crude oil pipelines was located as follows (in percent of the total length of the crude oil pipeline network):

Urals-Volga region	38.8
Siberia and the Far East	22.7
North and central oblasts of the RSFSR	4.4
North Caucasus	6.9
Central Asia and Kazakhstan	6.0
Transcaucasus	4.6
Ukraine	0.3
"Druzhba" oil pipeline	16.3

Thus, about two-thirds of the oil pipeline network (in length) are located in the regions of the Volga, Urals, Siberia, and the Far East. The network of product pipelines is distributed as follows: about 60 percent in regions of the Urals and West Siberia, about 23 percent in the central oblasts of the RSFSR, and about 17 percent in regions of the North Caucasus, the southern part of the RSFSR, and the Ukraine.

Considering the major scale of consumption of light oil products in the Ukraine, Tselinnyy Kray (new Lands), and the North Caucasus, the network of product pipelines here is obviously insufficiently developed.

Crude oil and oil products pipelines were built in 1958-64 mostly from large diameter pipe. By early 1965, the length of crude oil and oil product pipelines with a diameter of 500 millimeters and larger constituted 73 percent of the total length of pipelines, while in the US the length of pipelines of such diameters constituted little more than 10 percent.

The structure of the oil product pipelines by diameter changed essentially during these years, as is evident in Table 4.

Table 4

Distribution of Trunk Crude Oil and
Oil Product Pipeline System by Diameters*

Diameter in Millimeters	1958 Kilometers	1958 Percent of Total	1961 Kilometers	1961 Percent of Total	1964 Kilometers	1964 Percent of Total
219-273	1,564.5	13.7	1,673.5	9.8	1,731.5	7.4
325-377	4,257.5	37.2	4,503.5	26.3	4,503.5	19.2
529	4,640.0	40.5	6,740.0	39.4	8,408.0	35.9
630	--	--	--	--	441.0	1.9
720	989.0	8.6	3,680.0	21.5	4,948.0	21.1
820	--	--	498.0	3.0	2,111.0	9.0
1,020	--	--	--	--	1,275.0	5.5
Total	11,451.0	100.0	17,095.0	100.0	23,418.0	100.0

*Excluding pipelines of mixed diameters, and other small pipelines.

Thus, the length of large diameter crude oil and oil product pipelines increased from 5,629 kilometers (49 percent of the length of all pipelines) in 1958 to 17,183 kilometers (73.4 percent) in 1964.

With the development of the network of trunk pipelines and especially large diameter pipelines, the average length of haul for crude oil and oil products increased and, correspondingly, the average length of haul of oil by rail decreased, as is evident in Table 5.

Table 5

Average Length of Haul of Crude Oil
and Oil Products in 1958-1964
(in kilometers)

Average Length of Haul	1958	1960	1962	1964
For all types of transport	880	901	913	924
Including:				
Rail	1,369	1,360	1,325	1,309
Pipeline	356	391	451	527
Crude oil	337	350	416	493
Products	474	745	746	840

The average length of haul along products pipelines increased particularly in connection with the completion of the new trunk lines Kuybyshev-Penza-Bryansk and Omsk-Novosibirsk.

Of the main technical-economic indicators of oil pipeline transport in the USSR for the period under study, two are of considerable interest: the cost of pumping and the use of productive capacities.

The cost of pumping is the most important economic indicator which characterizes the productive-economic activities of transport. The cost of pumping crude oil and oil products by pipeline--this is the relation of annual operational expenditures to the amount of ton-kilometers performed.

According to averaged data, the cost of transport of oil on the whole for the USSR is presented in Table 6.

Table 6

Cost of Hauling Crude Oil and
Oil Products, by Type of Transport
(kopecks per 10 ton-kilometers)

Transport	1959	1960	1961	1962	1963	1964
Rail	3.00	3.03	2.81	2.67	2.70	2.70
River	1.69	1.70	1.62	1.67	1.70	1.76
Pipeline	1.26	1.20	1.28	1.22	1.15	1.06
Maritime	1.19	1.125	1.052	1.030	1.062	--

The cost of transporting oil by rail is 2.5 times more than by pipeline, and by river it is 1.6 times more than by pipeline. It is only with maritime transport that pipelines are not always competitive.

The structure of operational expenditures along pipelines of Glavneftesnab (the Main Administration of Oil Supply) of the RSFSR for 1964 is presented as follows (in percentage):

Salary and deductions for social security	14.1
Amortization of fixed assets	51.9
Electric power	15.1
Fuel, steam, and water	1.3
Current repair	0.8
Losses of oil and oil products	3.3
Administrative-managerial expenses	0.6
Other expenses	12.9
Total	100.0

Thus, the main items of expenses are amortization, electric power, and wages.

In analyzing the structure of operational expenditures, one can reveal the reserves for reducing the cost of pumping and draw conclusions on the growth of labor productivity and technical progress, the improvement of the organization of the transport process, etc.

The cost of pumping among individual oil pipeline administrations of Glavneftesnab, RSFSR for 1958-64 is presented in Table 7.

Table 7

Cost of Pumping Crude Oil and Oil Products by Oil Pipeline Administrations of Glavneftesnab RSFSR (in kopecks per 10 ton-kilometers)

Oil Pipeline Administration	1958	1959	1960	1961	1962	1963	1964
Urals-Siberia	1.09	0.98	0.93	0.98	0.93	0.95	0.92
Northwest	1.03	0.89	0.93	1.02	0.94	0.88	0.65
Kuybyshev	1.33	1.48	1.52	1.57	1.42	1.32	1.35
West Siberia	2.26	2.71	1.55	2.04	1.81	1.32	1.11
Grozny	1.93	1.81	1.73	1.84	1.85	1.98	1.85*
Volgograd	1.40	1.05	1.06	0.93	0.83	1.06	1.05
Krasnodar	2.85	2.64	2.61	2.88	2.71	3.18	2.09
Dagestan	1.74	1.65	1.62	3.53	4.32	8.23	--
Average	1.22	1.14	1.09	1.18	1.12	1.06	0.97

*Along with Dagestan-Oil Pipeline Administration.

The cost of pumping crude oil and oil products on the whole for Glavneftesnab RSFSR during the past 7 years has decreased by 20.5 percent. This is explained by the operation of the new large diameter pipelines (Al'met'yevsk-Gor'kiy II, Kuybyshev-Penza-Bryansk, Gor'kiy-Ryazan', Gor'kiy-Yaroslavl', etc.) and by the growth in labor productivity. For example, the cost for the Urals-Siberian Oil Pipeline Administration in 1964 had decreased by 1.47 percent in comparison with 1958.

Table 8

Coefficient of Intensive Use of Equipment

Oil Pipeline Administration	1962	1963	1964
Urals-Siberian crude oil			
pipelines	0.841	0.850	0.893
product pipelines	0.814	0.839	0.829
Northwest crude oil pipelines	0.823	0.834	0.852
Kuybyshev crude oil pipelines	0.915	0.919	0.890
product pipelines	0.833	0.834	0.880

The use of productive capacities is characterized by a number of indicators, one of which is the coefficient of the intensive use of equipment. This coefficient is formed from the ratio of the average actual capacity of the pipeline to the rated capacity. The degree of use of the productive capacities directly affects the cost of pumping. The overall coefficient for intensive use of equipment for three of the largest oil pipeline administration is characterized by data of Table 8.

This general indicator, being characterized by a certain conventionality, gives an idea of the use of standard equipment on the whole for the oil pipeline administrations although it will differ essentially for individual pipelines. Thus, in the system of the Northwest Administration (Tatar) for the Al'met'yevsk-Gor'kiy I and II pipelines, this coefficient reached 1.000 in 1964, and for other (Gor'kiy-Yaroslavl', Al'met'yevsk-Perm', Karabash-Bavly II, etc.) it decreased to 0.376 to 0.535. For the Urals-Siberian Administration, the coefficient of use for individual pipelines fluctuates from 0.606 to 0.993.

The highest coefficient of use of capacities was in the Kuybyshev Oil Pipeline Administration, but in 1964, a slight decrease in the amount of pumping of oil affected the coefficient and led to an increase in the cost of pumping in comparison with 1963. The coefficient for use of equipment on the Urals-Siberian and northwest oil pipeline administrations increased during these years by 6.2 and 3.5 percent respectively.

The main reasons for incomplete use of pipeline capacities for the last 3 years were lags in expansion of capacities of individual oil refineries; changes in crude oil flows as the result of shifts in previously proposed production rates for individual oil deposits and regions; the necessity of separate pumping of oil of individual grades to retain its quality; and incomplete filling of product pipelines on individual sectors because of offtake at oil bases and the rail loading points.

Obviously, with the elimination of these reasons and an improvement in planning of operation of individual pipelines, the coefficient of use of productive capacities will increase.

Conclusions

1. With the help of the operational system of trunk crude oil pipelines, more than 80 percent of the crude oil is delivered from the wells to the oil refineries. About one-fourth of the light oil products are transported by product pipelines.

2. At a number of trunk pipelines there are considerable reserves of unused productive capacities, which makes it possible to expand the volume of handling of crude oil and oil products. With the planned increase in the extraction of oil, as well as in its processing at refineries located on the routes of some pipelines, an increase in the coefficient of use of the standard capacities is of great importance. Of no less importance is an improvement in the use of capacities for reducing the cost of pumping.

3. The insufficient development of the network of product pipelines is unfavorable. Under conditions of an annually increasing consumption of light oil products, the share of transport by product pipelines has virtually not increased. This has resulted in an increase in the role of transport expenditures in the overall cost of oil products to the consumer.

4. With the established high rates of growth in the oil industry and of consumption of oil products in the USSR in the new Five Year Plan (1966-70), the importance of pipeline transport will increase still more. Therefore the network of oil pipelines and especially of product pipelines must grow sharply in those regions which link the new oil centers and the oil refineries being built, as well as the major centers of consumption of light oil products.

"Technical-Economic Foundations for Development of the System of Crude Oil and Petroleum Product Pipelines" Markova, A.N. and Smirnov, M.F. Tekhniko-ekonomicheskoye obosnovaniye razvitiya seti nefteproduktoprovodov, (Moscow, 1966), pp. 6-10

The current level of development and utilization of pipeline transport (excluding gas) in the unified transport system may be characterized by its share (Table 1) in the total volume of transport work (expressed in ton-kilometers) and in the volume of freight handled (tons-originated).

Table 1

The Share of the Various Means of Transport in the Total Volume of Transport Work and in the Volume of Hauling (Percent)

Means of Transport	1958	1964	1965 Plan
As a Share of Ton-Kilometers Performed			
Railroads	81.2	73.5	70.0
Maritime	6.6	11.8	14.4
River	5.3	5.0	4.9
Automobile	4.8	5.1	5.2
Pipeline	2.1	4.6	5.5
Total	100.0	100.0	100.0
As a Share of Tons Hauled			
Railroads	19.2	17.9	17.4
Maritime	0.6	0.9	1.0
River	2.2	2.0	1.9
Automobile	76.9	77.5	77.9
Pipeline	1.1	1.7	1.8
Total	100.0	100.0	100.0

Crude oil and petroleum products are a basic part of the fuels-energy resources and pipelines are utilized for their transport. At the present time the share of pipeline transport in the total volume of transport work (ton-kilometers) and in the total volume of freight hauled (tons-originated) reach 20 percent and 42 percent, respectively (Table 2).

Table 2

The Share of the Various Means of Transport in the Movement of Oil Freight (Percent)

Means of Transport	1958	1959	1960	1961	1962	1963	1964	1965 Plan
As a Share of Ton-Kilometers Performed								
Railroads	61.1	62.3	61.3	59.4	57.8	56.1	50.2	43.7
Maritime	19.3	17.5	17.9	19.8	20.4	20.5	24.0	27.1
River	6.2	6.0	5.5	5.3	4.8	4.8	5.5	5.6
Pipeline	13.4	14.2	15.3	15.5	17.0	18.6	20.3	23.6
Total	100.0	100.0	100.0	100.0	100.0	100.0	100.0	100.0
As a Share of Tons Hauled								
Railroads	43.9	45.0	45.4	45.6	45.7	45.3	43.3	40.9
Maritime	12.5	10.0	9.8	9.5	9.3	9.3	9.9	10.3
River	6.3	6.0	5.6	5.6	5.0	4.9	4.8	4.9
Pipeline	37.3	39.0	39.2	39.3	40.0	40.5	42.0	43.9
Total	100.0	100.0	100.0	100.0	100.0	100.0	100.0	100.0

Crude oil is moved chiefly by pipeline, but petroleum products are handled by a variety of transport means, depending upon prevailing conditions: by pipeline, by rail, maritime, river, and also by automobile and even by air. Distribution of the handling of petroleum products among the various means of transport is illustrated in Table 3.

Table 3

The Share of the Various Means of Transport
in the Movement of Petroleum Products (Percent)

Means of Transport	1961	1962	1963	1964
As a Share of Ton-Kilometers Performed				
Railroads	77.9	78.6	79.0	77.8
Coastal shipping	5.8	5.8	5.2	5.4
River	10.2	9.4	8.8	9.4
Pipeline	6.1	6.2	7.0	7.4
Total	100.0	100.0	100.0	100.0
As a Share of Tons Hauled				
Railroads	72.7	73.2	71.8	72.0
Coastal shipping	8.0	6.9	9.1	9.3
River	10.5	10.5	9.6	9.1
Pipeline	8.8	9.4	9.5	9.6
Total	100.0	100.0	100.0	100.0

Total expenditures by the national economy for the transport of petroleum products currently reach almost 500 million rubles and the average length of haul is about 1,000 kilometers. Characteristics of these indexes are given, by means of transport, in Table 4.

Table 4

Distribution of Petroleum Product Expenditures and
Average Lengths of Haul, According to Means of Transport

Means of Transport	Average Length of Haul (Kilometers)					Transport Expenditures (Percent)				
	1960	1961	1962	1963	1964	1960	1961	1962	1963	1964
Railroads	1,252	1,247	1,217	1,181	1,153	92.3	90.0	91.0	89.5	88.7
Coastal Shipping	800	842	946	602	625	1.7	2.6	2.5	2.4	2.4
River	1,000	1,133	1,010	990	1,093	3.1	4.4	2.9	4.1	4.5
Pipeline	793	809	746	792	875	2.9	3.0	3.6	4.0	4.4
Total	1,190	1,164	1,132	1,092	1,085	100.0	100.0	100.0	100.0	100.0

In 1950 there were only two petroleum product pipelines in operation in the Soviet Union: Groznyy-Trudovaya, length of 879 kilometers, designed for the delivery of light products to the Donbass, and the Astrakhan-Saratov pipeline, length of 655 kilometers, used for the movement of kerosine from Baku to the region of the PoVolga, where it was transferred to rail for further shipment to the central, northwest, west, and eastern regions of the country.

During 1950-64, the length of the petroleum product pipeline system increased by more than three times (Table 5).

Table 5

The System of Trunk Petroleum Product Pipelines of the USSR*
(Kilometers)

Pipeline	1950	1955	1960	1961	1962	1963	1964
Groznyy-Trudovaya	879	879	879	879	879	879	879
Astrakhan-Saratov	655	655	-**	-	-	-	-
Ishimbay-Ufa	-	-	168	168	168	168	168
Ufa-Petropavlovsk	-	912	912	912	912	912	912
Ufa-Omsk	-	1,189	1,189	1,189	1,189	1,189	1,189
Omsk-Novosibirsk	-	-	699	699	699	699	699
Kuybyshev-Penza-Michurinsk-Bryansk	-	-	400	708	1,161	1,161	1,161
Total	1,534	3,635	4,247	4,555	5,008	5,008	5,008***

* According to data of Glavneftesnaab RSFSR (Main Administration for Oil Supply).
** Transferred to the handling of crude oil.
*** In addition, 174 kilometers of the Ishimbay-Ufa pipeline (crude pipeline shifted to the movement of petroleum products).

In these 14 years the volume of movement of petroleum products increased by 8 times, with a somewhat higher growth in work performed (ton-kilometers). Despite these noted successes, there are prevailing shortcomings in the development of pipeline movement, reflecting the lag in growth of the pipeline system compared with the growth in production of petroleum products.

In this period the output of light products increased by more than 5 times while the length of the product pipeline system increased by only 3 times. In addition, existing pipelines are not utilized at their rated capacity. Along individual sections and directions the coefficient of utilization of the planned capacity of the product pipelines as of January 1, 1964 reached 60 to 70 percent. At the same time, light products moved in directions parallel to these pipelines, hauled by railroads. This comes about because of inadequate introduction of batch operation of the pipelines, the uneven consumption of light products in the summer and winter periods, and also from the shortcomings in construction and completion of pumping stations.

APPENDIXES

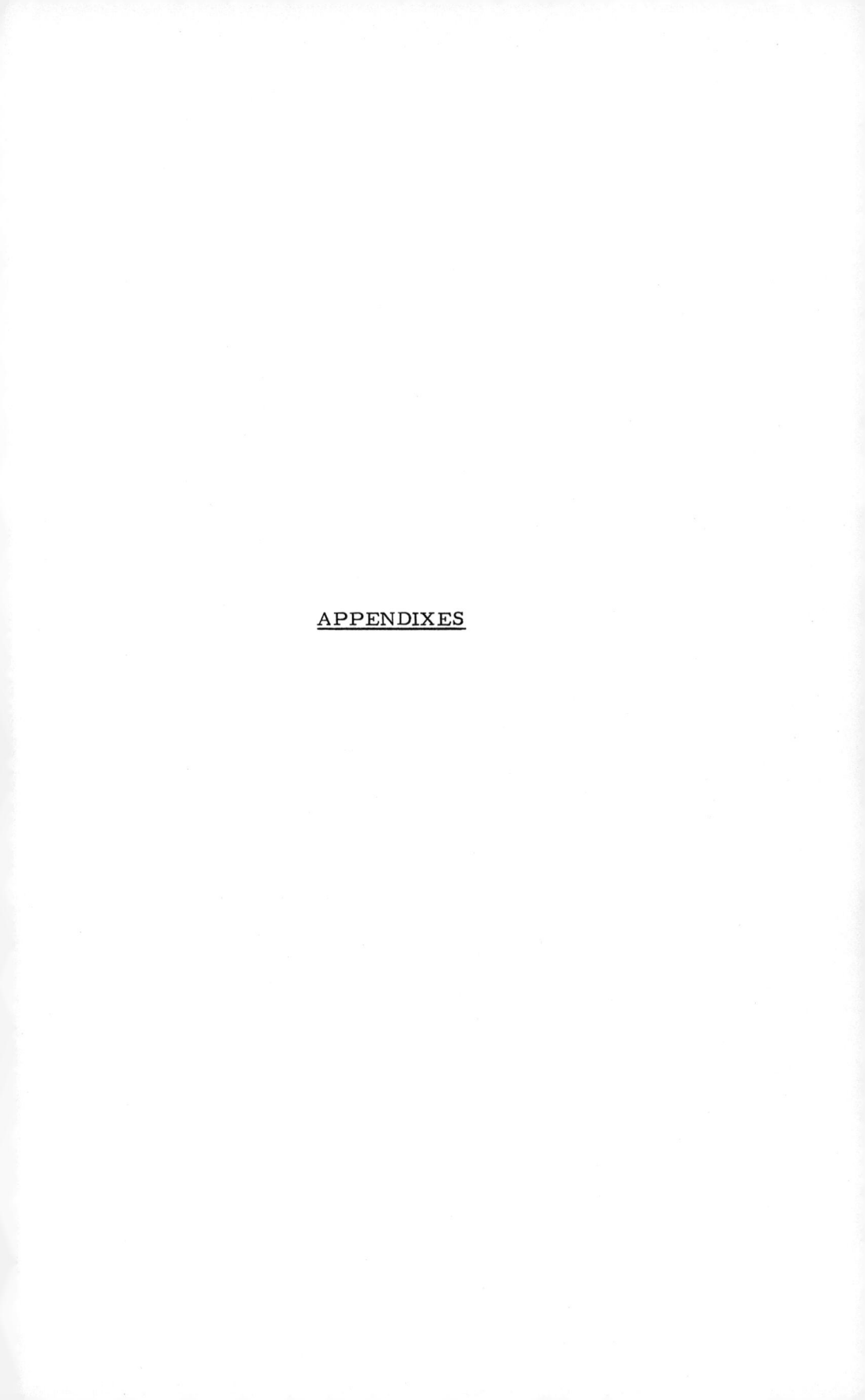

APPENDIX A   SELECTED STATISTICS

Data contained in Tables 53 through 59, which cover trade in petroleum (crude oil and petroleum products) by the Soviet Union for the years 1955 through 1967, have been derived from the various annual trade handbooks issued by the USSR Ministry of Foreign Trade. For the years 1955 through 1963 only Free World sales of total petroleum by country is indicated. For the subsequent years, sales both to Communist and non-Communist countries are listed, the export of petroleum products is broken down by type of product, and values are stated as well.

TABLE 53

Soviet Export of Oil to the Free World, by Country, 1955-63

(in thousand metric tons)

Destination	1955	1956	1957
Western Hemisphere			
Argentina	636.9	-	-
Brazil	-	-	-
Uruguay	-	-	-
Total Western Hemisphere	636.9	-	-
Free Europe			
Austria	37.4	26.1	57.4
Belgium	30.3	30.5	0.7
Denmark	2.1	0.6	22.7
Finland	612.5	1,011.8	1,214.0
France	269.3	408.9	551.3
West Germany	5.3	142.7	797.4
Greece	94.5	224.1	302.5
Iceland	283.3	258.9	299.8
Italy	183.3	500.4	502.3
Netherlands	10.3	15.1	0.2
Norway	35.5	26.1	146.8
Portugal	-	-	-
Sweden	725.6	694.2	536.4
Switzerland	0.1	1.2	128.6
United Kingdom	37.4	26.1	57.4
Yugoslavia	208.6	331.9	407.4
Total Free Europe	2,535.5	3,698.6	5,024.9
Other Eastern Hemisphere			
Afghanistan	21.1	27.4	34.3
Algeria	35.3	93.2	166.3
Ethiopia	-	-	-
Ghana	-	-	-
Guinea	-	-	-
India	-	-	-
Iran	0.5	0.2	0.3
Japan	-	-	-
Lebanon	-	-	11.2
Morocco	-	3.0	33.3
Syria	-	26.0	30.2
Tunisia	-	-	-
Turkey	0.6	5.7	4.7
United Arab Republic	329.5	920.7	1,072.4
Total other eastern hemisphere	387.0	1,076.2	1,352.7
Total free world	4,039.2	5,150.9	6,386.1

(continued)

TABLE 53 (Continued)

Destination	1958	1959	1960
Western hemisphere			
Argentina	911.3	444.8	-
Brazil	-	59.4	161.4
Cuba	-	-	2,164.8
Uruguay	214.4	500.7	71.4
Total western hemisphere	1,125.7	1,004.9	2,397.6
Free Europe			
Austria	60.5	526.7	605.2
Belgium	72.9	194.2	203.1
Denmark	38.7	96.5	153.4
Finland	1,233.7	1,856.3	2,127.9
France	710.7	807.6	785.2
West Germany	561.7	1,086.7	2,007.0
Greece	362.0	424.0	947.5
Iceland	332.2	365.4	339.2
Italy	1,082.0	3,035.9	4,702.5
Netherlands	103.0	47.9	40.1
Norway	158.0	263.3	249.1
Portugal	49.4	-	62.7
Sweden	870.4	1,451.4	1,968.3
Switzerland	-	39.4	28.5
United Kingdom	37.8	101.8	283.4
Yugoslavia	382.9	438.2	456.1
Unknown	-	-	-
Total Free Europe	6,055.9	10,734.9	14,959.2
Other eastern hemisphere			
Afghanistan	41.8	47.6	48.3
Algeria	38.0	25.6	61.5
Ethiopia	-	-	0.5
Ghana	-	-	0.1
Guinea	-	0.1	28.9
India	-	-	23.0
Iran	0.3	0.3	0.5
Japan	11.1	155.1	1,403.5
Lebanon	-	66.0	86.3
Morocco	30.0	64.6	43.5
Syria	270.8	448.9	273.3
Tunisia	17.9	13.7	-
Turkey	0.2	0.7	6.6
United Arab Republic	1,941.8	1,903.0	1,335.7
Total other eastern hemisphere	2,351.9	2,725.6	3,312.3
Total Free World	9,581.4	14,553.5	20,723.4

(continued)

TABLE 53 (Continued)

	1961	1962	1963
Destination			
Western hemisphere			
Argentina	-	-	-
Brazil	439.4	155.9	824.2
Uruguay	23.9	-	-
Total western hemisphere	463.3	155.9	824.2
Free Europe			
Austria	503.5	409.3	502.1
Belgium	231.4	393.9	368.1
Denmark	242.9	184.1	380.8
Finland	2,046.4	2,913.5	3,463.9
France	650.8	769.7	1,119.0
West Germany	2,656.9	3,040.5	3,225.2
Greece	989.8	826.3	1,177.9
Iceland	332.0	310.1	338.9
Italy	6,165.8	7,076.7	7,712.6
Netherlands	11.8	49.8	48.0
Norway	228.1	206.2	253.9
Portugal	-	-	-
Sweden	2,369.9	2,621.6	3,040.8
Switzerland	15.5	11.2	66.6
Spain	22.9	21.0	-
United Kingdom	157.8	200.9	204.9
Total Free Europe	16,625.5	19,034.8	21,902.7
Other eastern hemisphere			
Afghanistan	72.5	94.1	103.0
Algeria	54.6	27.6	-
Burma	21.9	56.5	34.4
Ceylon	-	175.6	264.9
Ethiopia	0.2	0.1	0.1
Guinea	86.8	129.0	74.8
India	174.8	452.0	708.3
Iran	1.4	0.9	0.1
Japan	3,042.7	2,979.2	3,086.7
Lebanon	11.8	12.8	12.4
Morocco	121.6	189.4	284.8
Syria	-	0.6	34.7
Pakistan	189.7	-	-
Tunisia	37.7	25.4	52.9
Turkey	47.1	8.6	19.4
United Arab Republic	1,793.7	1,331.5	974.5
Yemen	3.7	10.8	7.2
Total other eastern hemisphere	5,660.2	5,491.1	5,658.2
Total Free World	22,749.0	24,681.8	28,385.1

TABLE 54

Exports of Petroleum from the Soviet Union, 1964

Destination/Products	Quantity (Thousand Metric Tons)	Value (Thousand Rubles)
Total		
Crude oil	36,690.7	475,636
Petroleum products	19,930.1	373,176
Gasoline	2,747.2	87,506
Kerosine	899.5	24,351
Diesel fuel	6,851.4	142,775
Mazut	9,056.3	86,700
Lubes	288.2	23,342
Greases	6.5	1,756
Asphalt	23.5	693
Parrafin	33.4	4,592
Austria		
Crude oil	535.3	5,878
Petroleum products	19.7	315
Gasoline	13.5	251
Mazut	6.0	36
Paraffin (tons)	240	28
England		
Petroleum products	55.9	1,452
Diesel fuel	0.3	6
Mazut	27.9	322
Lubes	27.2	1,072
Paraffin	0.5	52
Belgium		
Petroleum products	168.6	1,870

(Continued)

TABLE 54 (Continued)

Destination/Products	Quantity (Thousand Metric Tons)	Value (Thousand Rubles)
Diesel fuel	62.7	1,016
Mazut	105.9	854
Bulgaria		
Crude oil	1,798.6	28,997
Petroleum products	1,327.0	28,360
Gasoline	214.8	8,315
Kerosine	10.3	319
Diesel fuel	230.2	5,961
Mazut	808.3	8,225
Lubes	40.2	4,380
Greases	0.3	158
Asphalt	20.3	520
Paraffin	2.3	299
Hungary		
Crude oil	1,757.8	34,204
Petroleum products	390.8	11,592
Gasoline	107.2	3,865
Diesel fuel	158.4	4,195
Mazut	98.4	1,597
Lubes and greases	26.8	1,933
East Germany		
Crude oil	3,936.4	67,645
Petroleum products	478.0	15,113
Gasoline	187.1	6,365
Diesel fuel	235.2	6,181
Lubes and greases	21.7	1,775
Petroleum coke	11.6	363

Destination/Products	Quantity (Thousand Metric Tons)	Value (Thousand Rubles)
The Netherlands		
Petroleum products	25.3	482
Diesel fuel	24.9	460
Mazut	0.3	4
Paraffin (tons)	155	18
Greece		
Crude oil	464.7	5,603
Petroleum products	690.9	7,833
Gasoline	11.1	165
Diesel fuel	176.2	3,658
Mazut	503.6	4,010
Denmark		
Petroleum products	459.5	4,181
Mazut	459.4	4,169
Paraffin (tons)	100	12
Iceland		
Petroleum products	330.3	6,006
Motor gasoline	46.6	1,074
Diesel fuel	191.6	3,705
Mazut	101.1	1,227
Italy		
Crude oil	7,017.3	56,470
Petroleum products	677.7	5,560

(Continued)

TABLE 54 (Continued)

Destination/Products	Quantity (Thousand Metric Tons)	Value (Thousand Rubles)
Diesel fuel	4.8	59
Mazut	669.1	5,062
Paraffin	3.8	439
Residual fuel for refining	236.8	1,867
Norway		
Petroleum products	233.8	3,734
Diesel fuel	165.6	3,150
Mazut	67.8	546
Paraffin	0.4	38
Poland		
Crude oil	1,702.6	33,666
Petroleum products	2,112.6	62,905
Gasoline	818.2	28,195
Kerosine	64.5	2,005
Diesel fuel	617.1	18,266
Mazut	564.5	10,122
Lubes and greases	19.4	1,437
Paraffin	16.6	2,449
Petroleum coke	12.3	419
Rumania		
Petroleum products	0.6	121
Lubes and greases	0.6	121
Lube additives (tons)	2	1
Spain		
Crude oil	114.2	1,162

Destination/Products	Quantity (Thousand Metric Tons)	Value (Thousand Rubles)
West Germany		
Crude oil	2,958.7	27,008
Petroleum products	777.1	10,994
Gasoline	137.5	1,563
Diesel fuel	634.5	9,260
Finland		
Crude oil	1,872.5	22,952
Petroleum products	2,326.6	39,527
Gasoline	4.4	140
Kerosine	12.5	326
Diesel fuel	1,483.0	30,204
Mazut	817.5	8,247
Lubes	7.9	445
Paraffin	1.3	164
France		
Crude oil	223.2	2,089
Petroleum products	973.5	14,530
Gasoline	42.2	802
Diesel fuel	491.0	9,610
Mazut	439.7	4,041
Paraffin (tons)	599	62
Czechoslovakia		
Crude oil	4,759.7	96,715
Petroleum products	378.7	12,533
Gasoline	108.1	3,629
Kerosine	236.1	7,552
Paraffin	1.8	255

(Continued)

TABLE 54 (Continued)

Destination/Products	Quantity (Thousand Metric Tons)	Value (Thousand Rubles)
Switzerland		
Crude oil	177.8	1,452
Petroleum products	68.1	638
Gasoline	68.1	638
Sweden		
Crude oil	28.7	281
Petroleum products	2,741.1	25,918
Gasoline	7.4	160
Kerosine	4.1	84
Diesel fuel	305.1	5,070
Mazut	2,424.0	20,549
Paraffin	0.5	55
Yugoslavia		
Crude oil	436.1	4,674
Petroleum products	413.2	6,289
Gasoline	21.6	947
Diesel fuel	131.4	2,565
Mazut	255.3	2,279
Paraffin (tons)	1,107	160
Lube oils	3.4	247
Afghanistan		
Petroleum products	107.5	4,667
Gasoline	75.7	2,952
Kerosine	2.7	183
Diesel fuel	24.8	934
Lubes	2.2	397
Greases	0.4	81
Asphalt	1.8	120

Destination/Products	Quantity (Thousand Metric Tons)	Value (Thousand Rubles)
Yemen		
Petroleum products	8.3	410
Kerosine	7.5	370
Diesel fuel	0.8	40
North Vietnam		
Petroleum products	101.3	4,099
Gasoline	37.6	1,262
Kerosine	21.1	660
Diesel fuel	34.1	992
Lubes	7.5	960
Greases	0.3	82
India		
Petroleum products	689.3	12,324
Kerosine	242.5	5,109
Diesel fuel	350.4	6,348
Mazut	96.4	853
Iran		
Petroleum products	0.1	17
Lubes	0	0
Paraffin (tons)	102	17
Communist China		
Petroleum products	504.9	19,422
Gasoline	269.9	10,509
Kerosine	139.1	4,396

(Continued)

TABLE 54 (Continued)

Destination/Products	Quantity (Thousand Metric Tons)	Value (Thousand Rubles)
Diesel fuel	79.5	2,670
Lubes	15.2	1,514
Greases	1.0	288
Paraffin	0.1	16
North Korea		
Petroleum products	437.5	15,980
Gasoline	290.1	6,472
Kerosine	6.2	197
Diesel fuel	168.7	5,106
Mazut	2.1	49
Lubes	47.3	3,580
Greases	1.4	304
Paraffin	1.7	220
Lebanon		
Petroleum products	5.6	82
Diesel fuel	0.6	14
Mazut	5.0	68
Mongolia		
Crude oil	39.7	912
Petroleum products	160.0	6,746
Gasoline	89.2	3,357
Kerosine	0.1	3
Diesel fuel	58.4	2,115
Lubes	11.7	1,177
Greases (tons)	392	87
Asphalt (tons)	38	2

Destination/Products	Quantity (Thousand Metric Tons)	Value (Thousand Rubles)
Turkey		
Petroleum products	27.6	626
Kerosine	26.5	612
Mazut	1.1	14
Ceylon		
Petroleum products	623.6	10,314
Gasoline	63.6	1,326
Kerosine	116.3	2,535
Diesel fuel	269.7	4,852
Mazut	173.7	1,558
Japan		
Crude oil	2,482.6	19,258
Petroleum products	1,106.7	12,645
Diesel fuel	509.7	7,377
Mazut	589.9	5,175
Guinea		
Petroleum products	55.8	1,308
Gasoline	32.5	802
Diesel fuel	23.3	498
Morocco		
Crude oil	300.7	3,116
United Arab Republic		
Crude oil	705.7	8,010
Petroleum products	43.5	1,532

(Continued)

TABLE 54 (Continued)

Destination/Products	Quantity (Thousand Metric Tons)	Value (Thousand Rubles)
United Arab Republic (cont'd)		
Gasoline	24.8	757
Diesel fuel	14.5	370
Mazut	1.6	21
Lubes	2.4	308
Syria (SAR)		
Petroleum products	130.3	2,627
Diesel fuel	130.3	2,627
Argentina		
Petroleum products	69.2	1,024
Diesel fuel	69.2	1,024
Brazil		
Crude oil	1,876.7	19,639
Petroleum products	20.2	439
Gasoline	20.2	439
Cuba		
Crude oil	3,426.6	35,147
Petroleum products	1,132.7	18,427
Gasoline	136.8	3,521
Diesel fuel	192.1	4,022
Mazut	752.0	6,601
Lubes	48.5	3,672
Paraffin	1.2	163
Greases	2.1	445

Destination/Products	Quantity (Thousand Metric Tons)	Value (Thousand Rubles)
<u>Burma</u>		
Crude oil	75.1	758
<u>Cyprus</u>		
Petroleum products	28.7	227
Mazut	28.7	227

TABLE 55

Exports of Petroleum from the Soviet Union, 1965

Destination/Products	Quantity (Thousand Metric Tons)	Value (Thousand Rubles)
Total		
Crude oil	43,432.0	550,841
Petroleum products	20,986.6	348,301
Gasoline	2,352.8	65,264
Kerosine	1,165.1	25,976
Diesel fuel	7,360.8	137,987
Mazut	9,709.5	89,298
Lubes	270.3	21,842
Greases	5.5	1,627
Asphalt	23.3	677
Paraffin	27.9	3,690
Austria		
Crude oil	470.1	5,162
Petroleum products	12.6	266
Gasoline	12.2	214
Paraffin (tons)	434.0	52
England		
Petroleum products	24.2	959
Lubes	23.8	922
Paraffin	0.4	37
Belgium		
Petroleum products	137.0	1,237
Diesel fuel	21.9	293
Mazut	115.1	944

Destination/Products	Quantity (Thousand Metric Tons)	Value (Thousand Rubles)
Bulgaria		
Crude oil	2,145.9	34,589
Petroleum products	1,304.2	26,957
Gasoline	141.0	5,391
Kerosine	11.3	341
Diesel fuel	288.3	6,917
Mazut	794.0	8,108
Lubes	45.9	4,872
Greases	0.5	237
Asphalt	20.3	516
Paraffin	2.4	333
Hungary		
Crude oil	2,046.3	40,741
Petroleum products	445.8	13,375
Gasoline	153.3	5,187
Diesel fuel	151.8	4,074
Mazut	101.5	1,601
Lubes and greases	28.0	2,176
East Germany		
Crude oil	4,922.9	75,436
Petroleum products	505.4	14,566
Gasoline	130.2	4,106
Diesel fuel	336.1	8,486
Lubes and greases	15.9	1,364
Petroleum coke	10.6	333
Greece		
Crude oil	421.8	4,859
Petroleum products	730.3	7,904

(Continued)

TABLE 55 (Continued)

Destination/Products	Quantity (Thousand Metric Tons)	Value (Thousand Rubles)
Greece (cont'd)		
Gasoline	32.0	458
Diesel fuel	205.2	3,829
Mazut	493.1	3,617
Denmark		
Petroleum products	386.3	3,510
Mazut	386.2	3,496
Paraffin (tons)	135.0	14
Iceland		
Petroleum products	401.6	6,412
Motor gasoline	48.5	1,037
Diesel fuel	242.5	4,149
Mazut	110.6	1,226
Italy		
Crude oil	6,588.2	52,217
Petroleum products	757.2	6,491
Mazut	700.6	5,575
Norway		
Petroleum products	303.6	4,095
Diesel fuel	191.0	3,164
Mazut	119.0	853
Paraffin	0.7	78

Destination/Products	Quantity (Thousand Metric Tons)	Value (Thousand Rubles)
Poland		
Crude oil	3,213.1	54,975
Petroleum products	1,489.5	41,411
Gasoline	510.9	17,329
Kerosine	30.0	912
Diesel fuel	450.9	12,079
Mazut	455.5	7,758
Lubes and greases	24.5	1,856
Paraffin	8.1	1,069
Petroleum coke	9.5	300
Rumania		
Petroleum products	0.6	94
Lubes and greases	0.6	94
Lube additives (tons)	2.0	1
Spain		
Crude oil	381.0	3,862
Petroleum products	51.6	499
Mazut	51.6	499
West Germany		
Crude oil	2,583.0	23,727
Petroleum products	521.0	5,763
Gasoline	139.6	1,209
Diesel fuel	381.4	4,554

(Continued)

TABLE 55 (Continued)

Destination/Products	Quantity (Thousand Metric Tons)	Value (Thousand Rubles)
Finland		
Crude oil	1,938.2	24,206
Petroleum products	2,546.3	39,641
Gasoline	1.1	35
Kerosine	10.0	261
Diesel fuel	1,580.8	29,478
Mazut	946.8	9,346
Lubes	6.1	326
Paraffin	1.5	194
France		
Crude oil	778.9	6,459
Petroleum products	833.2	10,834
Gasoline	67.9	1,048
Diesel fuel	410.9	6,760
Mazut	354.3	3,013
Paraffin (tons)	50.0	5
Ozokerite (tons)	36.0	8
Czechoslovakia		
Crude oil	5,694.0	107,739
Petroleum products	399.2	12,839
Gasoline	138.2	4,190
Kerosine	237.5	7,301
Paraffin	2.9	398
Diesel fuel	16.1	440
Switzerland		
Crude oil	916.9	7,614
Petroleum products	246.5	2,285

Destination/Products	Quantity (Thousand Metric Tons)	Value (Thousand Rubles)
Switzerland (cont'd)		
Gasoline	197.7	1,862
Diesel fuel	48.8	423
Sweden		
Petroleum products	2,803.6	24,827
Diesel fuel	277.7	4,236
Mazut	2,525.6	20,561
Paraffin	0.3	30
Yugoslavia		
Crude oil	592.4	6,292
Petroleum products	418.8	5,313
Gasoline	20.4	882
Diesel fuel	31.4	485
Mazut	359.5	3,069
Paraffin (tons)	3.163	475
Lube oils	3.9	300
Afghanistan		
Petroleum products	119.5	4,468
Gasoline	79.3	2,777
Kerosine	5.6	367
Diesel fuel	33.0	1,111
Lubes	0.8	142
Greases	0.1	19
Asphalt	0.8	52
Yemen		
Petroleum products	8.0	395

(Continued)

TABLE 55 (Continued)

Destination/Products	Quantity (Thousand Metric Tons)	Value (Thousand Rubles)
Yemen (cont'd)		
Kerosine	7.5	371
Diesel fuel	0.5	24
North Vietnam		
Petroleum products	121.0	4,790
Gasoline	42.1	1,381
Kerosine	26.8	873
Diesel fuel	43.2	1,258
Lubes	7.7	1,005
Greases	0.5	139
India		
Petroleum products	1,420.7	22,702
Kerosine	646.5	11,429
Diesel fuel	530.7	9,046
Mazut	243.4	2,199
Communist China		
Petroleum products	37.9	2,005
Gasoline	31.7	1,599
Kerosine	2.0	106
Diesel fuel	3.6	184
Lubes	0.4	79
Greases	0.1	36
North Korea		
Petroleum products	391.8	14,311

Destination/Products	Quantity (Thousand Metric Tons)	Value (Thousand Rubles)
North Korea (cont'd)		
Gasoline	172.9	5,417
Kerosine	1.6	51
Mazut	2.3	54
Lubes	35.9	2,805
Greases	1.9	447
Paraffin	1.2	162
Diesel fuel	175.0	5,332
Mongolia		
Crude oil	42.7	979
Petroleum products	158.1	6,810
Gasoline	96.0	3,627
Kerosine	0.1	1
Diesel fuel	49.2	1,795
Lubes	11.6	1,174
Greases (tons)	667.0	171
Asphalt (tons)	490.0	41
Turkey		
Petroleum products	82.0	1,065
Diesel fuel	45.4	778
Mazut	36.6	287
Ceylon		
Petroleum products	598.8	10,167
Gasoline	71.7	1,494
Kerosine	162.9	3,382
Diesel fuel	249.9	4,263
Mazut	114.3	1,019

(Continued)

TABLE 55 (Continued)

Destination/Products	Quantity (Thousand Metric Tons)	Value (Thousand Rubles)
Japan		
Crude oil	2,316.7	18,397
Petroleum products	1,564.2	17,427
Diesel fuel	645.0	9,222
Mazut	872.8	7,624
Petroleum coke	46.4	581
Guinea		
Petroleum products	52.4	1,188
Gasoline	30.1	724
Diesel fuel	22.3	464
Morocco		
Crude oil	417.0	4,180
United Arab Republic		
Crude oil	688.1	7,960
Petroleum products	84.5	2,902
Gasoline	50.4	1,605
Diesel fuel	18.0	464
Lubes	3.4	452
Syria (SAR)		
Petroleum products	155.2	3,170
Diesel fuel	154.8	3,092

Destination/Products	Quantity (Thousand Metric Tons)	Value (Thousand Rubles)
Argentina		
Crude oil	402.8	3,476
Petroleum products	547.1	7,711
Diesel fuel	547.1	7,711
Brazil		
Crude oil	2,354.1	23,674
Cuba		
Crude oil	3,512.6	36,948
Petroleum products	1,214.0	18,899
Gasoline	132.5	3,167
Diesel fuel	208.1	3,864
Mazut	811.0	7,381
Lubes	59.8	4,017
Paraffin	1.8	251
Greases	0.8	194
Burma		
Crude oil	140.0	1,396
Cyprus		
Petroleum products	110.7	877
Mazut	110.7	877
Ghana		
Crude oil	595.3	5,953

TABLE 56

Exports of Petroleum from the Soviet Union, 1966

Importing Country	Quantity (Thousand Metric Tons)	Value (Thousand Rubles)
Austria		
Crude oil	681.3	7,481
Products:		
Gasoline	15.3	261
Wax	0.380	44
Total	15.7	305
United Kingdom		
Products:		
Lube oils	13.7	491
Wax	0.5	45
Total	14.2	536
Belgium		
Products:		
Diesel fuel	109.1	1,464
Mazut	33.8	276
Total	142.9	1,740
Bulgaria		
Crude oil	2,622.8	42,621
Products:		
Gasoline	155.0	6,294
Kerosine	9.8	306
Diesel fuel	213.1	5,167
Mazut	932.8	9,690
Lube oils	50.6	5,352
Grease	0.8	343
Asphalt	25.5	673
Wax	3.0	412
Total	1,391.1	28,465

Importing Country	Quantity (Thousand Metric Tons)	Value (Thousand Rubles)
Hungary		
Crude oil	2,473.4	43,847
Products:		
Gasoline	57.8	1,890
Diesel fuel	145.6	3,630
Mazut	253.5	3,563
Lubes and greases	26.7	2,013
Total	492.1	11,518
East Germany		
Crude oil	6,114.0	82,543
Products:		
Gasoline	138.2	3,893
Diesel fuel	43.8	1,148
Lubes and greases	4.3	492
Petroleum coke	12.0	378
Total	208.7	6,112
Netherlands		
Products:		
Gasoline	10.8	140
Total	11.6	154
Greece		
Crude oil	467.5	5,345
Products:		
Gasoline	56.0	827
Diesel fuel	166.5	3,086
Mazut	353.7	2,654
Total	576.2	6,567

(Continued)

TABLE 56 (Continued)

Importing Country	Quantity (Thousand Metric Tons)	Value (Thousand Rubles)
Denmark		
Products:		
Mazut	384.9	3,095
Wax	0.055	6
Total	387.0	3,135
Iceland		
Products:		
Gasoline	53.4	1,075
Diesel fuel	251.5	4,188
Mazut	122.0	1,355
Total	426.9	6,618
Italy		
Crude oil	8,030.7	64,537
Products:		
Gasoline	172.7	1,933
Mazut	670.3	5,320
Total	846.3	7,590
Norway		
Products:		
Diesel fuel	196.1	2,872
Mazut	166.1	1,229
Wax	0.5	54
Petroleum coke	3.6	50
Total	366.3	4,205
Poland		
Crude oil	3,346.9	49,187
Products:		
Gasoline	657.2	17,733
Kerosine	35.1	1,026

Importing Country	Quantity (Thousand Metric Tons)	Value (Thousand Rubles)
Poland (cont'd)		
Diesel fuel	530.8	12,478
Mazut	453.3	6,576
Lubes and greases	21.9	1,910
Wax	8.6	998
Petroleum coke	6.5	189
Total	1,713.5	40,983
Natural gas (million M^3)	827.9	5,696
Rumania		
Products:		
Lubes and greases	0.4	77
Total	0.4	84
Spain		
Crude oil	163.6	1,646
West Germany		
Crude oil	3,337.3	30,634
Products:		
Gasoline	233.9	2,216
Diesel fuel	948.1	12,050
Total	1,182.0	14,266
Finland		
Crude oil	2,568.1	31,084
Products:		
Gasoline	0.9	29
Kerosine	11.9	328
Diesel fuel	2,352.6	42,565
Mazut	1,075.0	10,836
Lube oils	9.5	447
Wax	1.6	210
Total	3,451.6	54,416

(Continued)

TABLE 56 (Continued)

Importing Country	Quantity (Thousand Metric Tons)	Value (Thousand Rubles)
France		
Crude oil	1,658.9	13,647
Products:		
Gasoline	119.2	1,781
Diesel fuel	500.8	7,055
Mazut	401.9	3,011
Wax	.050	5
Ozokerite	.036	7
Total	1,021.9	11,860
Czechoslovakia		
Crude oil	6,396.2	98,769
Products:		
Gasoline	226.5	6,292
Kerosine	250.2	7,305
Diesel fuel	17.5	479
Wax	1.6	200
Total	503.7	14,995
Switzerland		
Crude oil	598.7	5,293
Products:		
Gasoline	103.8	969
Diesel fuel	35.9	328
Total	139.7	1,297
Sweden		
Products:		
Diesel fuel	472.4	6,472
Mazut	3,233.9	25,477
Wax	0.1	11
Total	3,706.4	31,960

Importing Country	Quantity (Thousand Metric Tons)	Value (Thousand Rubles)
Senegal		
Products:		
Wax	0.553	64
Total	0.553	64
Yugoslavia		
Crude oil	662.5	7,035
Products:		
Gasoline	21.0	905
Diesel fuel	69.3	1,030
Mazut	376.8	3,199
Lube oils	7.7	638
Wax	3.227	485
Total	498.3	6,694
Afghanistan		
Products:		
Gasoline	84.5	3,022
Kerosine	4.8	300
Diesel fuel	29.3	995
Lube oils	1.8	310
Grease	0.1	33
Asphalt	3.1	173
Total	123.6	4,833
Yemen		
Products:		
Kerosine	7.3	291
Diesel fuel	3.3	95
Total	11.6	447

(Continued)

TABLE 56 (Continued)

Importing Country	Quantity (Thousand Metric Tons)	Value (Thousand Rubles)
North Vietnam		
Products:		
Gasoline	26.4	990
Kerosine	12.1	391
Diesel fuel	42.4	1,284
Lube oils	4.1	470
Grease	0.3	61
Total	85.7	3,270
India		
Products:		
Kerosine	699.9	13,231
Diesel fuel	247.0	4,221
Mazut	266.8	2,423
Total	1,213.7	19,889
Iran		
Products:		
Wax	0.150	24
Total	0.150	24
Communist China		
Products:		
Gasoline	28.0	1,583
Kerosine	4.0	211
Diesel fuel	7.0	364
Lube oils	0.9	123
Grease	0.1	30
Total	40.0	2,314
North Korea		
Products:		
Gasoline	181.2	5,608

Importing Country	Quantity (Thousand Metric Tons)	Value (Thousand Rubles)
North Korea (cont'd)		
Kerosine	11.0	348
Diesel fuel	142.6	4,357
Mazut	13.5	270
Lube oils	31.0	2,485
Grease	1.5	346
Wax	1.9	271
Total	383.9	13,763
Mongolia		
Crude oil	45.0	1,033
Products:		
Gasoline	96.1	3,664
Kerosine	0.2	8
Diesel fuel	44.5	1,621
Lube oils	11.7	1,052
Grease	0.439	140
Asphalt	0.238	20
Total	155.5	6,577
Turkey		
Products:		
Diesel fuel	36.8	629
Mazut	11.0	87
Total	47.8	716
Ceylon		
Products:		
Gasoline	71.9	1,503
Kerosine	177.1	3,810
Diesel fuel	256.6	4,334
Mazut	139.0	1,238
Total	644.6	10,885

(Continued)

TABLE 56 (Continued)

Importing Country	Quantity (Thousand Metric Tons)	Value (Thousand Rubles)
Japan		
Crude oil	2,786.2	22,448
Products:		
Diesel fuel	647.0	8,519
Mazut	644.8	5,591
Petroleum coke	71.7	882
Total	1,363.5	14,992
Ghana		
Crude oil	603.5	5,096
Guinea		
Products:		
Gasoline	34.3	804
Diesel fuel	28.6	587
Total	62.9	1,391
Morocco		
Crude oil	525.2	5,267
Egypt (UAR)		
Crude oil	963.5	11,318
Products:		
Gasoline	49.8	1,671
Diesel fuel	19.0	488
Lube oils	3.9	521
Total	73.1	2,871
Syria		
Products:		
Diesel fuel	226.9	4,699
Total	227.0	4,711

Importing Country	Quantity (Thousand Metric Tons)	Value (Thousand Rubles)
Argentina		
Products:		
Diesel fuel	241.9	3,284
Total	241.9	3,284
Brazil		
Crude oil	2,195.2	20,865
Cuba		
Crude oil	3,840.2	39,792
Products:		
Gasoline	100.0	2,610
Diesel fuel	247.6	4,167
Mazut	825.6	7,535
Lube oils	74.1	5,251
Grease	1.0	185
Wax	1.9	256
Total	1,250.2	20,008
Burma		
Crude oil	214.8	1,982
Cyprus		
Products:		
Mazut	134.4	1,064
Total	134.4	1,064
Somalia		
Products:		
Gasoline	7.7	160
Kerosine	2.5	55
Diesel fuel	12.9	222
Total	23.1	437

(Continued)

TABLE 56 (Continued)

Importing Country	Quantity (Thousand Metric Tons)	Value (Thousand Rubles)
Colombia		
Products:		
Wax	0.350	36
Total	0.350	36
Uruguay		
Crude oil	18.8	155

TABLE 57

Imports of Petroleum by the Soviet Union, 1966

Origin	Quantity (Thousand Metric Tons)	Value (Thousand Rubles)
Rumania		
Products:		
Gasoline	648.1[a]	18,122
Kerosine	46.6[b]	1,383
Diesel fuel	116.4	2,820
Mazut	36.8	533
Lube oils	106.5	6,459
Grease	12.6	1,734
Asphalt	40.7	945
Wax	10.0	1,421
Naphthenic acids	0.6	139
Total	1,018.3	18,122
United Kingdom		
Products:		
Special purpose lube oils	0.512	128
Total	0.512	218
Hungary		
Products:		
Grease	13.4	1,844
Total	29.9	2,336
East Germany		
Products:		
Gasoline	230.0	6,555
Diesel fuel	27.6	739
Lubes and greases	0.3	47
Ozokerite	0.7	299
Total	259.6	7,671

(Continued)

TABLE 57 (Continued)

Origin	Quantity (Thousand Metric Tons)	Value (Thousand Rubles)
Netherlands		
Products:		
Demulsifier	0.400	173
Total	0.400	173
West Germany		
Products:		
Demulsifier	7.4	3,155
Total	7.4	3,155
Syria		
Products:		
Gasoline	39.2	806
Total	39.2	806

[a] Of which 589,200 tons were motor gasoline.
[b] Of which 45,900 tons were lamp kerosine.

TABLE 58

Exports of Petroleum from the Soviet Union, 1967

Destination/Products	Quantity (Thousand Metric Tons)	Value (Thousand Rubles)
Total		
Crude oil	54,117.3	641,724
Petroleum products	24,692.1	395,195
Gasoline	3,276.4	78,172
Kerosine	1,180.7	26,406
Diesel fuel	8,615.6	153,536
Mazut	11,166.2	105,540
Lube oils	260.3	22,873
Grease	6.5	1,686
Asphalt	32.7	1,021
Wax	24.5	2,925
Petroleum coke	97.6	1,875
Natural gas	1,290.5[a]	18,097
Austria		
Crude oil	698.2	7,667
Products	21.0	350
Gasoline	0.1	3
Diesel fuel	-	-
Mazut	-	-
Wax	0.175	19
United Kingdom		
Products	30.0	693
Diesel fuel	-	-
Mazut	-	-
Lube oils	11.3	393
Wax	-	5
Mogas	18.7	295

[a] Million cubic meters.

(Continued)

TABLE 58 (Continued)

Destination/Products	Quantity (Thousand Metric Tons)	Value (Thousand Rubles)
Belgium		
Crude oil	-	-
Products	339.1	5,260
Diesel fuel	226.6	3,758
Mazut	94.0	1,087
Bulgaria		
Crude oil	2,685.9	34,863
Products	1,788.1	35,499
Gasoline	211.3	6,988
Kerosine	8.7	260
Diesel fuel	265.0	6,303
Mazut	1,184.8	14,062
Lube oils	55.9	5,821
Grease	0.6	192
Asphalt	28.3	751
Wax	2.8	389
Hungary		
Crude oil	2,788.1	42,575
Products	617.2	13,392
Gasoline	56.9	1,792
Diesel fuel	158.2	3,707
Mazut	344.4	4,452
Lubes and grease	-	-
Lube oils	29.3	2,399

Destination/Products	Quantity (Thousand Metric Tons)	Value (Thousand Rubles)
East Germany		
Crude oil	6,137.1	82,542
Products	125.8	4,066
Gasoline	59.0	1,698
Diesel fuel	31.8	821
Lubes and grease	7.8	883
Petroleum coke	15.6	490
Netherlands		
Products	154.9	2,901
Diesel fuel	136.7	2,478
Gasoline	-	-
Greece		
Crude oil	672.3	7,733
Products	564.5	6,615
Gasoline	63.0	998
Diesel fuel	156.9	2,866
Mazut	344.6	2,751
Denmark		
Products	534.6	4,755
Diesel fuel	44.9	762
Mazut	489.7	3,993
Wax	0.002	-

(Continued)

TABLE 58 (Continued)

Destination/Products	Quantity (Thousand Metric Tons)	Value (Thousand Rubles)
Iceland		
Products	408.3	6,721
Mogas	50.2	1,167
Diesel fuel	248.8	4,337
Mazut	109.3	1,217
Italy		
Crude oil	10,576.4	101,354
Products	1,372.7	14,376
Gasoline	378.6	4,653
Mazut	978.8	9,135
Wax	-	-
Norway		
Products	538.9	6,676
Diesel fuel	311.1	4,475
Mazut	181.8	1,595
Wax	0.6	65
Petroleum coke	0.9	12
Gasoline	44.5	529
Poland		
Crude oil	3,609.2	52,737
Products	1,917.9	45,800
Gasoline	778.3	20,810
Kerosine	29.3	849
Diesel fuel	581.2	13,761

Destination/Products	Quantity (Thousand Metric Tons)	Value (Thousand Rubles)
Poland (cont'd)		
Mazut	487.1	7,085
Lubes and grease	24.7	2,189
Asphalt	-	-
Wax	6.8	774
Petroleum coke	10.4	303
Natural gas[a]	1,025.3	14,484
Rumania		
Products	0.704	164
Lubes and grease	0.6	115
Lube additives	0.104	49
Spain		
Crude oil	456.8	5,365
Products	-	-
West Germany		
Crude oil	4,209.0	38,639
Products	1,378.1	20,163
Mogas	473.2	6,039
Diesel fuel	904.6	14,119
Mazut	-	-
Finland		
Crude oil	3,968.6	46,150
Products	2,492.4	41,651

[a] Million cubic meters.

(Continued)

TABLE 58 (Continued)

Destination/Products	Quantity (Thousand Metric Tons)	Value (Thousand Rubles)
Finland (cont'd)		
Gasoline	44.2	1,306
Kerosine	11.9	316
Diesel fuel	1,731.2	31,813
Mazut	695.9	7,667
Lube oils	7.5	337
Wax	1.7	209
Ireland		
Products	295.4	2,392
Mazut	295.4	2,392
France		
Crude oil	1,773.6	15,091
Products	1,170.9	15,214
Gasoline	103.3	1,646
Diesel fuel	712.6	10,868
Mazut	354.9	2,684
Wax	-	-
Ozokerite	-	-
Czechoslovakia		
Crude oil	7,425.3	114,165
Products	498.3	14,189
Gasoline	185.6	4,857
Kerosine	249.5	7,235
Diesel fuel	33.8	923
Asphalt	-	-

Destination/Products	Quantity (Thousand Metric Tons)	Value (Thousand Rubles)
Czechoslovakia (cont'd)		
Wax	1.6	205
Natural gas[a]	265.2	3,913
Switzerland		
Crude oil	-	-
Products	177.8	2,848
Mogas	-	-
Diesel fuel	147.7	2,208
Mazut	-	-
Kerosine	30.1	640
Sweden		
Crude oil	-	-
Products	3,809.7	34,463
Gasoline	-	-
Kerosine	-	-
Diesel fuel	722.2	9,957
Mazut	3,074.0	24,269
Wax	0.1	9
Senegal		
Products	0.615	67
Wax	0.615	67
Yugoslavia		
Crude oil	733.2	7,914

[a] Million cubic meters.

(Continued)

TABLE 58 (Continued)

Destination/Products	Quantity (Thousand Metric Tons)	Value (Thousand Rubles)
Yugoslavia (cont'd)		
Products	755.3	10,490
Gasoline	10.2	442
Diesel fuel	236.5	4,245
Mazut	417.3	3,660
Lube oils	6.7	584
Wax	0.350	48
Kerosine	84.1	1,461
Afghanistan		
Products	116.0	4,678
Gasoline	77.1	2,790
Kerosine	6.6	403
Diesel fuel	28.3	964
Lube oils	1.8	355
Grease	0.1	34
Asphalt	1.9	132
Yemen		
Products	7.3	244
Gasoline	-	-
Kerosine	5.8	218
Diesel fuel	1.5	26
North Vietnam		
Products	214.4	8,127
Gasoline	92.3	3,450
Kerosine	22.0	694
Diesel fuel	92.1	2,901

Destination/Products	Quantity (Thousand Metric Tons)	Value (Thousand Rubles)
North Vietnam (cont'd)		
Lube oils	6.6	813
Grease	1.0	205
India		
Products	468.7	8,418
Kerosine	468.6	8,399
Diesel fuel	-	-
Mazut	-	-
Iran		
Products	0.410	51
Wax	0.410	51
Communist China		
Crude oil	-	-
Products	6.9	467
Gasoline	3.9	225
Kerosine	0.4	18
Diesel fuel	1.9	98
Lube oils	0.6	103
Grease	0.1	21
Wax	-	-
North Korea		
Products	446.3	15,106
Gasoline	186.1	5,780
Kerosine	12.5	397

(Continued)

TABLE 58 (Continued)

Destination/Products	Quantity (Thousand Metric Tons)	Value (Thousand Rubles)
North Korea (cont'd)		
Diesel fuel	177.9	5,423
Mazut	40.4	766
Lube oils	25.6	2,200
Grease	1.3	275
Wax	1.5	211
Mongolia		
Crude oil	48.0	1,101
Products	179.0	7,681
Gasoline	103.4	4,038
Kerosine	-	2
Diesel fuel	10.7	2,201
Lube oils	10.7	1,138
Grease	0.513	153
Asphalt	0.709	59
Mazut	3.0	89
Turkey		
Products	174.9	2,745
Gasoline	-	-
Kerosine	22.6	407
Diesel fuel	123.5	2,111
Mazut	28.8	227
Ceylon		
Crude oil	-	-
Products	564.8	9,180

Destination/Products	Quantity (Thousand Metric Tons)	Value (Thousand Rubles)
Ceylon (cont'd)		
Gasoline	54.5	1,129
Kerosine	161.5	3,421
Diesel fuel	199.2	3,302
Mazut	149.6	1,328
Japan		
Crude oil	1,798.1	16,781
Products	1,474.7	16,279
Kerosine	-	-
Diesel fuel	639.0	8,490
Mazut	772.2	6,900
Petroleum coke	63.4	889
Ghana		
Crude oil	452.6	3,565
Products	-	-
Guinea		
Products	61.9	1,436
Mogas	30.3	784
Kerosine	-	-
Diesel fuel	31.6	652
Mazut	-	-
Morocco		
Crude oil	582.9	6,002
Products	-	-

(Continued)

TABLE 58 (Continued)

Destination/Products	Quantity (Thousand Metric Tons)	Value (Thousand Rubles)
Egypt (UAR)		
Crude oil	1,003.5	11,892
Products	155.7	5,086
Gasoline	122.7	3,674
Diesel fuel	29.5	707
Mazut	-	-
Lube oils	3.3	553
Grease	0.199	149
Syria		
Products	96.3	1,510
Diesel fuel	96.2	1,479
Argentina		
Crude oil	-	-
Products	99.1	1,357
Diesel fuel	99.1	1,357
Brazil		
Crude oil	581.8	5,167
Products	-	-
Cuba		
Crude oil	3,837.7	39,759
Products	1,448.5	21,802

Destination/Products	Quantity (Thousand Metric Tons)	Value (Thousand Rubles)
Cuba (cont'd)		
Gasoline	104.2	2,624
Diesel fuel	320.9	5,261
Mazut	954.6	8,720
Lube oils	66.2	4,787
Grease	1.3	232
Wax	1.3	175
Burma		
Crude oil	79.0	662
Products	-	-
Nigeria		
Products	10.6	280
Diesel fuel	10.6	280
Cyprus		
Products	132.6	1,058
Mazut	132.6	1,058
Somalia		
Products	39.1	751
Gasoline	11.5	243
Kerosine	4.4	96
Diesel fuel	23.2	412

(Continued)

TABLE 58 (Continued)

Destination/Products	Quantity (Thousand Metric Tons)	Value (Thousand Rubles)
Colombia		
Products	0.025	3
Wax	0.025	3

TABLE 59

Imports of Petroleum by the Soviet Union, 1967

Origin/Products	Quantity (Thousand Metric Tons)	Value (Thousand Rubles)
Total		
Crude oil	61.3	796
Petroleum products	1,358.7	46,238
Gasoline	753.2	20,789
Kerosine	12.2	362
Diesel fuel	242.1	5,370
Mazut	160.8	1,833
Lube oils	110.9	7,008
Greases	27.0	3,758
Other products	52.4	7,019
Asphalt	25.6	595
Wax	8.0	1,135
Naphthenic acids	0.2	42
De-emulsifier	10.7	4,667
Algeria		
Crude oil	61.3	796
Rumania		
Products	776.6	26,859
Gasoline	445.4	12,659
Motor gasoline	393.1	10,308
Special gasoline	52.3	2,351
Kerosine	10.8	320
Diesel fuel	152.3	3,725
Mazut	11.2	160
Lube oils	107.4	6,317
Greases	13.3	1,821

(Continued)

TABLE 59 (Continued)

Origin/Products	Quantity (Thousand Metric Tons)	Value (Thousand Rubles)
Rumania (cont'd)		
Other products	36.2	1,857
Asphalt	25.6	595
Wax	9.0	1,133
Naphthenic acids	0.2	42
United Kingdom		
Products	0.661	165
Special purpose lube oils	0.661	165
Hungary		
Products	31.1	2,241
Greases	12.9	1,749
East Germany		
Products	271.9	8,255
Gasoline	236.2	6,757
Diesel fuel	32.9	887
Lubes and greases	0.8	187
Ozokerite	0.7	359
Netherlands		
Products	1.102	475
De-emulsifier	1.0	396
Lube oil additives	0.102	79

Origin/Products	Quantity (Thousand Metric Tons)	Value (Thousand Rubles)
West Germany		
Products	6.7	2,987
De-emulsifier	6.7	2,987
Syria		
Products	28.9	593
Gasoline	28.9	593
Pakistan		
Mazut	149.6	1,673
United States		
De-emulsifier	3.0	1,278

TABLE 60

Apparent Consumption of Petroleum in the Soviet Union,
Selected Years 1955-75

(in million metric tons)

Year	Quantity[a]
1955	60.4
1956	70.8
1957	79.3
1958	87.9
1959	95.9
1960	104.9
1961	113.2
1962	126.7
1963	140.0
1964	150.0[b]
1965	160.1
1966	170.6[b]
1967	186.8[b]
1968	203.[b]
1970	231
1975	304

[a] A summation of the estimated refinery yields, the production of natural gas liquids and synthetics and net trade in petroleum products. Refinery yields based on crude oil production less crude losses (3 percent of production) and net trade in crude oil as constituting the charge to refining, and yields estimated at 92 percent of the charge.

[b] To illustrate, the derivation of the apparent consumption of petroleum in the USSR for these years is indicated in Table 61.

TABLE 61

Derivation of Apparent Consumption of Petroleum
in the Soviet Union, 1964 and 1966-68

(in million metric tons)

Indicator	1964	1966	1967	1968[a]
Production of crude oil	223.6	265	288	309
Imports of crude oil	0	0	0.1	0
Total crude oil	223.6	265	288.1	309
Exports of crude oil	36.7	50.3	54.1	57
Crude oil losses	6.7	8	8	9
Crude oil available for charge to refining[b]	180.2	206.7	226	243
Charge to refining	180.2	206.7	226	243
Yield from refining[c]	165.8	190.2	207.9	223.5
Production of natural gas liquids, synthetics	2	2	2.2	3
Total petroleum products	167.8	192.2	210.1	226.5
Exports of petroleum products	19.9	23.3	24.7	25
Subtotal	147.9	168.9	185.4	201.5
Imports of petroleum products	2.1	1.7	1.4	1.5
Total petroleum products available for domestic consumption	150.0	170.6	186.8	203

[a] Estimated imports and exports.

[b] A reasonable estimate of capacity to refine crude oil in the Soviet Union is that refineries are operated at about 87 percent of capacity. Thus, to derive refinery capacity for these years, divide the charge to refining by 0.87.

[c] It is estimated that petroleum product yield from refining averages about 92 percent of the charge.

APPENDIX B PERTINENT CURRENT SOURCE MATERIALS ON THE COMMUNIST OIL INDUSTRY

The choice of the Western observer in selecting sources covering the oil and gas industries of the Communist world is quite broad, if he is willing to confine his study to the Soviet Union. If, however, he should attempt to broaden his scope to include the socialist countries of Eastern Eurpoe and Communist China as well, not to mention Albania, Cuba, Yugoslavia, North Korea, and North Vietnam (all part of the Communist world by our definition today), he will find that his efforts will be much less rewarding. Indeed, perhaps the best sources for information as to what is going on in the "other Communist countries" will again be the Russian-language publications. For, with the exception of a very limited selection of oil trade publications from Poland, Rumania, and Yugoslavia, there is very little available outside the Soviet Union. As for Communist China, the practice of divulging pertinent statistics on economic and industrial developments met an untimely death in the late 1950's. One now has to resort to piecing together scraps of information picked out of the air, so to speak, in the hope of perhaps being able to construct a series of link relatives long enough to reach back in time to an acceptable base year.

For the researcher who may not have access to Russian-language material or who may be handicapped by an inability to read such material, two English-language oil trade magazines--The Oil and Gas Journal (a weekly) and Petroleum Press Service (a monthly)--offer by far the most complete coverage of the subject of Communist oil and gas. There are a variety of other trade publications which carry occasional articles of value: World Oil (a monthly), World Petroleum (a monthly), Petroleum Times (a weekly), Petroleum Intelligence Weekly (a weekly newsletter), and Platt's Oilgram (a daily newsletter).

APPENDIX B 443

There are also a limited number of English-language studies on the various aspects of energy and oil in the Soviet Union, the most recent of which is Robert W. Campbell, The Economics of Soviet Oil and Gas (Baltimore: published for Resources For the Future by the Johns Hopkins Press, 1968). Professor Campbell, an economist, examines the oil industry of the Soviet Union against the background of his discipline, analyzing its expansion from the middle 1950's through 1965. His work, the product of painstaking research and examination of a multitude of primary sources, is particularly important for its insight into decision-making in the Soviet petroleum industry.

A ready reference for those concerned with the primary energy resource base of the Soviet Union is Jordan A. Hodgkins, Soviet Power: Energy Resources, Production, and Potential (New York: Prentice-Hall, 1961). Soviet Power is useful for its accumulation of data covering reserves and production of coal, oil shale, crude oil, and natural gas, but suffers from overemphasis on coal. The bulk of his statistics end with the middle 1950's.

What must be considered to be the most important and certainly the most authoritative study of oil in the Communist world is that undertaken by the National Petroleum Council, an industry advisory body to the secretary of the interior. In 1962 the council responded to a request by the secretary to make a factual study of the effects on the Free World of the exports of petroleum from the Soviet bloc, together with such comments and conclusions as were deemed appropriate. The council completed its assignment upon the release in October, 1962 of a two-volume study by the National Petroleum Council, Impact of Oil Exports From the Soviet Bloc (Washington, D.C., 1962). This was followed by an updating, issued in 1964, of the original study, which took into consideration new information which had become available in the intervening months and which projected probable Soviet bloc trade in oil through 1970.

In 1962 Demitri B. Shimkin, then with the U.S. Department of Commerce, completed a study, The Soviet Mineral-Fuels Industries, 1928-1958: A Statistical Survey, U.S. Bureau of the Census, International Population Statistics Reports, Series P-90, No. 19 (Washington, D.C.: U.S. Government Printing Office, 1962). To gather the data, which at that time was a particularly difficult and often frustrating task, Shimkin undertook an extraordinary search of original source material, and the results of this search are fully annotated in this book. The notes and sources to the tables in the appendix and to the various chapters in the text offer a valuable reference aid to the researcher.

In the summer of 1960 a team of oil experts, selected and sponsored by the American Petroleum Institute, toured a number of oil fields, refineries, and petroleum research institutes of the Soviet Union at the request of the U.S. Department of State. This tour was made under the U.S.-USSR Exchange Agreement of November 21, 1959 which provided for exchanges in scientific, technical, educational, and cultural fields between the two countries. The findings of this team, together with the results of additional research by the author, were assembled in a report to its sponsor, the American Petroleum Institute, which subsequently published these findings in book form as Robert E. Ebel, The Petroleum Industry of the Soviet Union (Washington, D.C.: American Petroleum Institute, 1961).

Perhaps the best-known work of Shimkin is his Minerals: A Key to Soviet Power (Cambridge, Mass.: Harvard University Press, 1953), a systematic survey of the mineral resources, production, and consumption position of the Soviet Union: the concluding chapter is an attempt to analyze the potential development of the mining industries. At the time of its publication, this study was considered a landmark among the early research efforts to construct a reasonable picture of the Soviet economy. Today this publication is noteworthy largely for its collection of pre-World War II data.

The first English-language study of the oil industry in the Soviet Union to be published after World War II was Heinrich Hassman's Oil in the Soviet Union (Princeton, N.Y.: Princeton University Press, 1953), translated from the German by Alfred M. Leeston. With the addition of a number of maps, and of many notes, the latest statistics, and a bibliography by Leeston, Oil in the Soviet Union found a ready and interested audience around the world, and it stood as the reference work on Soviet oil for a number of years. For those who wish an introduction to the development of the Russian and the Soviet oil industry, from the initial appearance of oil in the country up to the beginning of World War II, this book is quite rewarding.

APPENDIX B

The rapidly expanding technology in the oil industry, the ever-enlarging search for new supplies of oil and gas (which takes explorers and producers into areas heretofore known only to the local inhabitants), and changing political climates have brought forth a proliferation of special-interest journals and newsletters, each presented to the harried executive as the answer to his struggle to keep up with the changing oil scene. This deluge has also been extended to the Soviet Union in recent years, a most welcome change from the days of Stalin, when Western research on Communism was unencumbered by readily available statistics.

The extent of the regularly published titles of varying applicability to Soviet oil and gas is illustrated very clearly by the following list of 41 daily newspapers, general economic journals, and oil industry periodicals which, in the author's opinion, represent an acceptable minimum for any library which seeks to support serious research. In addition, seven East European titles are noted; unfortunately, comparable coverage is not available on each of the East European nations.

RUSSIAN-LANGUAGE PUBLICATIONS

Newspapers

Pravda	Truth
Pravda vostoka	Pravda of the East
Sovetskaya Belorussiya	Soviet Belorussia
Izvestiya	News
Bakinskiy rabochiy	Baku Worker
Vyshka	Derrick
Turkmenskaya iskra	Turkmen Spark
Stroitelnaya gazeta	Construction Newspaper
Pravda Ukrainy	Pravda of the Ukraine
Ekonomicheskaya gazeta	Economic Newspaper
Kazakhstanskaya pravda	Kazakhstan Pravda
Sovetskaya Rossiya	Soviet Russia

General Economic Journals

Voprosy ekonomiki	Questions of Economics
Vneshnyaya torgovlya	Foreign Trade
Planovoye khozyaystvo	Planned Economy
Vestnik statistiki	Herald of Statistics
Kommunist	Communist
Ekonomika promyshlennosti	Economics of Industry
Narodnoye khozyaystvo Kazakhstana	National Economy of Kazakhstan
Ekonomika i zhizn	Economics and Life
Promyshlennost Belorussiy	Industry of Belorussia
Ekonomika Sovetskoy Ukrainy	Economics of the Soviet Ukraine

Oil Industry Journals

Neftyanoye khozyaystvo	Oil Economy
Geologiya nefti i gaza	Geology of Oil and Gas
Gazovaya promyshlennost	Gas Industry
Stroitelstvo truboprovodov	Pipeline Construction
Neftyanik	Oil Worker
Ekonomika neftedobyvayushchey promyshlennosti	Economics of the Oil Extracting Industry
Neftegazovaya geologiya i geofizika	Oil and Gas Geology and Geophysics
Bureniye	Drilling
Neftepromyslovoye delo	Oil Field Economy
Transport i khraneniye nefti i nefteproduktov	Transport and Storage of Crude Oil and Petroleum Products
Gazovoye delo	Gas Economy
Mashiny i neftyanoye oborudovaniye	Machinery and Oil Equipment
Khimiya i tekhnologiya topliv i masel	Chemistry and Technology of Fuels and Lubricants
Neftyanaya i gazovaya promyshlennost	Oil and Gas Industry
Razvedka i okhrana nedr	Exploration and Conservation of Natural Resources
Neftepererabotka i neftekhimiya	Oil Refining and Petrochemistry
Neft i gaz	Oil and gas

Azerbaydzhanskoye　　　　　　Azerbaydzhan Oil Economy
　　neftyanoye khozyaystvo
Za tekhnicheskiy progress　　　After Technical Progress

EAST EUROPEAN PUBLICATIONS

Petrol si gaze (Rumania)　　　　Oil and Gas
Banyaszati lapok (Hungary)　　　Miners' Bulletin
Gospodarka paliwami i　　　　　Fuel and Energy Economics
　　energia (Poland)
Nafta (Poland)　　　　　　　　　Oil
Waidomosci naftowe (Poland)　　Petroleum News
Nafta (Yugoslavia)　　　　　　　Petroleum
Privredni pregled (Yugoslavia)　 Economic Review

ABOUT THE AUTHOR

Robert E. Ebel is a Foreign Oil Specialist with the Office of Oil and Gas of the U.S. Department of the Interior. He has held various positions with the U.S. Government since 1956, providing guidance to policy makers on matters of concern in the international oil industry.

A graduate of the USAFIT Russian language program at Syracuse University, he has long been an observer of developments in the oil and gas industries of Communist nations. He has traveled extensively in the Soviet Union; in 1960 as a member of the first American oil delegation to that country, and in 1969 as a member of a delegation of natural gas experts who were the first Americans to inspect the prolific new gas fields in the Soviet Arctic. In 1962 he served as the Government Co-Chairman of the National Petroleum Council Working Group which prepared a two-volume study on the impact of oil exports from the Soviet bloc.

Mr. Ebel received his bachelor's degree in petroleum geology from Texas Technological College and did postgraduate work at Syracuse University in the field of international relations.